Dahlem Workshop Reports
Life Sciences Research Report 47
Exchange of Trace Gases between Terrestrial Ecosystems and the Atmosphere

Goal of this Dahlem Workshop:
to analyze the biological and physicochemical regulation of trace gas exchange for modeling on local to global scales, and to evaluate the importance of trace gas exchange to ecology, climate, and atmospheric chemistry

Life Sciences Research Reports

Series Editor: Silke Bernhard

Held and published on behalf of the
Stifterverband für die Deutsche Wissenschaft

Sponsored by:
Senat der Stadt Berlin
Stifterverband für die Deutsche Wissenschaft
Stiftungsfonds Unilever im Stifterverband für die Deutsche Wissenschaft

Exchange of Trace Gases between Terrestrial Ecosystems and the Atmosphere

M.O. Andreae and D.S. Schimel, Editors

Report of the Dahlem Workshop on
Exchange of Trace Gases between Terrestrial
Ecosystems and the Atmosphere
Berlin 1989, February 19–24

Rapporteurs:
G.P. Robertson, T. Rosswall, J.W.B. Stewart, P.M. Vitousek

Program Advisory Committee:
M.O. Andreae and D.S. Schimel, Chairpersons
R. Conrad, D. Fowler, B.H. Svensson

A Wiley–Interscience Publication

John Wiley & Sons 1989
Chichester · New York · Brisbane · Toronto · Singapore

Copy Editors: J. Lupp, K. Klotzle

Photographs: E.P. Thonke

With 4 photographs, 48 figures, 32 tables

Library of Congress Cataloging-in-Publication Data

Dahlem Workshop on Exchange of Trace Gases between Terrestrial Ecosystems and the Atmosphere (1989 : Berlin, Germany)
Exchange of trace gases between terrestrial ecosystems and the atmosphere : report of the Dahlem Workshop on Exchange of Trace Gases between Terrestrial Ecosystems and the Atmosphere, Berlin 1989, February 19–24 / M.O. Andreae and D.S. Schimel, editors ; rapporteurs: G.P. Robertson . . . [et al.].
p. cm. — (Life Sciences Research report ; 47) (Dahlem workshop reports)
"A Wiley–Interscience publication."
Includes bibliographical references.
ISBN 0 471 92551 9
1. Biogeochemical cycles—Congresses. 2. Atmospheric chemistry—Congresses. 3. Gases, Rare—Measurement—Congresses. 4. Ecology—Congresses. I. Andreae, M.O. II. Schimel, David Steven. III. Robertson, G.P. IV. Title. V. Series. VI. Series: Dahlem workshop reports.
QH344.D34 1989
574.5'222—dc20 89–22459
CIP

British Library Cataloguing in Publication Data

Dahlem workshop on Exchange of Trace Gases between Terrestrial Ecosystems and the Atmosphere : 1989 : Berlin, Germany.
Exchange of trace gases between terrestrial ecosystems and the atmosphere.
1. Land ecosystems. Effects of atmospheric gases. Atmosphere. Gases. Effects on land ecosystems
I. Title II. Andreae, M.O. III. Schimel, D.S.
IV. Series
574.5'264

ISBN 0 471 92551 9

Typeset by Photo·graphics, Honiton, Devon
Printed and bound in Great Britain by Biddles Ltd, Guildford, Surrey

Table of Contents

The Dahlem Konferenzen

Founders

Recognizing the need for more effective communication between scientists, the Stifterverband für die Deutsche Wissenschaft*, in cooperation with the Deutsche Forschungsgemeinschaft**, founded Dahlem Konferenzen in 1974. The project is financed by the founders and the Senate of the City of Berlin.

Name

Dahlem Konferenzen was named after the district of Berlin called *Dahlem*, which has a long-standing tradition and reputation in the sciences and arts.

Aim

The task of Dahlem Konferenzen is to promote international, interdisciplinary exchange of scientific information and ideas, to stimulate international cooperation in research, and to develop and test new models conducive to more effective communication between scientists.

The Concept

The increasing orientation towards interdisciplinary approaches in scientific research demands that specialists in one field understand the needs and problems of related fields. Therefore, Dahlem Konferenzen has organized workshops, mainly in the Life Sciences and the fields of Physical, Chemical, and Earth Sciences, of an interdisciplinary nature.

The Dahlem Workshops provide a unique opportunity for posing the right questions to colleagues from different disciplines who are encouraged to state what they do not know rather than what they do know. The aim is not to solve problems or to reach a consensus of opinion, the aim is to define and discuss priorities and to indicate directions for further research.

---

* The Donors Association for the Promotion of Sciences and Humanities, a foundation created in 1921 in Berlin and supported by German trade and industry to fund basic research in the sciences.

** German Science Foundation.

Topics

The topics are of contemporary international interest, timely, interdisciplinary in nature, and problem oriented. Dahlem Konferenzen approaches internationally recognized scientists to suggest topics fulfilling these criteria. Once a year, the topic suggestions are submitted to a scientific board for approval.

Program Advisory Committee

A special Program Advisory Committee is formed for each workshop. It is composed of 6–7 scientists representing the various scientific disciplines involved. They meet approximately one year before the workshop to decide on the scientific program and define the workshop goal, select topics for the discussion groups, formulate titles for background papers, select participants, and assign them their specific tasks. Participants are invited according to international scientific reputation alone. Exception is made for younger German scientists. Invitations are not transferable.

Dahlem Workshop Model

Since no type of scientific meeting proved effective enough, Dahlem Konferenzen had to create its own concept. This concept has been tested and varied over the years. It is internationally recognized as the *Dahlem Workshop Model*. Four workshops per year are organized according to this model. It provides the framework for the utmost possible interdisciplinary communication and cooperation between scientists in a period of $4\frac{1}{2}$ days.

At Dahlem Workshops 48 participants work in four interdisciplinary discussion groups. Lectures are not given. Instead, selected participants write background papers providing a review of the field rather than a report on individual work. These papers, reviewed by selected participants, serve as the basis for discussion and are circulated to all participants before the meeting with the request to formulate written questions and comments to them. During the workshop, each of the four groups prepares reports reflecting their insights gained through the discussion. They also provide suggestions for future research needs.

Publication

The group reports written during the workshop together with the revised background papers are published in book form as the Dahlem Workshop Reports. They are edited by the editor(s) and the Dahlem Konferenzen staff. The reports are multidisciplinary surveys by the most internationally distinguished scientists and are based on discussions of advanced new concepts, techniques, and models. Each report also reviews areas of priority interest and indicates directions for future research on a given topic.

The Dahlem Workshop Reports are published in two series:

1) Life Sciences Research Reports (LS), and
2) Physical, Chemical, and Earth Sciences Research Reports (PC).

Director

Silke Bernhard, Dr med., Dr phil. h.c.

Address

Dahlem Konferenzen
Tiergartenstr. 24–27
D-1000 Berlin (West) 30

Tel.: (030) 262 50 41

THE DAHLEM WORKSHOP MODEL

MONDAY	TUESDAY	WEDNESDAY	THURSDAY	FRIDAY
A. Opening (P) B. Introduction (P) C. Selection of Problems for the Group Agendas (S) 1 2 3 4	1 2 3 4	1 3	F. Report Session 1 2 3 4	G. Distribution of the Reports H. Reading Time I. Discussion of the Group Reports (P)
D. Presentation of Group Agendas (P) E. Group Discussions (S) 1 2		2 4		J. Groups Meet to Revise their Reports (S) 1 2 3 4

Key: (P) = Plenary Session;
(S) = Simultaneous Sessions;
◯ = one discussion group

Explanation of the Dahlem Workshop Model

A. Opening
Background information is given about Dahlem Konferenzen and the Dahlem Workshop Model.

B. Introduction
The goal and the scientific aspects of the workshop are explained.

C. Selection of Problems for the Group Agenda
Each participant is requested to define priority problems of his choice to be discussed within the framework of the workshop goal and his discussion group topic. Each group discusses these suggestions and compiles an agenda of these problems for their discussions.

D. Presentation of the Group Agenda
The agenda for each group is presented by the moderator. A plenary discussion follows to finalize these agendas.

E. Group Discussions
Two groups start their discussions simultaneously. Participants not assigned to either of these two groups attend discussions on topics of their choice.
The groups then change roles as indicated on the chart.

F. Report Session
The rapporteurs discuss the contents of their reports with their group members and write their reports, which are then typed and duplicated.

G. Distribution of Group Reports
The four group reports are distributed to all participants.

H. Reading Time
Participants read these group reports and formulate written questions/comments.

I. Discussion of Group Reports
Each rapporteur summarizes the highlights, controversies, and open problems of his group. A plenary discussion follows.

J. Groups Meet to Revise their Reports
The groups meet to decide which of the comments and issues raised during the plenary discussion should be included in the final report.

Exchange of Trace Gases between Terrestrial Ecosystems and the Atmosphere
eds. M.O. Andreae and D.S. Schimel, pp. 1–5
John Wiley & Sons Ltd

Introduction

M.O. Andreae[1] and D.S. Schimel[2]

[1] *Abt. Biogeochemie*
Max-Planck-Institut für Chemie
Postfach 3060
6500 Mainz, F.R. Germany

[2] *NASA-Ames Research Center*
MS 239–12
Moffett Field, CA 94035, U.S.A.

The composition of the global atmosphere is predominantly a product of the activity of the biosphere. This has long been recognized to be the case for the main constituent gases of the atmosphere: Molecular oxygen is a "side product" of plant photosynthesis, and nitrogen is released by soil microbes during denitrification. Without the continuous regeneration of molecular oxygen and nitrogen by the biota, the atmosphere would proceed towards thermodynamic equilibrium, represented by carbon dioxide and nitrate.

In recent years, we have become increasingly aware of how strongly the physical and chemical properties of the Earth's atmosphere are influenced by emissions from the biosphere and by uptake of trace gases by the biota. Relevant examples are the emission of sulfur gases by marine phytoplankton, the production of methane by methanogenic bacteria, and the exchange of ammonia and nitrogen oxides between the atmosphere and the bacteria in soils. Plants also emit numerous trace gases, e.g., ammonia, hydrogen sulfide, and hydrocarbons; but on the other hand, their surfaces are major sinks for many atmospheric constituents, including pollutants such as sulfur dioxide, nitric acid, and ozone. Conversely, such deposition to plant surfaces appears to be a major cause of plant damage by air pollutants.

We have also begun to appreciate that the interactions between the biosphere and the atmosphere are part of a complex, interconnected system: The emission and uptake of atmospheric constituents by the biota influence chemical and physical climate through interactions with atmospheric photochemistry and the Earth's radiation budget. In turn, climate change and atmospheric pollution alter the rates and sometimes even the direction of chemical exchange between biosphere and atmosphere through influences both at the level of the individual organism and at the ecosystem level.

In spite of the indisputable importance of atmosphere/biosphere exchange for the composition of the atmosphere, the processes which control the flux rates and the magnitudes of the fluxes from many ecosystems remain poorly known. This is due both to the complexity of the biological and physico-chemical systems involved and to the difficulty of actually measuring exchange fluxes in the field. To improve our understanding of atmosphere/biosphere interactions, a close collaboration between chemists, biologists, and atmospheric scientists is required: Analytical methods to determine the relevant gas fluxes at the concentrations and against the background levels found in the natural atmosphere need to be developed and verified; ecosystems need to be characterized in relation to their potential for gas exchange with the atmosphere; physiological controls and biochemical processes involved in gas production and uptake have to be investigated; and the micrometeorological theory and field methods required to measure the exchange fluxes need to be developed. Finally, diagnostic and predictive models need to be designed at multiple spatial and temporal scales for extrapolation and analysis of scenarios of environmental change.

The multidisciplinary collaboration required to fulfill these tasks has begun to develop in the last few years, especially among atmospheric chemists and micrometeorologists. There still remains, however, a need for more intensive interaction between disciplines, including ecology, microbiology, and plant physiology. This appears particularly urgent at this time, in view of the ongoing planning activities in preparation for the International Geosphere/Biosphere Program, which is to take place in the 1990s.

To initiate the interactions of scientists across disciplines and national boundaries, essential to the anticipated research programs on atmosphere/biosphere exchange, a SCOPE Project on Trace Gas Exchange was established in 1987. The concept of a Dahlem Workshop on Trace Gas Exchange between Biosphere and Atmosphere evolved out of the discussions in this SCOPE project. In order to provide a sharp focus at this Dahlem Workshop, we limited the discussion to the trace gases CH_4, N_2O, and NO_x. These gases were selected because of their rich biogeochemical interactions and their importance for global climate and atmospheric chemistry. The fact that their atmospheric concentrations are increasing, apparently as a result of human activity, provides further complexity and interest to their study.

The following specific objectives were proposed in order to guide the discussions:

1. To critically assess our understanding of the controls and interactions in the plant–microorganism–soil–water system which regulate trace gas exchange with the atmosphere.
2. To assess the status of current techniques for deriving fluxes from production rates and from concentration measurements, and to explore the possibilities for the development of novel approaches to flux measurements. Also, to assess measurements on larger spatial scales such as aircraft-based techniques and approaches using remote sensing.
3. To develop a foundation for the design of diagnostic and predictive models which describe the exchange of trace gases between the biosphere and the world atmosphere. Such models form a critical link between the spatial and temporal scales at which measurements are conducted and the scales which are of interest for a scientific understanding of the exchange of gases between the biosphere and the atmosphere. These range from the microscopic scale, at which microbiological production and consumption of gases take place, to the regional and global scales relevant for atmospheric chemistry and climate.

In order to provide for optimal interaction of the workshop participants within, and particularly across, scientific disciplines, four discussion groups were formed and given the following topics:

1. *What are the relative roles of biological and environmental variables in regulating production and consumption of trace gases in ecosystems?*

To describe variations in rates of trace gas production and consumption between ecosystems and over time, and to analyze the response of systems to changing circumstances, we must understand physiological and ecosystem-level controls over turnover rates. Here, both inherently biological parameters (e.g., the number and species of microorganisms present) and environmental (physicochemical variables, e.g., temperature and soil moisture) function as controls over trace gas turnover. The critical variables and their relative importance should be identified and discussed.

2. *How should we extrapolate flux measurements to regional and global scales?*

Measurements of fluxes are inherently made at small scales and are samples in time, relative to the global atmosphere. Can ecosystem models be used to predict regional variations in source strength? Similarly, can physical models be used to describe variations in atmospheric behavior regulating emission and deposition? What role is there for aircraft measurements and remote sensing in extrapolation and model testing? How valid are current global extrapolations; what can be done to improve them?

3. *What are the processes controlling the fluxes of trace gases between terrestrial ecosystems and the atmosphere, and how do we best measure these fluxes?*

After trace gases are produced in the plant–soil–microorganism system, micrometeorological and photochemical processes may influence net emission to the atmosphere. Can variations in fluxes between systems be attributed primarily to the different rates of trace gas production, or are variations in photochemistry and micrometeorology due to vegetation structure dominant? How can the interactions of biological source and sink strength with physico-chemical processes be modeled? What measurement techniques are the most appropriate for understanding exchange processes and for obtaining representative flux values from key ecosystems?

4. *How does trace gas exchange interact with chemical and physical climate: Is there "geophysiological regulation"?*

How does trace gas exchange affect physical climate, and over what time scales? What feedback effects does modification of chemical climate through trace gas exchange have on biotic and chemical processes *in situ* and at distant locations? Can regional-scale transport of nutrients in the atmosphere affect the behavior and distribution of ecosystems? What is the current status of coupled atmosphere–ecosystem models on local and global scales? How adequate are current measurements programs worldwide for assessing the above issues, and what new experiments are feasible and of high priority? Finally, do current studies indicate significant stabilizing or destabilizing feedback loops between ecosystems and the atmosphere; is "geophysiology" a useful concept?

Following a week of discussions, most of the participants were impressed and surprised by the large number of fundamental gaps which still exist in our understanding of the exchange of trace gases between the biosphere and the atmosphere. This is true both for the absolute amounts of these exchange fluxes concerned and for the mechanisms which control them. This impression was particularly strong in the case of N_2O, where many participants entered the workshop thinking that the global cycle of this gas was reasonably well understood, and were surprised to find out that we can neither explain the magnitude of the flux of this gas to the atmosphere, nor account for the reasons for its increase in recent times.

The conferees came to the conclusion that, in view of the importance of biogenic trace gases for global climate and for the chemistry of the atmosphere, it is essential that scientific programs be designed to improve our understanding of their sources and sinks. The discussions yielded a substantial number of suggestions and recommendations for research towards this goal. These recommendations are contained in the group reports; in view of their timeliness and importance, some of them are discussed further in a separate chapter, which is intended to serve as an outline for forthcoming international scientific programs. Planning activities for such programs are already underway at this time, particularly within the working groups of the International Global Atmospheric Chemistry Program, which may come to be the atmospheric chemistry component of the International Geosphere/Biosphere Program. It is to be hoped that this Dahlem Workshop will provide an important contribution to the scientific basis for these programs.

Acknowledgements. We want to thank Dr. Silke Bernhard and her staff for the outstanding organization of this conference. In particular, we thank Julia Lupp for editorial assistance with the production of this book. Thanks also go to SCOPE for establishing the Project on Trace Gas Exchange, whose discussions stimulated us to propose and organize this Dahlem Workshop. Finally, we gratefully acknowledge the sponsors of the meeting, the Senate of Berlin, the Stifterverband für die Deutsche Wissenschaft, and the Stiftungsfonds Unilever, for their generosity.

Exchange of Trace Gases between Terrestrial Ecosystems and the Atmosphere
eds. M.O. Andreae and D.S. Schimel, pp. 7–21
John Wiley & Sons Ltd

Microbiological Basis of NO and N_2O Production and Consumption in Soil

M.K. Firestone and E.A. Davidson

Department of Plant and Soil Biology
University of California
Berkeley, CA 94720, U.S.A.

Abstract. Regulation of trace N-gas production via nitrification and denitrification occurs at two levels: (*a*) control of the rates of these processes and (*b*) control of the relative proportions of end products. At the cellular level nitrification rates are controlled primarily by O_2 and NH_4^+ availability. Similarly, denitrification is affected primarily by O_2, NO_3^-, and organic-C availability. The availability of each of these cellular controllers is affected by numerous physical, chemical, and biological properties of the ecosystem, many of which have been characterized for a number of ecosystems. In contrast, the relationship between ecosystem properties and factors affecting relative proportions of end products is less well understood. Production of N_2O by nitrifying bacteria results from reduction of NO_2 when O_2 is limiting, but the mechanism and factors affecting NO production during nitrification are not clear. Production of N_2O via denitrification is affected by relative availabilities of electron donors (organic-C) and electron acceptors (N-oxides). Any factor that slows the overall rate of denitrification may also cause N_2O to accumulate as a major end product. Production of NO via denitrification is more difficult to assess because control of cellular production and consumption is poorly understood. When NO diffusion is restricted by soil moisture, consumption by biological or abiological processes may be a dominant fate of this N-gas. Interaction of biological NO_2^- production and chemical NO_2^- decomposition (particularly in soil microsites) may also be an important source of NO.

INTRODUCTION

Microbial processes are important sources and sinks for N_2O and NO in the biosphere. Many groups of microorganisms are capable of producing N_2O and NO. In fact, it is likely that most biological processes involving the oxidation or reduction of N through the +1 or +2 state can produce trace amounts of these two relatively stable N-gases. However, the bacterial processes of denitrification and nitrification appear to be the dominant sources of N_2O in most natural systems. Only denitrification is recognized as a significant biological consumptive fate for N_2O and NO. While the origins of NO are not yet as clear, the involvement of nitrification, denitrification, and chemical decomposition of HNO_2 is well established. Considerable disagreement exists in the literature, however, as to the relative importance of nitrification, denitrification, biotic NO consumption, and abiotic production reactions in NO evolution. The following overview will address selected aspects of current knowledge and thought about N_2O and NO production and consumption in soil by nitrification, denitrification, other microbial processes, and chemical decomposition of NO_2^-.

A CONCEPTUAL MODEL OF TRACE N-GAS PRODUCTION BY NITRIFICATION AND DENITRIFICATION

This discussion will emphasize the production and consumption of N_2O and NO by microbial nitrification and denitrification. The environmental parameters which control the evolution of the trace N-gases by these two processes can be described by a "hole-in-the-pipe" model shown in Fig. 1. This simplistic model shows two levels of regulation for trace N-gas production by microbiological processes: (*a*) factors which control the rate of the overall process dictate the movement of N through the "process pipe"; and (*b*) factors which control the partitioning of the reacting N species to NO, N_2O, or a more oxidized/reduced product control the size of the holes in the pipe through which the N-gases "leak." The rate at which N moves through the process pipe determines the importance of the leaks—i.e., if denitrification rates are low, then trace N-gas production will be commensurately low regardless of the relative proportions of end products. When denitrification rates are high, factors that control the size of the "leaks" become critical for regulating N trace gas production. The first set of control parameters, those limiting the process rates, relate directly to control of the nitrogen cycle. That is, understanding the control of microbial N processing will provide a good starting point for predicting the conditions under which significant N-loss will occur. However, the factors controlling the portion of N being lost as NO and N_2O add another level of complication that must be understood to predict N trace gas production.

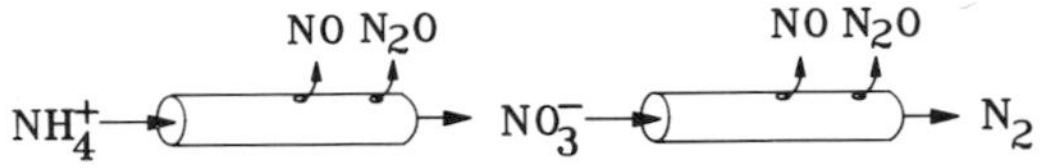

Fig. 1—A conceptual model of the two levels of regulation of N trace gas production via nitrification and denitrification: (*a*) flux of N through the process "pipes" and (*b*) holes in the pipes through which trace N-gases "leak."

NITRIFICATION

Process Regulation

Ammonium oxidation rates by chemoautotrophic nitrifying bacteria are affected by numerous factors, including NH_4^+, NO_2^-, PO_4^{3-}, O_2, acidity, temperature, water potential, and possible allelopathic compounds (Haynes 1986). Robertson (1989) has presented a conceptual model of environmental regulation of nitrification that terms the cellular controllers "proximal" and the environmental controllers "distal" (Fig. 2; Robertson 1989). In most soils, availability of NH_4^+ is the most important proximal factor controlling autotrophic nitrification rates. Oxygen is obligatory for the ATP-yielding process of NH_4^+ oxidation and, hence, nitrification rates generally decline as O_2 availability declines. Inhibition of nitrification by allelopathic inhibitors has been reported in some studies, but their effectiveness has not been verified (Bremner and McCarty 1988).

In the schematic diagram of Fig. 2, water is shown to control rates of nitrification by controlling the diffusional supply of NH_4^+ and O_2. Contemporary theories concerning water control of microbial processes in soil emphasize the importance of the diffusional impedance of substrate availability that occurs in thin water films of dry soil as a primary controller of microbial activity and hence process rates. Temperature is included in this diagram as a regulating factor throughout the continuum of distal controllers. The well documented temperature responses of soil microbial processes indicate that temperature should also be included as a proximal controller; however, the lack of response to temperature changes observed in some environments (Johansson and Sanhueza 1988) suggests that it be included at the bottom of the hierarchy.

Availability of NH_4^+ to nitrifying bacteria is controlled by several "distal" factors, including mineralization and immobilization rates, plant uptake, cation exchange, and diffusion (Fig. 2). Although nitrifying bacteria are often considered poor competitors for NH_4^+ relative to heterotrophic microorganisms and plants, microsite heterogeneity of NH_4^+ concentration, mineralization rates, and distribution of organisms and roots can account for significant nitrification rates in undisturbed grassland soils during seasons of active plant uptake and microbial immobilization (Jackson et al. 1989).

Nitrous Oxide Production via Nitrification

Nitrous oxide has been shown to be a product of a reductive process in which ammonium oxidizing bacteria use NO_2^- as an electron acceptor when oxygen is limiting (Poth and Focht 1985). This mechanism is consistent with observations that the ratio of N_2O/NO_3^- produced increases with decreasing availability of oxygen partial pressure (Goreau et al. 1980). The reduction

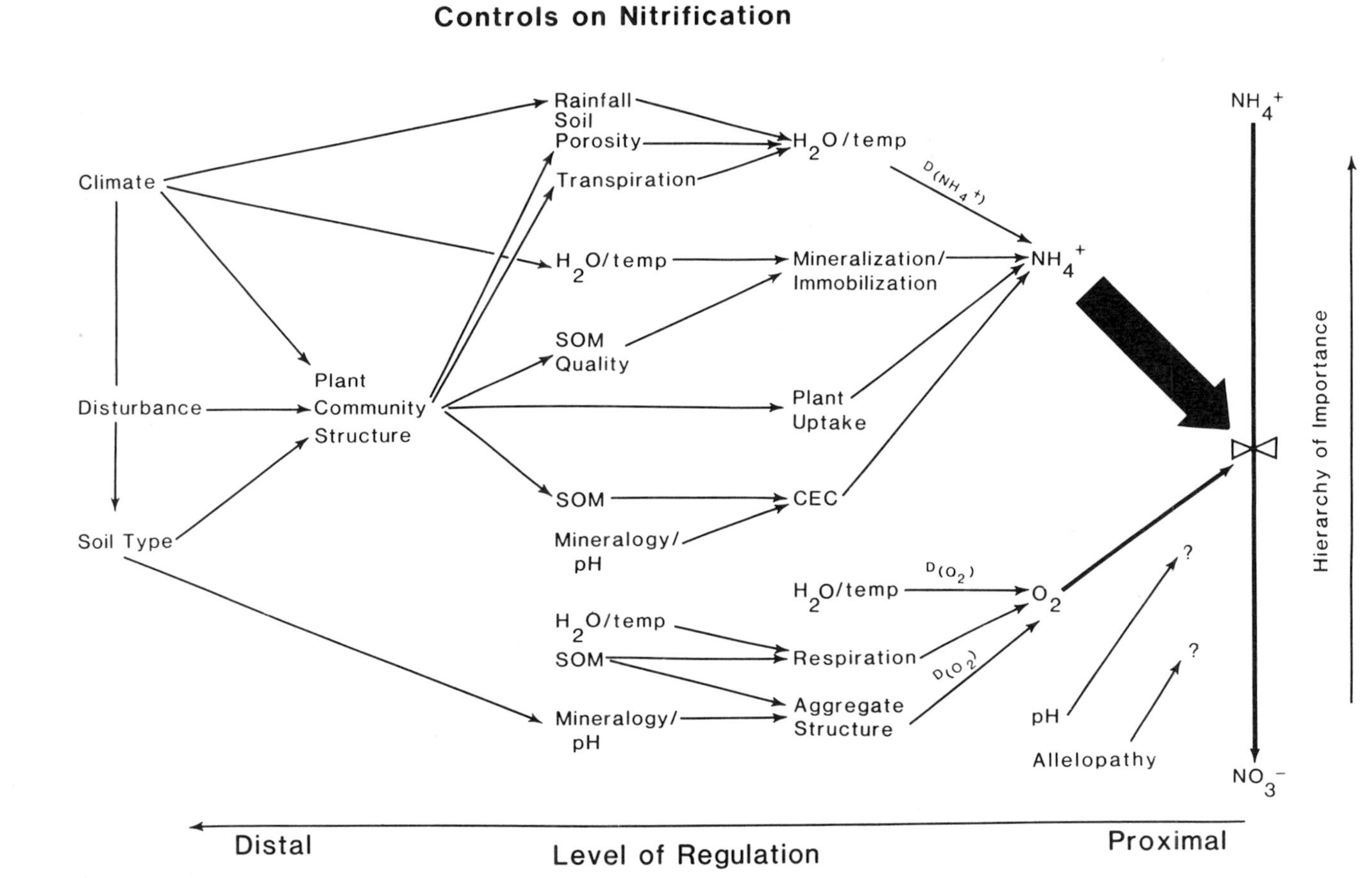

Fig. 2—A schematic diagram of the major factors regulating nitrification in soil (example given for rain forest soils). From Robertson (1989).

of NO_2^- to N_2O by nitrifiers also avoids accumulation of potentially toxic NO_2^-. The ratio of N_2O/NO_3^- produced increases with increasing acidity; the mechanism for this effect is unknown (Martikainen 1985). Although decreased O_2 and increased acidity increase the relative proportion of N_2O as an end product (the size of the hole in the pipe), these factors also generally retard rates of NH_4^+ oxidation (movement of N through the pipe) and, hence, the net effect on N_2O production can be difficult to predict. The ratio of N_2O/NO_3^- produced via nitrification has been reported as high as 20% (Martikainen 1985), but such ratios are usually below 1%. High rates of N_2O production are more commonly associated with denitrification rather than nitrification. Nevertheless, denitrification rates are spatially and temporally highly variable across ecosystem types, whereas nitrification is a relatively constant process in many ecosystems. Hence, a small N_2O/NO_3^- production ratio may be globally significant.

Nitric Oxide Production via Nitrification

Production of NO by nitrifying bacteria is poorly understood. The intermediates of +1 and +2 oxidation states in chemoautotrophic nitrification are not known with certainty (Hooper 1984; Fig. 3). Nitric oxide could be produced via oxidation of NH_2OH or reduction of NO_2^-. Recent studies of ammonium oxidizing bacteria (Conrad, pers. comm.) and aerobic soil (Tortoso and Hutchinson 1988) provide evidence for NO production via reduction of NO_2^-. Furthermore, the soil studies indicate that rates of NO production via nitrification are far greater than rates of N_2O production via nitrification. Conflicting evidence in the literature suggests that NO production via nitrification can be either inhibited (Lipschultz et al. 1981) or unaffected (Anderson and Levine 1986) by O_2 partial pressures.

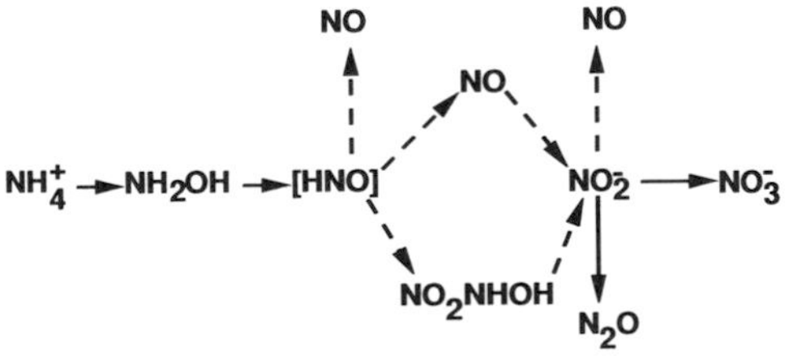

Fig. 3—Pathway of chemoautotrophic nitrification. Broken lines indicate unconfirmed pathways.

DENITRIFICATION

Process Regulation

In general terms, denitrification can be defined as a group of processes during which nitrate or nitrite are reduced to the gaseous nitrogen species NO, N_2O, or N_2. By this definition, several abiotic reactions, respiratory NO_3^- reduction by microbes, and nonrespiratory N_2O production would all be termed denitrification. Inclusion of such a variety of processes under one term may be useful for global budgets but will cause confusion when attempting to understand and predict the occurrence of nitrogen trace gas fluxes. For the purposes of this overview, *denitrification* identifies a form of anaerobic respiration in bacteria during which nitrogen oxide reduction is coupled to electron transport phosphorylation.

Denitrification is the only biotic or abiotic process capable of producing and consuming N_2O and NO. Hence this bacterial process plays a central role in global N trace gas dynamics. For this discussion, we will use the following sequence for N-oxide reduction:

$$\begin{array}{ccccccccc} & & & & NO & & & & \\ & & & & \uparrow\downarrow & & & & \\ NO_3^- & \rightarrow & NO_2^- & \rightarrow & [X] & \rightarrow & N_2O & \rightarrow & N_2. \end{array}$$

While the sequence of nitrogen species produced during denitrification has been known for a number of years, the exact reductive sequence and enzymology involved in the reduction of NO_2^- to N_2O is the topic of current controversy (Weeg-Aerssens et al. 1988; Zafiriou et al. 1989). However, in all of the pathways proposed, N_2O is an intermediate and NO behaves as if it is an intermediate in the reductive sequence.

The capacity to denitrify is widely spread among a number of taxonomic and physiological groups of bacteria; however, only a few genera seem to be numerically dominant in soil, marine, freshwater, and sediment environments. *Pseudomonas* species capable of denitrification are found in the greatest numbers in all of the environments listed, with *Alcaligenes* species commonly comprising the second most numerous denitrifying population (Tiedje 1988). The commonality of the taxonomic and physiological characteristics of the numerically dominant populations in a range of natural habitats is important in that it suggests that unifying concepts for the control of trace N-gas production by denitrification can be extrapolated across the major global environments.

The general requirements for denitrification to occur are: (*a*) the presence of bacteria possessing the metabolic capacity; (*b*) the availability of suitable reductants such as organic carbon; (*c*) the restriction of O_2 availability; (*d*) the availability of N oxides, NO_3^-, NO_2^-, NO, or N_2O. It is probably safe to assume that if the last three requirements (C, N-oxides, and limited O_2) are met in an environment, then bacteria with the appropriate metabolic capability will occupy the denitrification niche. Populations of denitrifying bacteria may be absent from environments where one or more of these three factors is always severely limiting (Davidson et al. 1989). It is interesting, however, that denitrification activity is high in anoxic soils/sediments from which NO_3^- pools are absent (Tiedje 1988); wetland soils may provide an environment in which net N_2O consumption can occur.

The cellular controllers of denitrification (O_2, NO_3^-, and carbon) are relatively simple to enumerate and conceptually encompass. However, in natural systems such as soil, the environmental factors which regulate O_2, NO_3^-, and carbon availability to microbial cells are numerous, interactive, and can impact more than one cellular controller. For example, in nonflooded terrestrial systems, plant roots can (*a*) create anaerobic zones by respiratory O_2 consumption and by supplying C for microbial respiration, (*b*) supply carbon as an electron donor for denitrification, (*c*) reduce NO_3^- availability by root uptake, and (*d*) increase rates of O_2 diffusion to the root zone by removing H_2O through evapotranspiration.

Robertson (1989) has presented a conceptual model of distal and proximal environmental regulation of denitrification (Fig. 4). This dissection of denitrification control into the component mechanisms assists in understanding and potentially modeling how a number of environmental factors interact to modulate rates of denitrification. It allows one to evaluate the significance of each potential environmental controller and imposes a hierarchy of controller importance for a specific environment. The diagram in Fig. 4 illustrates that O_2 most commonly limits denitrification. Soil water content, as controlled by precipitation and evapotranspiration, will be the dominant environmental controller of O_2 availability. Again, the effect of water on this microbial process is mechanistically expressed as it controls O_2 diffusion through air-filled pores and water films. Temperature effects are again included across all levels of distal regulation. While temperature can also control the cellular process rate, it should be noted that active denitrification has been observed in soil systems from 2–50°C (Keeney et al. 1979; Strauss and Firestone, unpubl. data).

The ordering of proximal and distal controllers varies among habitats. Tiedje (1988) proposes that the three cellular controllers of denitrification can be generally ranked in importance for habitat types based on gross characteristics of the habitats (Table 1). Oxygen availability is the dominant factor limiting denitrification in habitats (such as soil) exposed to atmosphere, while in dominantly anaerobic habitats (such as sediments), NO_3^- is the most important cellular controller. When anaerobic zones occur in nonfertilized soils, NO_3^- availability may control denitrification rates; however, in N-fertilized soils, C availability would limit denitrification.

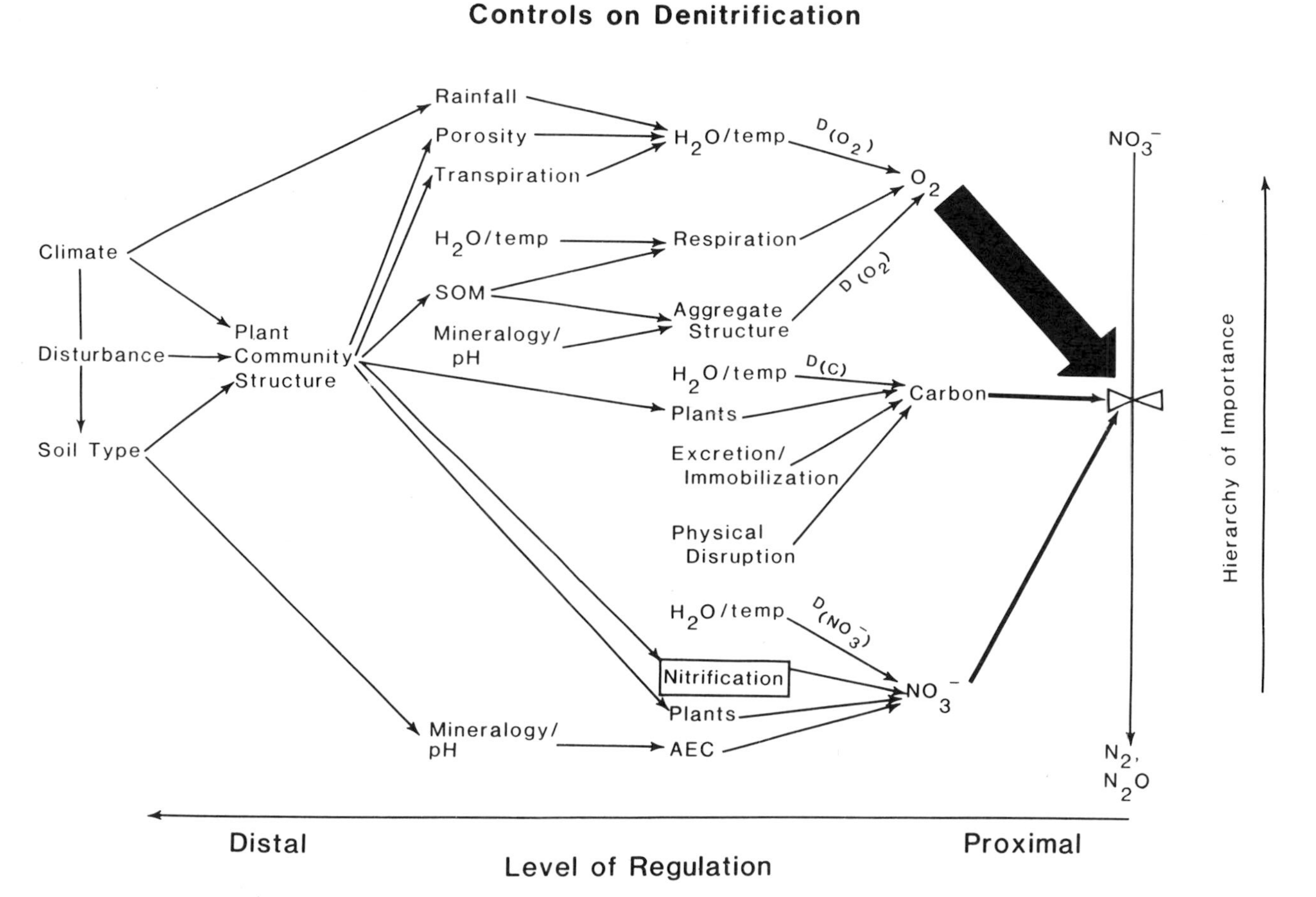

Fig. 4—A schematic diagram of the major factors regulating denitrification in soil (example given for rain forest soils). From Robertson (1989).

TABLE 1. Ranking of importance of the factors that limit denitrification in various habitats (from Tiedje 1988)[a].

Habitat	Oxygen	Regulator Nitrate	Carbon
Fertilized soil	1	3	2
Nonfertilized soil	1	2	3
Sediments	2	1	
Lake water column	1	2	
Ocean waters	1	3	2
Intestinal tract	2	1	
Anaerobic sludge		1	
Secondary waste treatment	1	2	

[a] A ranking of 1 indicates that this factor most commonly limits denitrification.

Understanding and analyzing the control of the process of denitrification is the first step in knowing when and where to look for trace N-gas fluxes, and in ultimately predicting magnitude, timing, and anthropogenic impact on these fluxes. As we become more sophisticated in our ability to predict the occurrence of denitrification, we can begin to superimpose another layer of controllers—those of trace N-gas production, i.e., those factors which control the production and consumption of NO, N_2O, or N_2.

Nitrous Oxide via Denitrification

It is probably valid to assume that all denitrifiers which are functionally significant in the natural environment are capable of the complete denitrification sequence. The reports of bacterial cultures which do not possess one or two steps in the reductive sequence generally reflect physiological changes which occur with retention on laboratory media. Why then do cultures of denitrifiers capable of reducing NO_3^- to N_2 sometimes produce large amounts of N_2O and sometimes produce none? Do denitrifiers ever produce quantities of NO, and if so, why?

Numerous factors have been reported to affect the proportion of N_2O produced relative to N_2 in denitrifying cells and soils (Table 2). Consideration of the relative availability of reductant vs. oxidant may help in conceptually clarifying the observations in Table 2. If the availability of oxidant (N-oxide) greatly exceeds the availability of reductant (most commonly organic carbon), then the oxidant may be incompletely utilized, i.e., N_2O will be produced. Hence recognition of which cellular regulator limits denitrification (NO_3^- or carbon) may be helpful in understanding the proportion of N occurring as N_2O in a given environment.

TABLE 2. Factors affecting the proportion of N_2O and N_2 produced during denitrification.

Factor	Will increase N_2O/N_2
$[NO_3^-]$ or $[NO_2^-]$	Increasing oxidant
$[O_2]$	Increasing O_2
Carbon	Decreasing C availability
pH	Decreasing pH
$[H_2S]$	Increasing sulfide
Temperature	Decreasing T
Enzyme status	Low N_2O reductase activity

Betlach and Tiedje (1981) presented a model to explain how a variety of factors may affect the accumulation of denitrification intermediates in culture solution. These workers applied a simple Michaelis-Menten model of enzyme kinetics to the reductive sequence. While it is intuitively obvious that the inhibition of any step in the reductive sequence could result in accumulation of the preceding intermediate, it is less apparent and more interesting that the model predicts accumulation of N_2O whenever a factor slows the rate of overall reduction. This interpretation provides a potential framework for understanding the effects of many environmental factors such as temperature, O_2 concentration, sulfide, acidity, etc., which have been observed to increase N_2O production in microbial culture.

Nitric Oxide Production via Denitrification

Nitric oxide functions as a freely exchangeable intermediate during denitrification (Firestone et al. 1979). Yet for many years very little, if any, NO was detected as a product of denitrifying cultures (Tiedje 1988). In part, this may reflect the fact that most microbiologists were not appropriately equipped to quantify NO; use of gas chromatographic systems to quantify NO is problematic at best. However, mass balance experiments confirmed the absence of NO as a major product of denitrifying cultures, indicating that the absence of NO as a major product of denitrification did not simply reflect an analytical deficiency. More recent work using chemiluminescence detectors to quantify NO production by several denitrifying cultures also indicated that this gas did not constitute a major denitrification product (Anderson and Levine 1986). Most of the work mentioned utilized dense bacterial cultures, the norm for batch culture microbiology. In the presence of a dense suspension of denitrifying cells, the rate of reutilization of any NO produced could have been very high. Recent work by Zafiriou et al. (1989) has demonstrated that in a very low density cell suspension, highly sparged to remove gaseous products, NO was the dominant product of denitrification. In a natural soil system, however, what density of denitrifying cells would be expected and how rapidly is NO removed from the zone of cell impact?

Work with soil columns incubated under anaerobic flow conditions indicates that NO can be a major product of denitrification in soil (Johansson and Galbally 1984). However, NO flux from the surfaces of soils in which denitrification should be occurring has generally been several orders of magnitude less than NO production by anaerobic flow columns. In field soils, anaerobiosis generally only occurs under conditions of high water content. Under conditions of high water content, only a small percentage of the NO produced may escape the soil surface (Fig. 5). If this line of reasoning is correct, then a significant amount of NO may be produced during denitrification in field soils and ultimately redistributed among soil microsites or within the soil profile.

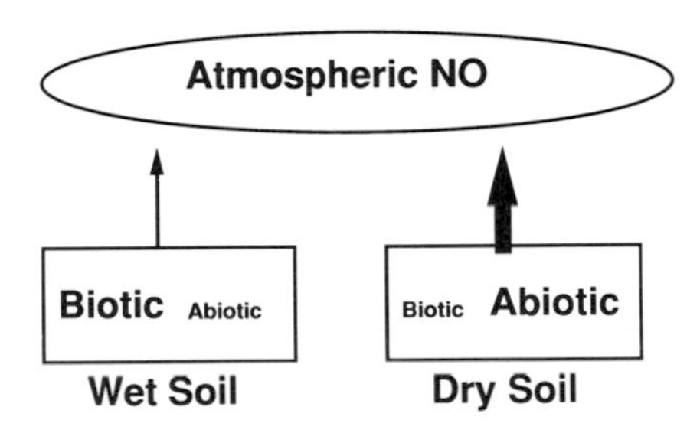

Fig. 5—A conceptual model of the relative importance of NO fluxes and of biotic and abiotic sources of NO production in wet and dry soil.

Abrupt changes in soil moisture are probably extremely important in affecting trace N-gas fluxes. Numerous studies have reported observations of soil NO fluxes with wet-up of dry soil (Johansson and Sanhueza 1988). Wet-up of soil causes major changes in all of the aforementioned proximal controllers of nitrification and denitrification. Significant increases in C and N availability as well as the physical impact of water on O_2 diffusion combine to produce a transient period when trace N-gas production by both denitrifiers and nitrifiers may be high. While the microbial capacity may limit process rates during these wet-up periods, denitrifying bacteria have been found to survive prolonged drought and to become metabolically active rapidly following wetting of soil (Smith and Parsons 1985). We have recently observed production of N_2O by denitrifiers within one hour of wetting very dry (<-9 MPa) soil (Rudaz, Davidson, and Firestone, unpubl. data).

N_2O and NO production via denitrification in the soil environments. The biological control of trace N-gas production in any environment must

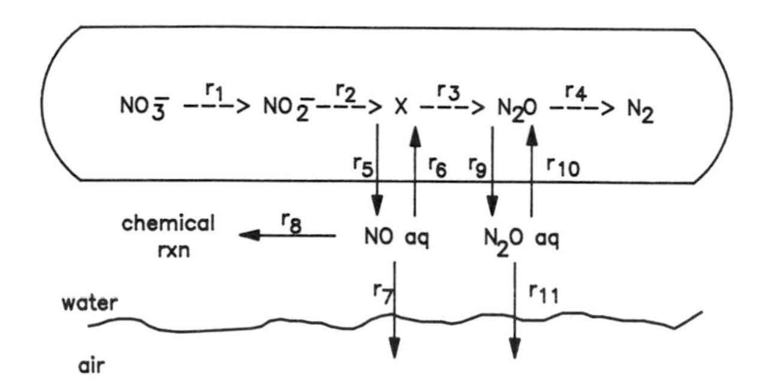

Fig. 6—A schematic diagram of production and consumption reactions of trace N-gases.

integrate control of production and consumption. This simple fact is diagrammed in Fig. 6. The preceding discussion addressed those biological factors which control production of N_2O_{aq} and NO_{aq}. However, whether biological reutilization of NO and N_2O will occur is dependent on the rate of consumption relative to the rate of escape of the gas from the biologically active site. This fact has been demonstrated by Zafiriou et al. (1989) for NO in a simple system; the amount of NO produced from a culture of denitrifying bacteria in solution was a function of the rates of production and consumption of NO and the rate of NO leaving the aqueous phase (r_5, r_6, and r_7 in Fig. 6).

The yield of NO and N_2O from an environment is a direct function of the rates r_5 through r_{11}:

- r_5 — rate of biological NO production. Control is described above.
- r_6 — rate of biological NO consumption. Control may be dependent on factors discussed above as well as cell density.
- r_7 — rate of NO movement out of solution. Controlled by diffusion rate and distance to air–water interface.
- r_8 — rate of NO chemical consumption. Controlled by NO concentration and reaction rate with O_2, organic groups, metals, etc. Work by Johansson and Galbally (1984) indicates that NO is relatively stable when present at low concentration in sterile soil.
- r_9 — rate of N_2O production. Control discussed above.
- r_{10} — rate of N_2O consumption. Control dependent on factors discussed above as well as cell density.
- r_{11} — rate of N_2O movement out of solution. Controlled by diffusion rate and distance to air–water interface.

The cell shown in Fig. 6 represents all microbial producers and consumers. The only microbial process known to consume significant amounts of NO and N_2O is denitrification. The possibility of NO consumption by assimilatory reduction to NH_4^+ is intriguing but, as of yet, totally untested. This simple model leads to several predictions:

1. Differences between NO and N_2O with respect to exchange between aqueous and gaseous phases and with respect to potential consumptive sinks could yield very different patterns of fluxes in response to environmental controllers.
2. While increasing water content might increase r_5 and r_9, it would decrease r_7 and r_{11} thus increasing the opportunity for biological consumption or chemical reaction. The net effects on trace N-gas flux from soils may appear complex without requisite understanding of biological production and consumption.
3. Concepts and models addressing trace N-gas flux from soils must integrate control of biological production/consumption with physical control of water–air transfer rates and with models of gaseous diffusion in soil.

BIOLOGICAL SOURCES OF N_2O OTHER THAN NITRIFICATION AND DENITRIFICATION

Numerous organisms are capable of producing N_2O, but the mechanisms of production by microorganisms other than nitrifying and denitrifying bacteria are unclear. The topic of N_2O production by other organisms has been recently reviewed by Tiedje (1988). Nitrous oxide production by nonnitrifying and nondenitrifying mechanisms has been observed in acid forest soils, and N_2O production has been observed in pure cultures of some fungi; hence, a fungal source appears plausible. The importance of this source of N_2O remains unknown. In one study, most of the N_2O produced in soil cores from an old-growth hardwood forest appears to have been from mechanisms other than nitrification or denitrification (Robertson and Tiedje 1987), but a study of the organic horizon (pH 4.0 to 4.7) of a pine forest indicated that autotrophic nitrification was the primary source of N_2O (Martikainen 1985). Although nitrification and denitrification appear to be the major processes producing N_2O in soils, further work on elucidating other biological sources of N_2O may prove useful.

CHEMODENITRIFICATION

There are a number of chemical reactions which may be termed chemodenitrification. In soil systems, the most relevant and significant reactions involve the spontaneous reactions of HNO_2 to form NO. Chemical decomposition of HNO_2 to form NO and HNO_3 is well documented to occur in acid soils, and high levels of organic material may also accelerate HNO_2 reaction to form NO (Nelson 1982). The most important factors controlling the significance of these mechanisms of NO generation are accumulation of NO_2^-, pH values, and the amount of organic matter present. Hence, in acid soils, particularly acid soils containing high organic matter, this mechanism of NO generation must be accounted. It appears that in neutral to alkaline soils this mechanism of N-gas production should be insignificant. However, analyses of bulk soil for NO_2^- concentrations and acidity may not be indicative of microsite characteristics. Dynamic soil physical processes, such as drying and freezing, may concentrate NO_2^- in thin water films, thus accelerating chemical decomposition of NO_2^-. Furthermore, biological and abiological processes may be linked at the microsite level, e.g., NO_2^- and H^+ might accumulate near a colony of NH_4^+ oxidizers, thus accelerating HNO_2 decomposition to NO. While the importance of chemical decomposition of NO_2^- in dominantly neutral soils has not been demonstrated, the role of microsite characteristics in abiotic and biotic/abiotic reactions may prove important in understanding mechanisms of NO generation in soil.

In summary, relating control of NO and N_2O production in the laboratory to fluxes in the field is made difficult by the complex interactions of productive and consumptive processes. Not only can NO be produced via nitrification, denitrification, and abiotic processes, it can also be consumed via further reduction by denitrifiers or by abiotic oxidation processes. The relative importance of these processes may vary with soil moisture.

Acknowledgements. Support for our work has been provided by the National Science Foundation Ecosystems Studies Program and by a McIntyre Stennis Project of the University of California Experiment Station. We thank G.P. Robertson for permission to use Figs. 2 and 5.

REFERENCES

Anderson, E.C., and J.S. Levine. 1986. Relative rates of nitric oxide and nitrous oxide production by nitrifiers, denitrifiers, and nitrate respirers. *Appl. Env. Microbiol.* 51:938–945.

Betlach, M.R., and J.M. Tiedje. 1981. Kinetic explanation for accumulation of nitrite, nitric oxide, and nitrous oxide during bacterial denitrification. *Appl. Env. Microbiol.* 42:1074–1084.

Bremner, J.M., and G.W. McCarty. 1988. Effects of terpenoids on nitrification in soil. *Soil Sci. Soc. Am. J.* 52:1630–1633.

Davidson, E.A., D.D. Myrold, and P.M. Groffman. 1989. Denitrification in forest ecosystems. In: Sustained Productivity of Forest Lands, ed. S.P. Gessel, D.S. Lacate, and R.F. Powers. Proc. Seventh North American Forest Soils Conference, Vancouver, BC, in press.

Firestone, M.K., R.B. Firestone, and J.M. Tiedje. 1979. Nitric oxide as an intermediate in denitrification: evidence from nitrogen-13 isotope exchange. *Biochem. Biophys. Res. Comm.* 91:10–16.

Goreau, T.J., W.A. Kaplan, S.C. Wofsy, M.B. McElroy, F.W. Valios, and S.W. Watson. 1980. Production of NO_3^- and N_2O by nitrifying bacteria at reduced concentrations of oxygen. *Appl. Env. Microbiol.* 40:526–532.
Haynes, R.J. 1986. Nitrification. In: Mineral Nitrogen in the Plant–soil System, ed. R.J. Haynes, pp. 127–165. New York: Academic.
Hooper, A.B. 1984. Ammonium oxidation and energy transduction in the nitrifying bacteria. In: Microbial Chemoautotrophy, ed. W.R. Strohl and O.H. Tuovinen, pp. 133–167. Columbus, OH: Ohio St. Univ. Press.
Jackson, L.E., J.P. Schimel, and M.K. Firestone. 1989. Short-term partitioning of ammonium and nitrate between plants and microbes in an annual grassland. *Soil Biol. Biochem.* 21:409–415.
Johansson, C., and I.E. Galbally. 1984. Production of nitric oxide in loam under aerobic and anaerobic conditions. *Appl. Env. Microbiol.* 47:1284–1289.
Johansson, C., and E. Sanhueza. 1988. Emission of NO from savanna soils during rainy season. *J. Geophys. Res.* 93:14193–14198.
Keeney, D.R., I.R. Fillery, and G.P. Marx. 1979. Effect of temperature on the gaseous nitrogen products of denitrification in a silt loam soil. *Soil Sci. Soc. Am. J.* 43:1124–1128.
Lipschultz, F., O.C. Zafiriou, S.C. Wofsy, M.B. McElroy, F.W. Valios, and S.W. Watson. 1981. Production of NO and N_2O by soil nitrifying bacteria. *Nature* 294:641–643.
Martikainen, P.J. 1985. Nitrous oxide emission associated with autotrophic ammonium oxidation in acid coniferous forest soil. *Appl. Env. Microbiol.* 50:1519–1525.
Nelson, D.W. 1982. Gaseous losses of nitrogen other than through denitrification. In: Nitrogen in Agricultural Soils, ed. F.J. Stevenson. *Agronomy* 22:327–364.
Poth, M., and D.D. Focht. 1985. ^{15}N kinetic analysis of N_2O production by *Nitrosomonas europaea*: an examination of nitrifier denitrification. *Appl. Env. Microbiol.* 49:1134–1141.
Robertson, G.P. 1989. Nitrification and denitrification in humid tropical ecosystems: potential controls on nitrogen retention. In: Mineral Nutrients in Tropical Forest and Savanna Ecosystems, ed. J. Procter. Oxford: Blackwell Scientific, in press.
Robertson, G.P., and J.M. Tiedje. 1987. Nitrous oxide sources in aerobic soils: nitrification, denitrification and other biological processes. *Soil Biol. Biochem.* 19:187–193.
Smith, M.S., and L.L. Parsons. 1985. Persistence of denitrifying enzyme activity in dried soils. *Appl. Env. Microbiol.* 49:316–320.
Tiedje, J.M. 1988. Ecology of denitrification and dissimilatory nitrate reduction to ammonium. In: Biology of Anaerobic Microorganisms, ed. J.B. Zehnder, pp. 179–244. New York: Wiley.
Tortoso, A.C., and G.L. Hutchinson. 1988. Abstract. *Am. Soc. Agron.*, p. 226.
Weeg-Aerssens, E., J.M. Tiedje, and B.A. Averill. 1988. Evidence from isotope labeling studies for a sequential mechanism for dissimilatory nitrite reduction. *J. Am. Chem. Soc.* 110:6851–6856.
Zafiriou, O.C., Q.S. Hanley, and G. Snyder. 1989. Nitric oxide and nitrous oxide production and cycling during dissimilatory nitrite reduction by *Pseudomonas perfectomarina*. *J. Biol. Chem.* 10:5694–5699.

Exchange of Trace Gases between Terrestrial Ecosystems and the Atmosphere
eds. M.O. Andreae and D.S. Schimel, pp. 23–37
John Wiley & Sons Ltd

Factors Controlling NO_x Emissions from Soils

I.E. Galbally

CSIRO Division of Atmospheric Research
Private Bag 1
Mordialloc, Victoria 3195, Australia

Abstract. Odd nitrogen oxides, NO_x, are produced in soils by the biological processes of nitrification and denitrification and by several chemical reactions involving nitrite. Current observations show that the dominant emission to the atmosphere is nitric oxide and there are lesser emissions of other nitrogen oxides (e.g., nitrogen dioxide). The two variables that frequently influence NO_x emission are soil temperature and soil moisture content. The wider scale factors that influence NO_x emission to the atmosphere include climate (through temperature and rainfall), plant growth and decay, biomass burning, and fertilization.

There have been no studies to trace the production of NO_x from its microbiological or chemical origin within the soil to its emission from the soil into the atmosphere. Thus, while some processes have been observed in the field, the observations are incomplete and understanding inadequate so that even empirical modeling of variations of NO_x emissions within or between ecosystems is not yet possible.

INTRODUCTION

This paper reviews the physical, chemical, and biological processes that control odd nitrogen oxide emission from soils. The odd nitrogen oxides, NO_x, are nitrogen oxides that do not have two nitrogen atoms bonded together as occurs in nitrous oxide. This review includes biological and chemical mechanisms for nitrogen oxide production and uptake in soils and the broader environmental factors that modulate both this production and the escape of these nitrogen oxides from the soil surface. Processes specifically excluded include the internal biochemistry of the microorganisms and also the micrometeorological processes that move NO_x from the soil/plant surface to the free atmosphere.

The nitrogen oxides considered include nitric oxide (NO), nitrogen dioxide (NO_2), nitrous acid (HNO_2), and methyl nitrite (CH_3ONO). Emissions from soil of NO and CH_3ONO have been identified (Galbally and Roy 1978; Magalhaes et al. 1985). Unidentified soil emissions of NO_x have been measured (Slemr and Seiler 1984); however, neither NO_2 nor HNO_2 have been positively identified as participant gases. A mechanism exists for thc HNO_2 generation in soils, and one has been proposed for NO_2 (Slemr and Seiler 1984) but this mechanism may not be appropriate to the composition of soils. In nearly all field studies NO is the major (approximately 90%) NO_x gas emitted.

Nitrogen can adopt a range of valency states varying from ammonia to nitrate, and several biological processes lead to oxidative and reductive transformation of fixed nitrogen. The states and species of relevance here are: −3, NH_3; −1, NH_2OH; 0, N_2; +1, N_2O; +2, NO; +3, HNO_2, NO_2^-, CH_3ONO; +4, NO_2; +5, NO_3^-. Obviously NO can be an intermediary in some of the biological processes and, if it is poorly bound to the biological material, it can leak from the microorganism to the soil atmosphere: this constitutes the biological production of NO in soils. A difficulty in quantifying these processes arises because of the abundance of types of nitrogen-transforming microorganisms and the fact that some of them do not necessarily use the same biochemical reactions to achieve a given end product, hence the role of intermediary compounds such as NO may vary from species to species.

MICROBIAL PRODUCTION AND UPTAKE OF NO_x IN SOILS AND PLANTS

The four biological processes of nitrogen transformations that may be relevant here are:

1. denitrification: anaerobic reduction of NO_3^- and NO_2^- to N_2,
2. nitrification: aerobic oxidation of NH_4^+ to NO_2^- and NO_3^-,
3. anaerobic reduction of NO_2^- to NH_4^+, and
4. uptake by plants.

There was another candidate mechanism in the dissimilatory reduction of NO_3^- to NH_4^+ in anaerobic conditions; however, available studies on two microbial species indicate that NO is not produced during this fermentative process (Bleakley and Tiedje 1982; Anderson and Levine 1986) so it is not discussed here.

Denitrification

Denitrification is an anaerobic process where, in the absence of oxygen, bacteria utilize nitrate and nitrite for growth and reduce them to nitrous oxide and molecular nitrogen. The accepted sequence for denitrification is

$$NO_3^- \rightarrow NO_2^- \rightarrow NO \rightarrow N_2O \rightarrow N_2$$

(Payne 1981). Denitrifying bacteria are widespread in nature and are generally present in all soil and freshwater samples analyzed. Denitrification is repressed by the presence of oxygen, O_2. Hence, denitrification is associated with either high soil water contents, which impedes the diffusion of O_2 into the soil, or large respiration and oxygen consumption rates in the soil (Parkin 1987). Most denitrifying bacteria are heterotrophs and thus require the presence of readily oxidizable organic carbon. The rate of denitrification is temperature dependent, roughly doubling for every 10°C rise (i.e., $Q_{10} = 2$). The maximum rate occurs at around 60–75°C and measurable denitrification occurs even in the range 0–5°C. The maximum rate of NO production occurs around 40°C (McKenny et al. 1982).

Studies of specific denitrifying bacteria show that they both produce and consume NO depending on the atmospheric concentration present in contact with the microorganisms and the degree to which the other available fixed nitrogen present has already been reduced to N_2. During part of the denitrification sequence these microorganisms maintain a near constant NO concentration in the atmosphere surrounding them (Betlach and Tiedje 1981). In one study the microorganisms preferentially reduced NO in the presence of NO_3^- (Nommik et al. 1984). This dual behavior of production and uptake of NO was observed also in studies on soil cores in anaerobic conditions (McKenny et al. 1982; Johansson and Galbally 1984).

The rates of NO production in anaerobic conditions for three denitrifying species have been determined to range over 1×10^{-13} to 5×10^{-13} mmol cell^{-1} day^{-1} (Anderson and Levine 1986). Denitrifying populations in agricultural soils can lie in the range 10^4 to 10^8 organisms per gram of soil. Information such as this can be combined to estimate the NO production rate per gram of soil, a rate that can be independently measured. The data above indicate NO production rates between 0.00001 and 0.5 ng N g^{-1} soil min^{-1}. Observed rates on anaerobic soil cores (McKenny et al. 1984; Johansson and Galbally 1984) range over 1 to 4 ng g^{-1} soil min^{-1} which is, at closest, only a factor of 2 larger production than observed with the pure cultures. Simultaneous measurements on the same soil of numbers of denitrifying bacteria per gram of soil and bulk NO production per gram of soil would establish whether the correct species are being studied to elucidate the microbial chemistry of NO production.

Current knowledge indicates that the NO flux from the soil will depend both on physical transfer processes from the site of the denitrification to the atmosphere and on the relative rates of production and consumption of NO. It appears that denitrification will flourish best in those warm moist soils which contain both NO_3^- and available organic carbon. In suitable conditions essentially all the NO_3^- in soils can be consumed, producing NO, N_2O, and N_2. Field experiments show that denitrification is in fact an episodic event and that N_2 is the main product. Presumably this occurs because most of the NO produced during denitrification is reconsumed by soil bacteria and little of the NO produced leaks out to the atmosphere.

Nitrification

Nitrification involves the biological oxidation of fixed nitrogen. The most common form of this involves the oxidation of ammonium (NH_4^+) to NO_3^- with NO_2^- as an intermediate (Bremner and Blackmer 1981). Separate bacteria oxidize NH_4^+ to NO_2^- and NO_2^- to NO_3^-. These bacteria are generally chemoautotrophic, requiring only CO_2, H_2O, O_2, and either NH_4^+ or NO_2^- for growth. Nitrification will occur in all well-aerated soils with NH_4^+ or urea ($(NH_2)_2CO$) present, except perhaps where the soil is very acidic with pH 4 or less (Rosswall 1982).

There has been a detailed study of the nitrogen transformations carried out by the autotrophic nitrifying bacterium, *Nitrosomonas*, which converts NH_4^+ to NO_2^- and which is also able to reduce NO_2^- (Ritchie and Nicholas 1972). These authors proposed NO as an intermediate prior to NO_2^- production. There has been some debate about whether or not during nitrification NO is an obligatory intermediate (presumably enzyme bound) because so far there has been no demonstration of an enzyme-catalyzed conversion of NO to NO_2^- (Hooper and Terry 1979; Hooper 1978).

Two other laboratory studies have shown NO production from the soil nitrifying bacteria *Nitrosomonas* in aqueous solution (Lipschultz et al. 1981; Anderson and Levine 1986). Nitrite was produced at the same time as NO, but sterile experiments showed that the NO was of biological rather than chemical origin. In the first study the molar fraction of NO produced compared with NO_2^- production varied from 3.8% at oxygen concentrations of 0.5% to 0.1% at the normal atmospheric level of 21% O_2. Thus only a small fraction of the nitrogen oxidized during nitrification is lost as NO. Johansson and Galbally (1984) in soil core studies measured NO production from soil which corresponded to about 2% of the rate of nitrification in aerobic conditions. This agrees well with the studies on individual species of nitrifiers.

NO_x loss rates from fertilized soils have been measured by Johansson and Granat (1984) and Slemr and Seiler (1984). At bare soil sites Slemr and

Seiler (1984) found the fractional losses of fertilizer as NO_x, typically during two weeks after the addition, to be 0.1 to 0.3% for nitrate, 1.8 to 2.7% for ammonium, and 5.4% for urea. These are consistent with the rates of NO production by nitrification.

The study of Anderson and Levine (1986) showed that nitrifiers under near-anaerobic conditions produce NO in proportion to the amount of NO_2^- in the surrounding solution. This is prima facie evidence of the production of NO by nitrifiers during reducing conditions where NO_2^- is converted to NH_4^+ as suggested by Ritchie and Nicholas (1972), Hooper (1978), and Hooper and Terry (1979). The importance of this mechanism for field production of NO is currently unknown.

Plant Uptake

The other biological process of importance in the exchange of NO and NO_2 at the Earth's surface involves the metabolism of plants (Farquhar et al. 1983). In plants nitrate is reduced to nitrite in the cytoplasm after which the nitrite is reduced to NH_3 in the chloroplasts (Farquhar et al. 1983). The effectiveness of these biological processes in absorbing NO_2 is shown in the ^{15}N tracer study of Rogers et al. (1979). In plants harvested immediately after a 3-hour exposure to $^{15}NO_2$ and then freeze-dried, it was found upon analysis that only 3% of the ^{15}N was present as $^{15}NO_3^-$, 35% as soluble reduced ^{15}N (amino acids, amides, lipids, and chlorophyll), and 63% as insoluble ^{15}N compounds (primarily proteins and nucleic acids). Thus the metabolic processes proceed very rapidly removing excess NO_2, NO_3^-, and NO_2^- from within the plant tissue in a matter of minutes.

Nitric oxide is taken up by plants much more slowly than NO_2 (Hill 1971). Traditionally this lower uptake has been associated with the lower solubility of NO versus NO_2 (Hill 1971), but in fact the slower rate of uptake may be due to the necessity of converting NO into some readily metabolized form (e.g., NO_2^-) before it follows the pathway of NO_2. There are inorganic chemical reactions which may cause NO uptake to be accelerated in the presence of NO_2 or HNO_3 (see next section). This has not been examined in the field.

CHEMICAL PRODUCTION OF NO_x IN SOILS

There are four known chemical processes that regulate the exchange of NO and NO_x at the Earth's surface:

1. the vaporization of HNO_2 from soil aqueous solution,
2. the decomposition of HNO_2 to NO and NO_2,

3. the reaction of HNO_2 with soil organic matter yielding CH_3ONO, and
4. the photolysis in aqueous solution of NO_2^- yielding NO.

All of these processes require the presence of nitrite in the soil. As nitrite is both produced and consumed by soil microorganisms, one must note that any overview of the role of these processes in the soil must take into account the biological processes that modulate the nitrite concentration in the soil.

Volatilization of HNO_2

One chemical process which drives NO_x exchange involves the direct volatilization of dissolved HNO_2 from the aqueous phase to the gas phase. Mixed phase equilibra for HNO_2 have recently been reviewed by Schwartz and White (1983) and we have the equilibrium

$$HNO_2\ (g) = H^+ + NO_2^-$$
$$K = 2.5 \times 10^{-2}\ M^2\ atm^{-1} \quad \Delta H = -7.0\ kcal/mol$$

This equilibrium can be used along with soil composition analyses to calculate the partial pressure of HNO_2 in the soil atmosphere. This was done using some of the soil analyses that accompany published NO flux measurements. In Table 1 some soil analyses from Johansson (1984) and Galbally et al. (1985) are presented. Two of the sites chosen represent those that will yield the largest partial pressures of HNO_2, and one is a more typical site which has conditions that could easily occur in any mildly acidic soil with an active nitrogen cycle. The partial pressures of HNO_2 are calculated for equivalent situations, that is above aqueous solutions that have the pH of the soil and the NO_2^- and NO_3^- concentrations equivalent to those occurring if all NO_2^- and NO_3^- in the soils were partitioned into the soil water phase.

TABLE 1. The composition of soil cores at sites of NO emission measurements and calculated partial pressure of HNO_2 above an aqueous solution (at 298 K) with composition similar to that observed in soil cores (+ = estimated).

Site	Rutherglen		Jädraås
Soil Moisture %	14.2	12.9	20$^+$
pH	6.9	4.8	4.0
NO_3^- (ppm)	6.1	60.1	< 0.1
NO_2^- (ppm)	0.4	2.8	0.94
P_{HNO_2} (atm)	1×10^{-9}	1×10^{-6}	1×10^{-6}

It appears from Table 1 that HNO_2 can have a significant partial pressure in the soil atmosphere under the more extreme conditions examined. This HNO_2 in the soil atmosphere could diffuse to the surface where, because of the nonspecific nature of current NO_x detectors, it would be detected as an emission of NO_x. There is every reason to believe that some of the present measurements of NO_x emission are in fact emissions of HNO_2.

Nitrite Decomposition

Another process that may lead to NO_x emissions from soil involves the decomposition of HNO_2. The thermodynamic basis of HNO_2 decomposition was examined by Van Cleemput and Baert (1976). These authors showed that under the relevant conditions HNO_2 preferentially decomposes to NO and NO_3^-, and that NO_2 formation is unlikely except in extremely acid conditions. The equilibria and kinetics of NO and NO_2 exchange from aqueous solutions to the air have been studied by Schwartz and co-workers (Schwartz and White 1983; Lee and Schwartz 1981). Here we are interested in the case of decomposition of NO_2^- in aqueous solutions yielding NO and NO_2 in the gas phase. The relevant chemical reactions are

$$2H^+ + 3NO_2^- = 2NO\ (g) + NO_3^- + H_2O$$

and

$$NO_2\ (g) + NO_2^- = NO_3^- + NO\ (g).$$

The first reaction involves the decomposition of soil NO_2^-. Calculations undertaken, using the same information as presented in Table 1 and the various equilibrium constants from Schwartz and White (1983), show that at equilibrium, nitrate and nitrite concentrations equivalent to those observed in acid soils maintain NO in the gas phase at partial pressures of 10^{-5} to 10^{-2} atm (see Table 2). The same equilibria predict that NO_2 will be present in concentrations of 10^4 to 10^8 times less than the abundance of NO. These calculations suggest that nitrite decomposition in soils may be a ready source of NO and that this NO will diffuse to the atmosphere to produce observable emissions of NO.

To further examine this proposal let us look at the rate of NO production from nitrite decomposition in soil. The kinetics of NO_2^- decomposition have been reviewed by Schwartz and White (1983). The decomposition rate has been calculated for the three solutions (appropriate to field conditions) presented in Table 1, and the production rate of NO is listed in Table 2. These rates are calculated for the decomposition reaction

$$3HNO_2 \rightarrow H^+ + NO_3^- + H_2O + 2NO\ (g)$$

TABLE 2. Equilibrium ratios, equilibrium partial pressures, and rates of NO production within the soil atmosphere for HNO_2 chemistry and soil conditions equivalent to those described in Table 1.

Site	Rutherglen		Jädraås
P_{NO}/P_{NO_2}	5×10^5	3×10^5	$>7 \times 10^7$
P_{NO} (atm)	3×10^{-6}	2.5×10^{-3}	$> 1.5 \times 10^{-2}$
Rate of NO Production (10^{-9} g (N)g solution/min)	4×10^{-7}	3×10^3	9×10^3

with the rate equation of

$$d[NO]/dt = 2\ k_{fwd}\ [HNO_2]^4/(P_{NO})^2$$

where $k_{fwd} = 2.6 \times 10^{-1}$ atm^2 M^{-3} s^{-1} (Schwartz and White 1983). The calculations require an assumption about the concentration of NO in the soil and so values of 0.1 ppmV for the first site and 1 ppmV for the second and third site were chosen based on the soil equilibrium NO concentrations determined by Johansson and Galbally (1984). The calculated NO production rates from nitrite decomposition presented in Table 2 may be compared with the observations of biological NO production rates in soil by Johansson and Galbally (1984) of 0.06 ng N g^{-1} soil min^{-1} in aerobic conditions and 3.7 ng N g^{-1} soil min^{-1} in anaerobic conditions at 298 K. Thus it appears that in the more usual conditions of the Rutherglen site, uncatalyzed NO_2^- decomposition is much less effective than microbial production. However, under the more acidic conditions observed at sites at Rutherglen and Jädraås, NO_2^- decomposition may be more effective than microbial production of NO. Under these conditions there would be significant NO emission from nitrite decomposition in the soil.

Reaction with Soil Organic Matter

The third mechanism that can lead to NO_x emission from soils involves the production of methyl nitrite from the reaction of nitrous acid with soil organic matter (Magalhaes et al. 1985). These authors showed that CH_3ONO was formed by the addition of HNO_2 to clay soils under moderately acidic and moist conditions: pH 4.8 and 6.3; H_2O, 410 and 480 g/kg dry wt. The addition of HNO_2 was constrained to give realistic soil concentrations of NO_2^- of 260 mg N kg^{-1} soil, a value reached in some situations after the application of anhydrous NH_3 fertilizer. The yield of CH_3ONO was less than 1% of the added NO_2^- in the more favorable case and perhaps a factor of 10 less in the second soil examined. A second study (Magalhaes and Chalk 1986) demonstrated that CH_3ONO rapidly decomposes in aqueous solution and in soils with a lifetime of less than 10 hours. The main products identified were NO_3^- in acidic and distilled water and NO_2^- in alkaline water and soil.

There are no field observations of CH_3ONO emission to the atmosphere, but it appears that such emission can occur.

Photolysis

The photolysis of nitrite to produce NO has been examined by Zafiriou and McFarland (1981). The suggested mechanism is

$$NO_2^- + H_2O + h\nu \rightarrow NO + OH + OH^-.$$

Zafiriou and McFarland (1981) have made measurements of the rate of NO_2^- disappearance in sterile seawater in sunlight. They found that the net NO_2^- photochemical loss rate for their "standard UV light day" (62 J cm^{-2} of 300–390 nm radiation) had a median value of 8% for 53 observations. They found that detectable NO occurred in seawater containing NO_2^- exposed to sunlight, but that in the dark or in the absence of NO_2^-, NO was undetectable. Within the seawater the NO had a half-life of 20 to 150 seconds when the sample was removed from sunlight. The observations showed that the partial pressure of NO in seawater is strongly dependent on both the UV radiation and NO_2^- concentration, yielding typical values of P_{NO}(sea) of 3×10^{-8} atm for $[NO_2^-] = 0.2$ mM and UV radiation = 1.5 mW cm^{-2}. Galbally et al. (1987) examined this process in an N-fertilized flooded rice field and showed that the loss of NO from the water to the atmosphere is around 10^{-5} of the NO_2^- photolysis rate. It appears that most of the NO produced reacts with various radical scavengers within the water, and it is these reactions, rather than loss to the atmosphere, which lead to the disappearance of NO. It appears that the NO fluxes from NO_2^- photolysis in natural waters present a negligibly small source of NO to the atmosphere. Galbally et al. (1987) found the NO flux from the fertilized rice field to the atmosphere to be 5×10^{-15} g N m^{-2} s^{-1}. It was possible that these fluxes arose from either a biological or photochemical source.

FACTORS THAT MODULATE THE PRODUCTION, UPTAKE, AND EMISSION OF NO_x FROM SOILS

So far this review has concentrated on the chemical and biological factors that produce and consume NO at the scale ranging from molecules to microbes. While this gives insight into the processes that produce NO, it is not so useful for modeling global soil NO_x emissions. Two other areas now to be discussed are the escape of NO from the soil and the larger-scale processes that modulate the environment affecting NO production and emission rates.

Soil Diffusion

The escape of NO from soil is driven by molecular diffusion within the soil pore spaces and by the advective and convective air flows through the soil due to pressure and thermal forcing.

Liquid water in soil pores can, through capillary forces, collect to form blockages in individual pores and so affect NO fluxes. When both the solubility and the different molecular diffusivities in water and air are taken into account, it appears that NO diffuses 10^5 times slower through water than through an equivalent thickness of air. Hence blockage of soil pores by water can drastically affect the escape of NO to the atmosphere. This is made clear in an analysis of effective gaseous diffusion coefficients in wet porous media (Currie 1970). As the porous medium is progressively wet, the effective gaseous diffusion coefficient decreases according to the fourth power of the quotient of the remaining air-filled porous space and the total porous space (Currie 1970).

In field studies of NO fluxes, it has been observed that as the soil water content approaches field capacity or is flooded, the NO emission decreases substantially (Johansson and Granat 1984; Galbally et al. 1987; Anderson and Levine 1987; Colbourn et al. 1987). This behavior is probably due to the water forming a substantial barrier to the NO escape from the soil; however, the simultaneous increase in N_2O emission (Anderson and Levine 1987) does complicate this interpretation. The role of diffusivity in regulating NO emissions is clearly illustrated in a model of NO emissions from soils (Galbally and Johansson 1989).

Larger-Scale Processes

The larger-scale processes that control NO fluxes from soils include climate (through temperature and rainfall), plant growth and decay, biomass burning, and fertilization.

The microbial and chemical processes of NO generation are temperature dependent as are the observed field emissions of NO from moist soils. The temperature dependence of NO emissions from field observations gives activation energies that lie in the range of 40 to 100 kJ/mol (Johansson and Granat 1984; Slemr and Seiler 1984). These activation energies lie in the same range as those observed for NO production in soil cores in the laboratory (McKenney et al. 1984).

Some field studies indicate a lack of, or very weak, temperature dependence of the NO emission under dry soil conditions. This behavior changes to a stronger temperature dependence when moisture is added to the soil (Johansson 1984; Anderson and Levine 1987; Johansson et al. 1988; Johansson and Sanhueza 1989). This weak temperature dependence can be translated into an activation energy of 10 to 18 kJ/mol for the increase in NO flux between 20 and 40°C for tropical savanna (Johansson et al. 1988). The reason for this unusually weak temperature dependence is unknown but it presumably arises because of water stress on the soil microbial community.

It is now clear that soil moisture has a profound effect on NO emissions. As discussed, NO emissions are low at low soil water content, rise to a maximum in moist aerated soils, and decrease again as soils become waterlogged.

The effect of rain or artificial irrigation on dry soils is to increase the NO emissions by anything between a factor of two at the least and twenty at the most, and this increase persists during the following one or more days after the water addition (Johansson and Granat 1984; Slemr and Seiler 1984; Johansson 1984; Anderson and Levine 1987; Williams et al. 1987; Johansson et al. 1988; Anderson et al. 1988; Johansson and Sanhueza 1989). In the first case, the period for which the enhanced flux persists may correspond to the period of drying out of the soil. However, after repeated watering most sites lose the capacity to produce enhanced NO emissions from new watering. This suggests that some substrate other than water has become limiting on the NO production within the soil. It should be noted that there is no knowledge of the mechanism that relates soil water to NO emissions in the dry soil situation. The coupling may in fact be through some third property of the soil/microbe system.

Johansson and Sanhueza (1989) have made the only systematic study of the variation of NO emission with soil water content. They found an increase factor of twenty in the NO flux when the soil water content of a tropical savanna soil increased from 2% to between 14 and 17%, these measurements being at near constant soil temperature.

The presence or absence of vegetation has an important effect on the magnitude of NO emissions. Johansson and Granat (1984) found an increase factor of five in NO emission from a fertilized field with a barley crop when the vegetation was removed, but they found much less of an increase when vegetation was removed from an unfertilized field. Slemr and Seiler (1984) found similar large differences between otherwise identical sites, one with bare soil and the other with grass cover. However, in the latter case the differences were observed for both fertilized and unfertilized soils. The difference in NO emissions in the presence or absence of plants can be explained by the uptake of NO and NO_x by plants as discussed earlier.

There have been no observations, so far, of the effect of plants on NO emission via the action of plants in sequestering the inorganic nitrogen from the soil in competition with the soil microorganisms. Presumably this is another way that plants can reduce NO emissions from the soil.

There has been a suggestion that senescing plants can emit NO_x compounds to the atmosphere (Wetselaar and Farquhar 1980), but so far this has not been confirmed by studies in the field.

The studies of Johansson et al. (1988) and Anderson et al. (1988) have shown that biomass burning has a substantial effect on NO emissions from soil. From a burnt site in a tropical savanna, Johansson et al. (1988) found that the NO emissions increased by a factor of ten for the four days during which the measurements were conducted after the burning. They suggest that this change comes about because of the effect of the fire temperatures on the soil composition at the site of the measurements. Given that much of the agricultural and grazing land of the world is burnt regularly, this effect probably has global significance. The processes that cause the increase are currently unknown.

The addition of inorganic nitrogen compounds to the soil is another way that NO emissions are stimulated. This addition can take place through artificial fertilizer addition or through the natural accession of inorganic nitrogen compounds in rainwater. Many of the field studies have deliberately or otherwise measured the fractional loss rate of NO_x from fertilizer addition (Johansson and Granat 1984; Slemr and Seiler 1984; Johansson 1984; Colbourn et al. 1987; Anderson and Levine 1987; Williams et al. 1987, 1988; Kaplan et al. 1988; Johansson et al. 1988; Johansson and Sanhueza 1989). At bare soil sites the fractional losses of fertilizer as NO_x, typically during two weeks after the addition, are 0.1 to 0.3% for nitrate, 1.8 to 2.7% for ammonium, and 5.4% for urea. Over vegetated surfaces the fractional losses are 0.15 to 0.4% for nitrate, 0.7% for ammonium, and roughly 1% for urea. (The data of Colbourn et al. [1987] are not included because of a large imbalance between added nitrogen and that observed in the soil during the first day after the addition.)

These observations of the wider influences on NO emissions can be illustrated by a generalized picture of the annual cycle of NO emissions from tropical savanna, based mainly on the work of Johansson et al. (1988) and Johansson and Sanhueza (1989). In tropical savanna the onset of the rainy season promotes NO emission from the previously dry and burnt soil. This is accentuated because of the higher levels of inorganic nitrogen in the rainwater at the start of the wet season. As the wet season progresses the growth of plants and the waterlogging of areas of soil probably diminishes the NO flux from at least part of the region. At around the end of the rainy season the soils start drying out and then the plants die. The emissions of NO may rise as the soils pass through the moist aerated phase and also as the plants senesce. Then during the hot dry period the NO emissions decrease to a minimum value. Finally, biomass burning stimulates soil processes and increases the fluxes over burnt soil for a period of time before the start of the rainy season when the cycle starts again. This gives a picture of the wider influences on the soil NO emissions. However, it should be noted that this picture is based on fragmentary evidence and will no doubt be modified as further measurements are made.

KEY QUESTIONS

1. What are the NO_x species, other than NO, that appear as NO_x emission?
2. Can the measured NO production rate per cell and the measured number of active microbes per unit of soil give the same NO production rate as measured on a soil core?
3. Are the NO emissions observed in the field due to nitrification or denitrification?
4. What processes actively remove NO from within the soil?
5. Why is the rate of NO emission depressed at high soil water contents?
6. How does the addition of water to dry soil stimulate NO emission?
7. How does biomass burning cause an increase in NO emissions?

REFERENCES

Anderson, I.C., and J.S. Levine. 1986. Relative rates of nitric oxide and nitrous oxide production by nitrifiers, denitrifiers and nitrate respirers. *Appl. Env. Microbiol.* 51:938–945.

Anderson, I.C., and J.S. Levine. 1987. Simultaneous field measurements of nitric oxide and nitrous oxide. *J. Geophys. Res.* 92:965–976.

Anderson, I.C., J.S. Levine, M.A. Poth, and P.J. Riggan. 1988. Enhanced biogenic emissions of nitric oxide and nitrous oxide following surface biomass burning. *J. Geophys. Res.* 93:3893–3898.

Betlach, M.R., and J.M. Tiedje. 1981. Kinetic explanation for accumulation of nitrite, nitric oxide and nitrous oxide during bacterial denitrification. *Appl. Env. Microbiol.* 42:1074–1078.

Bleakley, B.H., and J.M. Tiedje. 1982. Nitrous oxide production by organisms other than nitrifiers or denitrifiers. *Appl. Env. Microbiol.* 44:1342–1348.

Bremner, J.M., and A.M. Blackmer. 1981. Terrestrial nitrification as a source of atmospheric nitrous oxide. In: Denitrification, Nitrification and Atmospheric Nitrous Oxide, ed. C.C. Delwiche, pp. 151–170. New York: Wiley.

Colbourn, P., J.C. Ryden, and G.J. Dollard. 1987. Emission of NO_x from urine-treated pasture. *Envir. Poll.* 46:253–261.

Currie, J.A. 1970. Movement of gases in soil respiration. *S.C.I. Monog.* 37:152–169.

Farquhar, G.D., R. Wetselaar, and B. Weir. 1983. Gaseous nitrogen losses from plants. *Dev. Plant Soil Sci.* 9:159–180.

Galbally, I.E., J.R. Freney, W.A. Muirhead, J.R. Simpson, A.C.F. Trevitt, and P.M. Chalk. 1987. Emission of nitrogen oxides (NO_x) from a flooded soil fertilized with urea: relation to other nitrogen loss processes. *J. Atmos. Chem.* 5:343–365.

Galbally, I.E., and C. Johansson. 1989. A model relating laboratory measurements of rates of nitric oxide production and field measurements of nitric oxide emission from soils. *J. Geophys. Res.* 94:6473–6480.

Galbally, I.E., and C.R. Roy. 1978. Loss of fixed nitrogen from soils by nitric oxide exhalation. *Nature* 275:734–735.

Galbally, I.E., C.R. Roy, C.M. Elsworth, and H.A.H. Rabich. 1985. The measurement of nitrogen oxide (NO, NO_2) exchange over plant/soil surfaces. CSIRO Division of Atmospheric Research Technical Paper 8.

Hill, A.C. 1971. Vegetation: a sink for atmospheric pollutants. *J. Air Poll. Cont. Assn.* 21:341–347.

Hooper, A.B. 1978. Nitrogen oxidation and electron transport in ammonia-oxidizing bacteria. In: Microbiology, ed. D. Schlessinger, pp. 299–304. Washington, D.C.: Am. Soc. Microbiol.

Hooper, A.B., and K.R. Terry. 1979. Hydroxylamine oxidoreductase of Nitrosomonas production of nitric oxide from hydroxylamine. *Biochim. Biophys. Acta* 571:12–20.

Johansson, C. 1984. Field measurements of emission of nitric oxide from fertilized and unfertilized forest soils in Sweden. *J. Atmos. Chem.* 1:429–442.

Johansson, C., and I.E. Galbally. 1984. The production of nitric oxide in a loam under aerobic and anaerobic conditions. *Appl. Env. Microbiol.* 47:1284–1289.

Johansson, C., and L. Granat. 1984. Emission of nitric oxide from arable land. *Tellus* 36B:25–37.

Johansson, C., H. Rodhe, and E. Sanhueza. 1988. Emission of NO in a tropical savanna and a cloud forest during the dry season. *J. Geophys. Res.* 93:7180–7192.

Johansson, C., and E. Sanhueza. 1989. Emission of NO from savanna soils during rainy season. *J. Geophys. Res.*, in press.

Kaplan, W.A., S.C. Wofsy, M. Keller, and J.M. da Costa. 1988. Emission of NO and deposition of O_3 in a tropical forest system. *J. Geophys. Res.* 93:1389–1395.

Lee, Y.-N., and S.E. Schwartz. 1981. Evaluation of the rate of uptake of nitrogen dioxide by atmospheric and surface liquid water. *J. Geophys. Res.* 11971–11983.

Lipschultz, F., O.C. Zafiriou, S.C. Wofsy, M.B. McElroy, F.W. Valois, and S.W. Watson. 1981. Production of NO and N_2O by soil nitrifying bacteria. *Nature* 294:643.

Magalhaes, A.M.T., and P.M. Chalk. 1986. Decomposition of methyl nitrite in solutions and soils. *Soil Sci. Soc. Am. J.* 50:72–75.

Magalhaes, A.M.T., P.M. Chalk, A.B. Rudra, and D.W. Nelson. 1985. Formation of methyl nitrite in soil treated with nitrous acid. *Soil Sci. Soc. Am. J.* 49:623–625.

McKenney, D.J., G.P. Johnson, and W.I. Findlay. 1984. Effect of temperature on consecutive denitrification reactions in Brookston clay and sandy loam. *Appl. Env. Microbiol.* 47:919–926.

McKenney, D.J., K.F. Shuttleworth, J.R. Vriesacker, and W.I. Findlay. 1982. Production and loss of nitric oxide from denitrification in anaerobic Brookston Clay. *Appl. Env. Microbiol.* 43:534–541.

Nommik, H., D.J. Pluth, and J. Melin. 1984. Dissimilatory reduction of ^{15}N-labelled nitrate in the presence of nonlabelled NO or N_2O. *Can. J. Soil Sci.* 64:21–29.

Parkin, T.B. 1987. Soil microsites as a source of denitrification variability. *Soil Sci. Soc. Am. J.* 51:1194–1199.

Payne, W.J. 1981. The status of nitric oxide and nitrous oxide as intermediates in denitrification. In: Denitrification, Nitrification and Atmospheric Nitrous Oxide, ed. C.C. Delwiche, pp. 85–103. New York: Wiley.

Ritchie, G.A.F., and D.J.D. Nicholas. 1972. Identification of the sources of nitrous oxide produced by oxidative and reductive processes in Nitrosomonas europaea. *Biochim J.* 126:1181–1191.

Rogers, H.H., J.C. Campbell, and R.J. Volk. 1979. Nitrogen-15 dioxide uptake and incorporation by phaseolus vulgaris (L.). *Science* 206:333–335.

Rosswall, T. 1982. Microbiological regulation of the biogeochemical nitrogen cycle. *Plant Soil* 67:15–34.

Schwartz, S.E., and W.H. White. 1983. Kinetics of reaction dissolution of nitrogen oxides into aqueous solution. In: Advances in Environmental Science and Technology, ed. J.O. Nriagu, vol. 12, p. 151. New York: Wiley.

Slemr, F., and W. Seiler. 1984. Field measurements of NO and NO_2 emissions from fertilized and unfertilized soils. *J. Atmos. Chem.* 2 (1):1–24.

Van Cleemput, O., and L. Baert. 1976. Theoretical considerations on nitrite self-decomposition reactions in soils. *Soil Sci. Soc. Am. J.* 40:322–324.

Wetselaar, R., and G.D. Farquhar. 1980. Losses of nitrogen from the tops of plants. *Adv. Agron.* 23:263–302.

Williams, E.J., D.D. Parrish, M.P. Buhr, F.C. Fehsenfeld, and R. Fall. 1988. Measurement of soil NO_x emissions in Central Pennsylvania. *J. Geophys. Res.* 93:9539–9546.

Williams, E.J., D.D. Parrish, and F.C. Fehsenfeld. 1987. Determination of nitrogen oxide emissions from soils: results from a grassland site in Colorado, United States. *J. Geophys. Res.* 92:2173–2179.

Zafiriou, O.C., and M. McFarland. 1981. Nitric oxide from nitrite photolysis in the central equatorial Pacific. *J. Geophys. Res.* 86:3173–3182.

Exchange of Trace Gases between Terrestrial Ecosystems and the Atmosphere
eds. M.O. Andreae and D.S. Schimel, pp. 39–58
John Wiley & Sons Ltd

Control of Methane Production in Terrestrial Ecosystems

R. Conrad

Fakultät für Biologie
Universität Konstanz
7750 Konstanz, F.R. Germany

Abstract. This chapter gives an overview of the parameters and processes that may be important in the control of production and emission of methane from terrestrial ecosystems into the atmosphere. A large part of the produced CH_4 is oxidized before it reaches the atmosphere. The various controls of CH_4 production and emission depend on the structure of the ecosystem and of the microbial communities within. They further depend on the availability of organic substrates and electron acceptors and are affected by temperature and pH. Perturbation and fertilization can have severe effects which, however, need more research. Special emphasis is given to shallow vegetated methanogenic ecosystems (e.g., paddy fields) which seem to be most complex but especially important for the global CH_4 budget.

INTRODUCTION

Biogenic CH_4 production is exclusively accomplished by the so-called methanogenic bacteria that can metabolize and live only in the strict absence of oxygen and which need redox conditions of less than −200 mV. This means that CH_4 production and emission can be expected only in those terrestrial ecosystems which contain sufficiently anoxic and reduced sites. Thus, methanogenic ecosystems usually are aquatic environments (such as swamps, marshes, fens, bogs, paddies, lakes, river banks, etc.) where oxygen-deficient zones develop due to O_2 consumption by respiration and limitation of O_2 diffusion from the atmosphere (Fig. 1). Other ecosystems does not seem to be the case for the bulk of lignin, which apparently can only be decomposed in the pressence of oxygen; however, this is still controversially of atmospheric CH_4 when they dry up. Most of the aerobic soils do not support CH_4 production and emission; quite the contrary, they seem to be

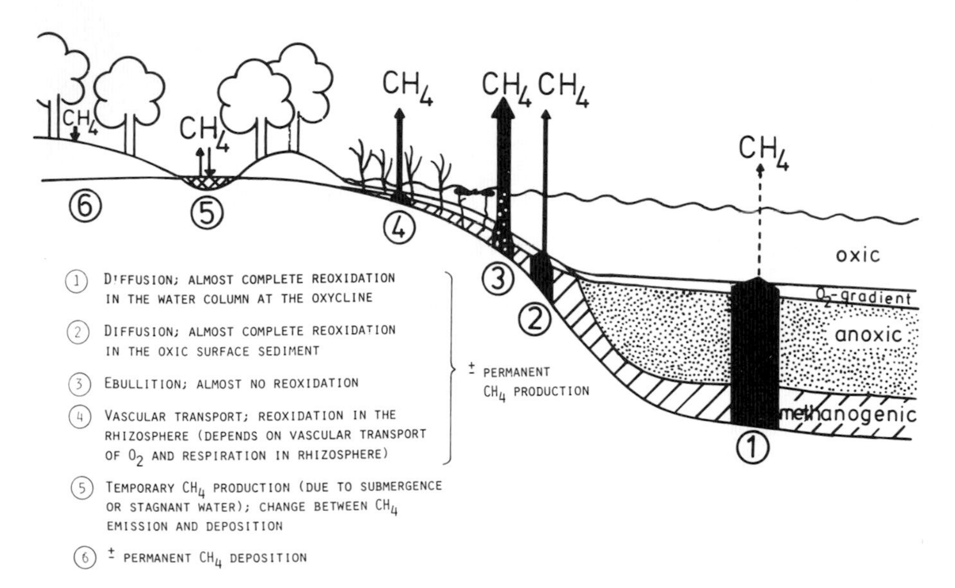

Fig. 1—Transport and oxidation processes involved in CH_4 cycling between terrestrial ecosystems and the atmosphere.

important sites for deposition and microbial oxidation of atmospheric CH_4 (Cicerone and Oremland 1988; Seiler and Conrad 1987). The guts of termites or the rumen of cattle are also anoxic methanogenic environments which are part of the terrestrial ecosystems, as are the anoxic cores of wetwood trees (e.g., cotton wood) or anthropogenic environments such as landfills and anaerobic digestors. Many of the regulatory principles apply to all of these ecosystems. They even apply to marine sediments, where methanogenesis is largely replaced by sulfate reduction.

Detailed reference to the literature can be found in recent reviews by Anthony (1986), Conrad and Schütz (1988), Cicerone and Oremland (1988), Seiler and Conrad (1987), Ward and Winfrey (1985), Whitman (1985), and in a recently published book on the biology of anaerobic bacteria (Zehnder 1988). The latter book contains excellent review articles on all aspects of anaerobic metabolism and biogeochemistry that are relevant to methanogenic ecosystems and quotes most of the relevant literature published until 1985.

TRANSPORT, OXIDATION, AND EMISSION OF THE PRODUCED METHANE

Methane emission from a particular ecosystem is basically controlled by two different microbial processes: by CH_4 production and CH_4 oxidation (Rudd and Taylor 1980). Only that part of CH_4 which is not oxidized will enter the atmosphere. Whereas CH_4-producing bacteria (the methanogens) require strictly anoxic conditions, the CH_4-oxidizing bacteria (the methanotrophs) require oxygen for metabolism. All the methanotrophs so far isolated and described need O_2 and are unable to use other electron acceptors, such as nitrate, ferric iron, or sulfate. Many scientists have attempted to isolate bacteria that would be able to oxidize CH_4 without O_2, so far without success. However, there is convincing geochemical and isotopic evidence that CH_4 is partially oxidized by anaerobic processes in anoxic marine sediments and hypersaline water (Alperin and Reeburgh 1984; Cicerone and Oremland 1988). In terrestrial, freshwater-dominated ecosystems, on the other hand, anaerobic CH_4 oxidation probably does not occur (Zehnder and Brock 1980). There, CH_4 oxidation is controlled by the availability of oxygen.

The major factors controlling the availability of O_2 and the oxidation of CH_4 are the structure of the particular ecosystem (see below) and the transport pathway by which the produced CH_4 reaches the atmosphere (Fig. 1). Whenever CH_4 production rates are low and/or diffusion paths are long, there is a big chance that most of the produced CH_4 is oxidized on its way to the atmosphere. However, when CH_4 production rates are high and/or diffusion paths are short, a substantial amount of the produced CH_4 may reach the atmosphere. For example, highly productive sediments often produce so much CH_4 that it can no longer escape by diffusion alone; instead gas bubbles are formed and pass through the overlying oxic sediment surface and water so rapidly that there is little chance for CH_4 to be reoxidized by methanotrophs. On the other hand, if CH_4 production rates are too low for bubble formation or if the ebullition is hindered by a high hydrostatic pressure, there is a good chance that the upward-diffusing CH_4 is completely oxidized at anoxic–oxic interfaces. Methanotrophic bacteria and CH_4 oxidation activity are usually concentrated in the oxic sediment surface layers or in the oxycline of stratified lakes (Rudd and Taylor 1980). Generally, it can be expected that CH_4 emission from shallow aquatic ecosystems is more important than from deep aquatic ecosystems.

Ecosystems with aquatic plants (e.g., littoral, paddy fields, fens, bogs) are especially interesting environments. There, the plant roots (in the rhizosphere) may build a network that traps gas bubbles at least temporarily. The root surface constitutes an additional anoxic–oxic interface within the CH_4 production zone of the sediment (Fig. 2). The ability of the aquatic plants to control oxidation of CH_4 depends largely on the vascular gas transport system which allows the diffusion of O_2 into the roots and from there eventually into the sediment. The same vascular system likewise allows the diffusion of CH_4 from the sediment into the atmosphere (Sebacher et al. 1985). The mechanism and efficiency of vascular transport seems to be specific to plant species and following molecular diffusion or Knudsen-

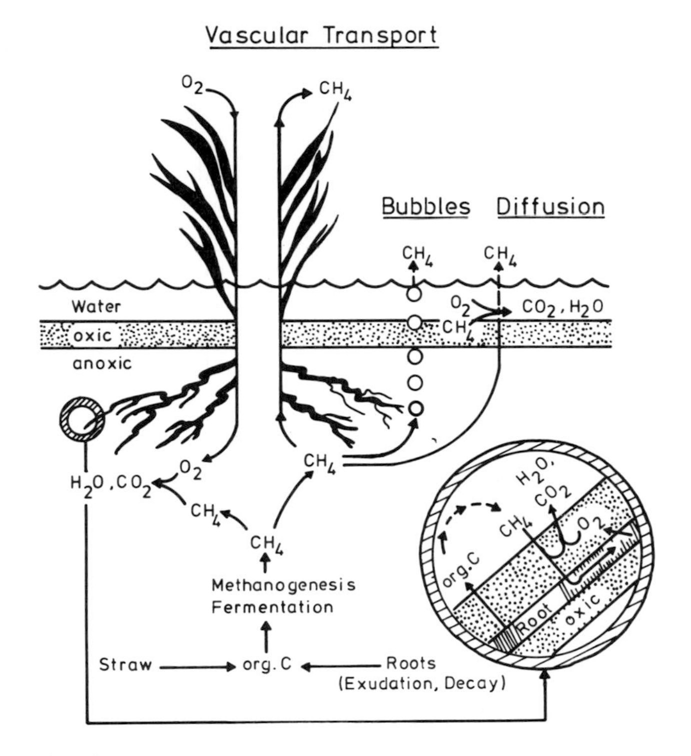

Fig. 2—The role of aquatic plants in production, reoxidation, and transport of CH_4.

transitional flow in pressurized leaves (Armstrong 1979). Availability of O_2 in the rhizosphere is controlled by respiration of the roots. Furthermore, aerobic heterotrophic microorganisms living in the rhizosphere and oxidizing organic substrates may compete with aerobic methanotrophic bacteria for the little O_2 available. The competition for O_2 depends on the availability of oxidizable organic matter (e.g., root exudates) or inorganic compounds (e.g., ammonium) (Watanabe and Furusaka 1980).

Rice paddy fields are a good example of the importance of aquatic plants and their influence on the gas transport for oxidation of CH_4. This is illustrated by results obtained in an Italian paddy field (Schütz et al. 1989; Table 1). During the early part of the season, before rice plants were visible, CH_4 was almost exclusively emitted by ebullition and most of the produced CH_4 actually reached the atmosphere. Later in the season when rice plant shoots developed and vascular transport versus ebullition dominated CH_4 emission increasingly, an increasingly larger percentage of the produced CH_4 was oxidized.

Although these results are a good example of what can happen, other vegetated aquatic ecosystems may behave differently. Oxygen transport into

TABLE 1. Seasonal change of CH_4 production and emission in a paddy field in Italy, 1985[a].

Date	Plant Height (cm)	CH_4 Production ($ml \cdot m^{-2}.h^{-1}$)	CH_4 Emission ($ml \cdot m^{-2}.h^{-1}$)	CH_4 Reoxidation (%)	Vascular Emission (%)	Bubble Emission (%)
6 June	5–10	20	11	45	0	100
9 July	50–60	67	24	64	48	51
31 July	60–80	192	40	79	90	9
27 August	60–80	373	33	91	97	2

[a] Data from Schütz et al. (1989)

the root zone may be smaller or higher, the extension of roots into the soil matrix may be different, and the microflora of the rhizosphere may be affected by the plant species. Thus, Holzapfel-Pschorn et al. (1986) observed a decrease of CH_4 production and an increase of CH_4 oxidation in soils which were vegetated with weed instead of rice plants.

METHANE-PRODUCING MICROBIAL COMMUNITIES

Under anaerobic conditions, organic matter (e.g., cellulose, hexose, etc.) is degraded to the gaseous end products CO_2 and CH_4:

$$C_6H_{12}O_6 \rightarrow 3\,CO_2 + 3\,CH_4. \tag{1}$$

However, no single microbial species is able to accomplish this reaction on its own. Methanogens, in particular, can only use a limited number of very simple compounds (Vogels et al. 1988; Whitman 1985; Table 2). Instead, many different microbial species contribute to this fermentation reaction by consecutively degrading complex organic molecules to simpler compounds (Fig. 3). This is done in a substrate food chain, where the fermentation end products excreted by one bacterium are utilized by another one until the organic matter is finally broken down to substrates which can be utilized by methanogens to form CH_4 as an end product (Zehnder 1978; Zeikus 1983).

The microbial communities and the pathways of electron and carbon flow through the substrate food chain can be different in different ecosystems. The type of organic matter, which requires different metabolic capabilities for degradation, is the determining factor. Polysaccharides such as cellulose, hemicellulose, and chitin, which originate from plant material and from lower animals (insects, crustaceans) are certainly the dominant organic materials in most ecosystems. Hence, microbial communities that degrade

TABLE 2. Substrates of methanogenic bacteria.

		$\Delta G^{\circ\prime}$ (kJ/mol CH_4)	Abundance[c]
Complete Degradation			
$4\ CO + 2\ H_2O$	$\rightarrow 3\ CO_2 + CH_4$	−186[a]	many species
$4\ H_2 + CO_2$	$\rightarrow 2\ H_2O + CH_4$	−136	almost every species
$4\ HCOOH$	$\rightarrow 3\ CO_2 + 2\ H_2O + CH_4$	−130	most species
$4\ CH_3OH$	$\rightarrow CO_2 + H_2O + 3\ CH_4$	−105[b]	*Methanosarcina*, *Methanococcus*
$4\ (CH_3)_3\ NH^+ + 6\ H_2O$	$\rightarrow 3\ CO_2 + 4\ NH_4^+ + 9CH_4$	−73[b]	
$2\ (CH_3)_2\ S + 2\ H_2O$	$\rightarrow CO_2 + 2\ H_2S + 3\ CH_4$	−49[b]	recently discovered
CH_3COOH	$\rightarrow CO_2 + CH_4$	−31	*Methanosarcina*, *Methanothrix*
Incomplete Degradation			
$2CH_3CH_2OH + CO_2$	$\rightarrow 2\ CH_3COOH + CH_4$	−116[a]	
$4\ CH_3CHOHCH_3 + CO_2$	$\rightarrow 4\ CH_3COCH_3 + H_2O + CH_4$	−36[a]	recently discovered (Widdel 1986)

[a] Ecological importance unknown
[b] Noncompetitive substrates (Oremland 1988)
[c] e.g., see Vogels et al. (1988)

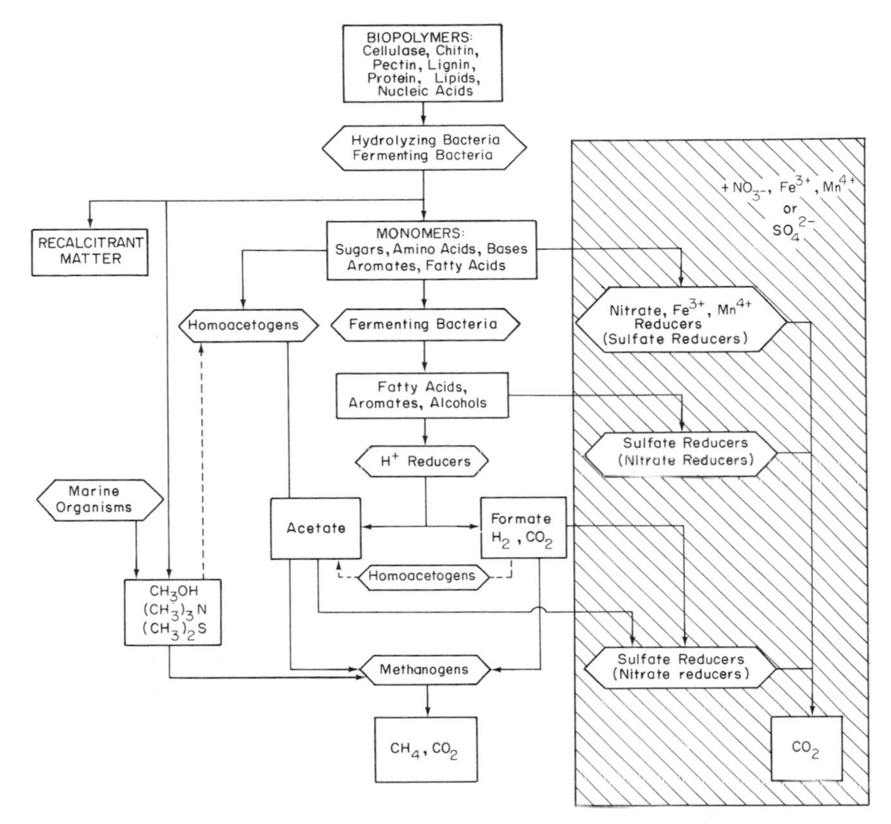

Fig. 3—Pathways of anaerobic degradation of organic matter including methanogenic (left and middle) and nonmethanogenic microbial communities (right). The latter are only operative in the presence of suitable and sufficient electron acceptors other than bicarbonate.

sugars and sugar acids are the most common and important. Proteins, lipids, and nucleic acids also occur in small amounts wherever biomass is degraded (McInerney 1988). After depolymerization to amino acids, long chain fatty acids, sugars, and purin or pyrimidine bases, the monomers are further fermented by specialized groups of bacteria. Most of the aromatic compounds that originate from plant matter can also be degraded by anaerobic microbial communities (Colberg 1988; Schink 1988; Tiedje 1988; Widdel 1988). This does not seem to be the case for the bulk of lignin, which apparently can only be decomposed in the presence of oxygen; however, this is still controversially discussed (Zeikus 1981; Colberg 1988; Schink 1988). Little is known about the degradation of "aged" organic carbon found in older sediment layers, bogs, etc. (Nedwell 1984; Cicerone and Oremland 1988).

Despite the differences in the organic materials and "specialist" microorganisms involved, there seems to be a general scheme of anaerobic methanogenic degradation which requires the cooperation of four types of bacteria within a substrate food chain (Fig. 3): (*a*) hydrolytic and fermenting bacteria, (*b*) H^+-reducing bacteria, (*c*) homoacetogenic bacteria, and (*d*) methanogenic bacteria. The first group hydrolyzes polymers (often by excreted enzymes) and ferments the resulting monomers to smaller molecules such as alcohols and short chain fatty acids, H_2, and CO_2. Only H_2/CO_2, formate, acetate, and a few other simple compounds (Table 2) can immediately be utilized by methanogens and converted to CH_4 and CO_2. The other fermentation products (fatty acids, alcohols, aromates, etc.), however, cannot be utilized by methanogenic bacteria. They are utilized by the so-called H^+ reducers which use H^+ as an electron acceptor forming H_2 while oxidizing their substrate to acetate and CO_2 (Dolfing 1988). This degradation step requires low H_2 partial pressures (1–100 Pa) because of thermodynamical reasons and, thus, is usually accomplished by syntrophic microbial associations (see below).

The situation is different if the methanogenic microbial community of an ecosystem is for some undetermined reason dominated by homoacetogens (Fig. 3). Homoacetogenic bacteria are very versatile bacteria which can use sugars, alcohols, fatty acids, purines, and aromatic compounds as well as H_2, CO, formate, and methanol as substrates and produce acetate as the sole fermentation product (Dolfing 1988). There are a few examples (e.g., termite guts, acidic lake sediments) where homoacetogens seem to dominate the fermentation of organic matter so that the fermenting and H^+-reducing bacteria are largely replaced by them (Breznak and Switzer 1986; Phelps and Zeikus 1984).

Finally, some organic polymers (e.g., pectin, lignin) contain methoxy groups which are released as methanol at a very early step in degradation (Schink 1984, 1988). Methanol can be used directly by methanogenic bacteria and converted to CO_2 and CH_4 (Fig. 3). The same is true for methylamines (e.g., trimethylamines) or dimethylsulfide which mainly originate from osmolytically active compounds (betaine, propiothetin) present in marine algae, marine plants, or marine animals. These substrates are noncompetitive substrates since they seem to be exclusively used by methanogenic bacteria even when sulfate (!) or nitrate (?) are present as alternative electron acceptors (Ward and Winfrey 1985; Oremland 1988).

The availability of electron acceptors other than bicarbonate allows the competition by nonmethanogenic microbial communities for common energy substrates (see below). In the presence of nitrate, denitrifying or nitrate-ammonifying bacteria can successfully compete with fermenting, H^+-reducing, and methanogenic bacteria for most of the soluble substrates including sugars, aromates, H_2, etc. (Tiedje 1988). The same is probably the case for iron- or manganese-reducing microbial communities, although our knowledge of these bacterial groups is still very limited (Lovley 1987; Ghiorse 1988). In the presence of sulfate, sulfate-reducing bacteria can successfully compete with the methanogens for their common substrates H_2 and acetate, and with H^+-reducing bacteria for alcohols and fatty acids (see below). Sulfate reducers usually also outcompete fermenting bacteria on common substrates, but many substrates (e.g., sugars, amino acids) cannot be utilized by most sulfate reducers (Widdel 1988).

HYDROGEN TURNOVER, SYNTROPHY, AND COMPETITION

Together with acetate, H_2 is the most important immediate precursor for methanogenesis. While only two genera of methanogens are able to use acetate, almost all are able to use H_2 (Table 2). Even more important, however, is the regulatory role of H_2 in the degradation of alcohols, fatty acids, and other compounds which can only be fermented if the H_2 partial pressure is low enough because of thermodynamical reasons. Buildup of high H_2 partial pressures is often diagnostic for disrupted degradation processes, observed for example in unbalanced digestors, pathogenic rumen, acidic wetwood, etc. (Zeikus 1983; Goodwin and Zeikus 1987). In balanced methanogenic ecosystems, however, H_2 concentrations are kept within a low concentration range (Conrad et al. 1986; Table 3). In fact, *in situ* H_2 concentrations are generally so low that the free energy available for H_2-dependent methanogenesis allows the synthesis of less than one mole ATP per mole of CH_4 produced, i.e., these types of methanogens are limited by energy substrate.

On the other hand, the low H_2 allows the fermentation of otherwise nonfermentable substrates. This fermentation is accomplished by the H^+-reducing bacteria which belong to different taxonomic groups, e.g., *Syntrophomonas, Clostridium, Pelobacter, Desulfovibrio* (Dolfing 1988). Evidence is accumulating that these bacteria not only take advantage of the low H_2 established in a methanogenic ecosystem, but also live in juxtaposed syntrophic association with H_2-utilizing methanogens (Conrad et al. 1985; Fig. 4). Syntrophic methanogenic associations may be found in the

TABLE 3. Energetics of H_2-dependent methanogenesis in different anaerobic environments *in situ*[a].

Environment	H_2 (nM)	H_2 (Pa)	Δ G *in situ* (kJ/CH_4)
Lake Mendota sediment	31	3.7	-27
Knaack Lake sediment	42	4.8	-35
Anoxic paddy soil	55	8.4	-33
Fetid liquid, Cottonwood	103	14.0	-42
Anaerobic sewage sludge	176	26.9	-33

[a] Data from Conrad et al. (1986) and Conrad, Schütz et al. (1987)

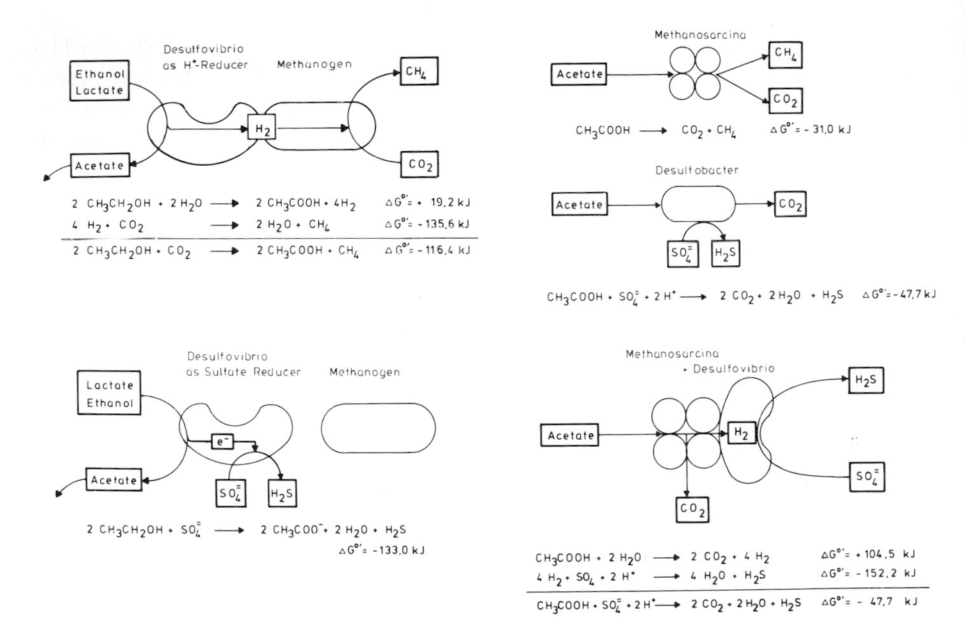

Fig. 4—Substrate utilization by H_2-syntrophic microbial associations in the absence or presence of sulfate: (*a*) utilization of lactate or ethanol; (*b*) utilization of acetate.

form of flocs or simply by sharing a microcompartment which is formed in soil or sediment pores. In these syntrophic associations, the methanogens practically act as a living electron acceptor reducing CO_2 to CH_4. The electrons are provided from the H^+ reducers via interspecies H_2 transfer. There are arguments that H_2 may sometimes be replaced by formate (Thiele and Zeikus 1988).

The function of methanogens may be substituted by other H_2-utilizing bacteria which differ in their capacity to scavenge traces of H_2 (Table 4). This capacity apparently increases with an increasing redox potential of the electron acceptor used and an increasing free energy yield of the H_2-consuming reaction (Cord-Ruwisch et al. 1988). Thus, sulfate reducers can outcompete methanogens for available H_2 if sulfate becomes available in sufficient amounts (usually > 20 μM) since they can use lower concentrations of H_2. In addition, they also have better kinetics (K_m, K_s) for H_2 consumption and growth (Ward and Winfrey 1985). Methanogens, on the other hand, should be able to outcompete homoacetogens because of similar reasons. However, this apparently is not always the case. The reason for the successful competition by homoacetogens is presently not understood but may be due to their ability for multiple substrate utilization and their better adaptation to low *in situ* temperatures (see below).

TABLE 4. Minimum H_2 concentrations and K_m values of different metabolic groups of anaerobic bacteria[a].

Reaction	$\Delta G^{\circ\prime}$ (kJ)	$E^{\circ\prime}$ (mV)	H_2 min. (nM)	K_m (μM)
Homoacetogens				
$4\ H_2 + 2\ CO_2 \rightarrow$ acetate $+ 2\ H_2O$	−104.4	−279	340–620	6–27
Methanogens				
$4\ H_2 + CO_2 \rightarrow CH_4 + 2\ H_2O$	−135.6	−238	16–65	2.5–13
Sulfate reducers				
$4\ H_2 + H_2SO_4 \rightarrow H_2S + 4\ H_2O$	−152.0	−217	5–12	0.7–1.9
Nitrate reducers				
$4\ H_2 + HNO_3 \rightarrow NH_3 + 3\ H_2O$	−599.6	+363	0.02	?

[a] Data from Breznak et al. (1988), Cord-Ruwisch et al. (1988), Phelps (1985), and Ward and Winfrey (1985)

The H_2-scavenging function of the methanogens may be substituted by suitable organic (e.g., fumarate) or inorganic (e.g., nitrate) electron acceptors, provided they can be used by the H^+-reducing bacteria alternatively to H^+. An important example is the functioning of sulfate reducers as H^+-reducing bacteria in a syntrophic methanogenic association (Fig. 4). In the presence of sulfate, the interspecies H_2 transfer will be disrupted and electron flow will be shifted away from CH_4 towards H_2S formation (Conrad, Lupton et al. 1987). In addition, the sulfate reducers may consume any H_2 originating from other fermenting bacteria and finally outcompete and replace H_2-utilizing methanogens completely (Ward and Winfrey 1985).

Acetate-utilizing sulfate reducers (e.g., *Desulfobacter, Desulfotomaculum*) are able to outcompete methanogens for their common substrate acetate (Ward and Winfrey 1985). However, acetate-utilizing sulfate reducers may not be present in every ecosystem and, as a result, acetate may be exclusively used by methanogens even when sulfate is available. In this respect it is important to note that only a few of the sulfate reducers are able to oxidize organic substrates to CO_2 completely and that most degrade their substrates just to acetate (Widdel 1988). Even then, however, more CO_2 versus CH_4 may be produced from acetate because of the syntrophic transfer of H_2 from an acetate-degrading methanogen to a nonacetotrophic but H_2-utilizing sulfate reducer (Phelps et al. 1985; Fig. 4).

Little is known about the role of syntrophy based on the transfer of acetate between fermenting acetogenic and methanogenic bacteria. Recently, bacterial culture studies have indicated that accumulation of acetate may result in thermodynamic inhibition of acetogenic fermentation similar to accumulation of H_2 results in inhibition of fermentative H_2 production (Platen and Schink 1987; Dolfing and Tiedje 1988; Ahring and Westermann 1988).

POPULATION DYNAMICS

The pathways and rates of organic matter degradation in general are dependent on the composition of the microbial community and the growth dynamics of the different bacteria. For many purposes it may be sufficient to know about the dominant species involved in the degradation of organic matter. However, we know virtually nothing about how fast the different physiological groups of bacteria multiply and die under *in situ* conditions. There is an urgent need for techniques to study individual microbial populations and physiological groups (Parkes 1987).

Some recent observations show that the production of CH_4 is not necessarily controlled by the population size of the methanogens. Using most probable number counting techniques, the numbers of H_2 and acetate-utilizing methanogens in paddy fields stayed constant during the season, although the rates of CH_4 production and emission changed significantly (Schütz et al. 1989). Moreover, the dry and oxic soils of unflooded paddy fields contained the same amount of methanogenic bacteria (Mayer and Conrad, submitted). CH_4 production was apparently controlled by the populations of the nonmethanogenic bacteria which establish the right redox conditions and supply the precursors for methanogenesis.

STRUCTURE OF AQUATIC METHANOGENIC ECOSYSTEMS

Methanogenic ecosystems have structure and may thus contain different microbial communities that may be individually controlled by the quality and quantity of organic substrates and the availability of electron acceptors. In steady state, undisturbed methanogenic ecosystems show a characteristic stratification (Fig. 5). The availability of O_2, nitrate, ferric iron, and sulfate typically decreases with the distance from the anoxic–oxic interface, so that in the end bicarbonate is the only available electron acceptor (Zehnder and Stumm 1988). This is the zone where CH_4 production becomes the dominant final step in organic matter degradation. These different redox zones establish in vertical strata (i.e., sediment layers) and around O_2-containing plant roots or pipes of benthic worms. The microbial activity within the individual redox zones is influenced by the availability of energy substrates and electron acceptors.

In paddy fields, for example, the kinetics of reduction processes are strongly affected by the composition and texture of soil and its content of

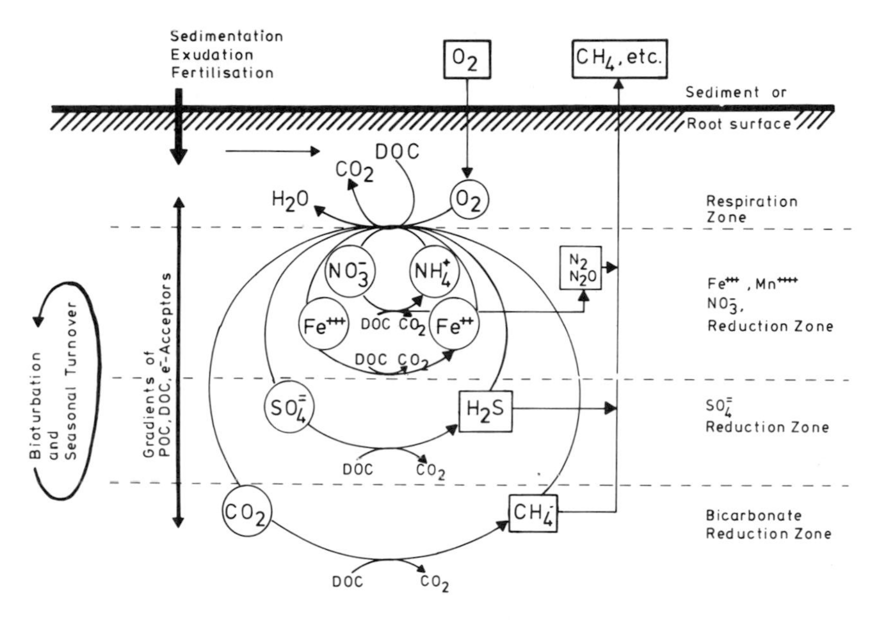

Fig. 5—Structure of methanogenic ecosystems with respect to redox zonation, regeneration of inorganic electron acceptors, and role of dissolved organic carbon (DOC). The scheme does not depict the reoxidation of CH_4 within the O_2 zone (see Fig. 2).

inorganic electron acceptors (Ponnamperuma 1981). The period between flooding of the soil and onset of methanogenesis can apparently be different for the various soils. It is unclear, however, whether the soil type also affects the rates of methanogenesis and CH_4 emission when steady-state conditions have been reached.

The availability of electron acceptors (e.g., nitrate, ferric iron, sulfate) may also be influenced by their regeneration, with O_2 as the ultimate electron acceptor. Hence, oxygen controls CH_4 not only by direct oxidation (Fig. 2) but also by structuring the ecosystem (Fig. 5). The latter ultimately affects the degradation process of organic matter and the formation of CH_4. Any disturbance by fertilization or mechanical perturbation (e.g., runoff, agricultural management) can have a great impact on CH_4 production and emission.

The structure of the ecosystem also affects the quality of organic substrates which reach the methanogenic microbial community (Nedwell 1984). In deep sediments, for example, fresh material is deposited as particulate organic carbon (POC) on top of the sediment. The bacteria of the deeper methanogenic layers have to rely on organic carbon which either has already undergone degradation processes (aged organic carbon) or diffuses as dissolved organic carbon (DOC) from the top sediments. This situation is influenced by benthic animals, which cause more or less strong bioturbation. In vegetated aquatic ecosystems, organic matter may be introduced deeply into the methanogenic zones by root exudation or root decay as suggested by Holzapfel-Pschorn et al. (1986). The different quality of the primary organic substrates may have a large effect on the degradation pathways and on the efficiency of CH_4 production.

CONTROL OF METHANE PRODUCTION BY SUBSTRATES

There are three classes of substrates which influence production and emission of CH_4: (*a*) organic matter, (*b*) electron acceptors (nitrate, iron, manganese, sulfate), and (*c*) nutrients (N, P, S, K, trace elements).

In general, organic matter stimulates CH_4 production, provided temperature and pH are not the major limiting factors for the whole year (see below). This stimulation is due to enhanced fermentative production of CH_4 precursors. This will immediately affect the CH_4 production since the methanogens are substrate-limited (H_2-limited) (see above). Hence, sediments with high organic load are more active in CH_4 production than oligotrophic sediments; CH_4 production may be stimulated by root exudation of aquatic plants or application of organic fertilizer (e.g., rice straw, manure), etc. In principle, CH_4 production could be expected to be more or less proportional to the input of organic carbon (Eq. 1). However, in ecosystems where sulfate reduction takes place as well, part of the organic carbon will be used for these processes and will result in production of CO_2 rather than CH_4 (Ward and Winfrey 1985; Conrad and Schütz 1988). Therefore, in most cases, the control of methanogenesis by organic substrates is nonlinear.

In general, electron acceptors such as nitrate or sulfate are preferred over bicarbonate and thus inhibit CH_4 production by competition (see above). In fact, this is the reason for the described stratification of methanogenic ecosystems (Fig. 5). Any input of oxidized compounds (e.g., sulfate- or nitrate-containing fertilizers) that may serve as electron acceptors will have a severe effect on CH_4 production. However, effects will also occur upon perturbation, e.g., bioturbation by benthic animals, turnover of stratified lakes, seasonal inundation, or growth of aquatic plants. I especially want to emphasize aquatic plants, since they possibly affect the availability of electron acceptors in two ways: (*a*) by providing organic root exudates which help to deplete electron acceptors (e.g., denitrification of nitrate or sulfate reduction); and (*b*) by providing oxygen which helps to regenerate electron acceptors (e.g., nitrification of ammonium to nitrate or oxidation of sulfide). Hence, cycling of nitrogen (Reddy and Patrick 1984) and sulfur (Freney et al. 1982) may affect CH_4 production in perturbed or vegetated ecosystems. This is especially relevant in fertilized paddy fields (Schütz and Seiler, this volume) but possibly also in natural wetlands with atmospheric input of N and S. There is definitely a need for more research in this field.

The effect of nutrients (e.g., fertilizer treatment) on CH_4 production is difficult to predict. Nutrients such as nitrogen, phosphate, etc., are necessary for biomass formation. The vegetation of ecosystems adapts to nutrient limitation, and a change of nutrient supply will change the species composition. Aquatic plants may have a pronounced effect on CH_4 production and emission (see above) so that nutrient-stimulated growth of plant biomass will affect CH_4 via (*a*) the provision of more organic substrates, (*b*) vascular transport of CH_4, (*c*) reoxidation of CH_4 by vascular O_2 transport, and (*d*) depletion and/or regeneration of nitrate, ferric iron, oxidized manganese, and sulfate. A good example is N-fertilized rice paddies, which may exhibit stimulated as well as inhibited CH_4 emission (Schütz et al., submitted). In contrast to plants, it is not known whether or how the bacterial species of methanogenic microbial communities adapt to limitation and changes of nutrients.

CONTROL OF METHANE PRODUCTION BY TEMPERATURE

Most isolates of methanogenic bacteria are mesophilic, i.e., they have temperature optima of 30–40°C (Vogels et al. 1988). However, there are also quite a number of thermophilic (40–70°C) and extremely thermophilic (up to 97°C) methanogens. Psychrophilic acetate-utilizing methanogens with temperature optimum below 20°C seem to occur in acidic peat, although they have not yet been isolated (Svensson 1984). However, in most environments methanogens seem to be limited by suboptimal temperatures. Since other members in the methanogenic microbial community are also temperature-sensitive, it is not immediately evident where the temperature gauntlet lies. Thus, the production of methanogenic substrates by fermenting bacteria can be more temperature-limited than the production of CH_4 itself (Conrad, Schütz et al. 1987). In contrast to methanogens, homoacetogens often seem to be better adapted to low *in situ* temperatures, i.e., they are psychrotrophic (Conrad et al. 1989). This may be one reason why they are so competitive in some ecosystems.

If the supply of organic matter is not limiting, increasing temperatures (T) generally stimulate CH_4 production (P) in most methanogenic environments. This stimulation can best be quantified by the logarithmic form of the Arrhenius equation:

$$\ln P = -(E_a/R)(1/T) + \text{const} \qquad (2)$$

where E_a is the apparent activation energy and R is the gas constant.

The value of E_a depends on the individual metabolic processes involved but in general reflects the temperature dependence of the overall process.

E_a is usually in a range of 60–90 kJ/mol corresponding to a 2.5- to 3.5-fold increase in CH_4 production with a temperature increase of 10°C (Conrad, Schütz et al. 1987). For CH_4 emission under field conditions, the variations can be substantially higher, since temperature may also affect CH_4 oxidation and, in addition, all processes related to the activity of aquatic plants, especially the production of organic carbon.

In principle, there is a fine balance between temperature control and control by substrate supply. Either usually limits degradation and thus CH_4 production. Temperature increase will stimulate degradation and deplete the substrates. Temperature decrease will reduce degradation and cause substrate accumulation somewhere in the degradation pathway. If temperature is permanently limiting (e.g., in deep sediments) it may preclude stimulation by organic matter. Hence, diurnal and seasonal temperature changes in shallow methanogenic ecosystems presumably play a great role in this balance. Observational data are scarce, however.

CONTROL OF METHANE PRODUCTION BY pH

Most methanogenic bacteria are neutrophilic i.e., they grow optimally in a narrow range around pH 7 (Conrad and Schütz 1988). Most of them tolerate pH values between 6 and 8. A few isolates have their optimum pH between 8 and 10, i.e., they are alkaliphilic. No acidophilic methanogen has so far been isolated, although one isolate was shown to be acid-tolerant and tolerated pH values of about pH 3 (but optimum at pH 6–7) (Williams and Crawford 1985). This strain was isolated from acidic peat, which in general shows substantial rates of CH_4 production (Svensson 1984). Hence, the methanogenic microbial communities of acidic environments may operate under suboptimal conditions. Recent observations (Delmas, pers. comm.) indicated that CH_4 emission rates from tropical forest swamps of the Congo were one order of magnitude lower at low pH (< 4) than at neutral sites, although the acidic sites contained significant numbers of methanogenic bacteria.

Fermentation processes and H_2 turnover apparently are also limited in acidic environments (Goodwin and Zeikus 1987; Phelps and Zeikus 1984). This actually seems to be an adaptation to the pH-limited CH_4 production to prevent accumulation of fermentation products (fatty acids) which would further acidify the environment. Increase of pH stimulates CH_4 production and organic matter degradation. As already mentioned, environments may occasionally turn sour because of imbalances in the degradation pathway. This is a fatal process that prevents any further anaerobic degradation. If the pH is permanently and generally too low or too high, it may exert a master control on CH_4 production similar to temperature. The velocity of anaerobic degradation is then reduced and, consequently, organic matter may accumulate unless inputs decline (Zeikus 1983).

So far, nothing is known about the effects of pH changes within microenvironments of methanogenic ecosystems. Localized pH changes may be caused by root exudation, by cation-H^+ or anion-OH^- exchange at roots or with mosses (e.g., *Sphagnum*), by photosynthesis, or by microbial oxidation and reduction reactions.

CONCLUDING REMARKS

This review is mainly based on our knowledge of CH_4 production in lake sediments and paddy fields. The latter are one of the most important terrestrial sources for the atmospheric CH_4 budget (Schütz and Seiler, this volume). Natural wetlands, e.g., high latitudinal wetlands (tundra, taiga) or tropical wetlands (big swamps such as the Pantanal, Brazil), may be equally important but have been investigated much less. For a better understanding of terrestrial sources for atmospheric CH_4, the following areas of research are emphasized:

1. Analysis of the structure and dynamics of methanogenic bacterial communities in different natural wetlands and rice paddies by using kinetic, tracer, and enrichment techniques (Conrad and Schütz 1988). CH_4 production is just the final step in a very complex chain of anaerobic biogeochemical processes which may be influenced by soil conditions and vegetation.
2. Analysis of the role of CH_4 oxidation processes which act as a biofilter for CH_4 diffusing into the atmosphere (Galchenko et al., this volume). CH_4 oxidation may be as important as CH_4 production to define the source strength of a terrestrial ecosystem. CH_4 oxidation in the root zone of aquatic plants deserves special attention.
3. Analysis of influences on CH_4 cycling in methanogenic ecosystems by diurnal and seasonal rhythms, by acidification, and by fertilizer inputs (N, P, S) from the atmosphere (Melillo et al., this volume) or man's activity; all these variables may have a significant influence on the complex methanogenic or methanotrophic microbial communities and may exert a feedback due to global changes.
4. Development of mechanistic models of the biogeochemistry and the microbial population dynamics involved in CH_4 cycling in the various methanogenic ecosystems (Lovley and Klug 1986); modeling efforts should parallel experimental research.

Acknowledgements. I thank B.H. Svensson, M.O. Andreae, A.F. Bouwman, R.A. Delmas, and M. Keller for helpful comments, and Ursula

Schmidt for drawing the figures. This work benefited from research grants by the Deutsche Forschungsgemeinschaft (Schwerpunkt Methanogene Bakterien and SFB 248).

REFERENCES

Ahring, B.K., and P. Westermann. 1988. Product inhibition of butyrate metabolism by acetate and hydrogen in a thermophilic coculture. *Appl. Env. Microbiol.* 54:2393–2397.

Alperin, M.J., and W.S. Reeburgh. 1984. Geochemical observations supporting anaerobic methane oxidation. In: Microbial Growth on C-1 Compounds, ed. R.L. Crawford and R.S. Hanson, pp. 282–289. Washington, D.C.: Am. Soc. Microbiol.

Anthony, C. 1986. The bacterial oxidation of methane and methanol. *Adv. Microb. Physiol.* 27:113–210.

Armstrong, W. 1979. Aeration in higher plants. *Adv. Bot. Res.* 7:226–332.

Breznak, J.A., and J.M. Switzer. 1986. Acetate synthesis from H_2 plus CO_2 by termite gut microbes. *Appl. Env. Microbiol.* 52:623–630.

Breznak, J.A., J.M. Switzer, and H.-J. Seitz. 1988. *Sporomusa termitida* sp. nov., an H_2/CO_2-utilizing acetogen from termites. *Arch. Microbiol.* 150:282–288.

Cicerone, R.J., and R.S. Oremland. 1988. Biogeochemical aspects of atmospheric methane. *Glob. Biogeochem. Cyc.* 2:299–327.

Colberg, P.J. 1988. Anaerobic microbial degradation of cellulose, lignin, oligolignols, and monoaromatic lignin derivatives. In: Biology of Anaerobic Microorganisms, ed. A.J.B. Zehnder, pp. 333–372. New York: Wiley.

Conrad, R., F. Bak, H.J. Seitz, B. Thebrath, H.P. Mayer, and H. Schütz. 1989. Hydrogen turnover by psychrotrophic homoacetogenic and mesophilic methanogenic bacteria in anoxic paddy soil and lake sediment. *FEMS Microbiol. Ecol.* 62:285–294.

Conrad, R., F.S. Lupton, and J.G. Zeikus. 1987. Hydrogen metabolism and sulfate-dependent inhibition of methanogenesis in a eutrophic lake sediment (Lake Mendota). *FEMS Microbiol. Ecol.* 45:107–115.

Conrad, R., T.J. Phelps, and J.G. Zeikus. 1985. Gas metabolism evidence in support of juxtapositioning between hydrogen producing and methanogenic bacteria in sewage sludge and lake sediments. *Appl. Env. Microbiol.* 50:595–601.

Conrad, R., B. Schink, and T.J. Phelps. 1986. Thermodynamics of H_2-consuming and H_2-producing metabolic reactions in diverse methanogenic environments under *in situ* conditions. *FEMS Microb. Ecol.* 38:353–360.

Conrad, R., and H. Schütz. 1988. Methods of studying methanogenic bacteria and methanogenic activities in aquatic environments. In: Methods in Aquatic Bacteriology, ed. B. Austin, pp. 301–343. Chichester: Wiley.

Conrad, R., H. Schütz, and M. Babbel. 1987. Temperature limitation of hydrogen turnover and methanogenesis in anoxic paddy soil. *FEMS Microbiol. Ecol.* 45: 281–289.

Cord-Ruwisch, R., H.J. Seitz, and R. Conrad. 1988. The capacity of hydrogenotrophic anaerobic bacteria to compete for traces of hydrogen depends on the redox potential of the terminal electron acceptor. *Arch. Microbiol.* 149:350–357.

Dolfing, J. 1988. Acetogenesis. In: Biology of Anaerobic Microorganisms, ed. A.J.B. Zehnder, pp. 417–468. New York: Wiley.

Dolfing, J., and J.M. Tiedje. 1988. Acetate inhibition of methanogenic, syntrophic benzoate degradation. *Appl. Env. Microbiol.* 54:1871–1873.

Freney, J.R., V.A. Jacq, and J.F. Baldensperger. 1982. The significance of the biological sulfur cycle in rice production. In: Microbiology of Tropical Soils and Plant Productivity, ed. Y.R. Dommergues and H.G. Diem, pp. 271–317. The Hague: Nijhoff.

Ghiorse, W.C. 1988. Microbial reduction of manganese and iron. In: Biology of Anaerobic Microorganisms, ed. A.J.B. Zehnder, pp. 305–331. New York: Wiley.

Goodwin, S., and J.G. Zeikus. 1987. Ecophysiological adaptations of anaerobic bacteria at low pH: analysis of anaerobic digestion in acidic bog sediments. *Appl. Env. Microbiol.* 53:57–64.

Holzapfel-Pschorn, A., R. Conrad, and W. Seiler. 1986. Effects of vegetation on the emission of methane from submerged paddy soil. *Plant Soil* 92:223–233.

Lovley, D.R. 1987. Organic matter mineralization with the reduction of ferric iron: a review. *Geomicrobiol. J.* 5:375–399.

Lovley, D.R., and M.J. Klug. 1986. Model for the distribution of sulfate reduction and methanogenesis in freshwater sediments. *Geochim. Cosmo. Acta* 50:11–18.

McInerney, M.J. 1988. Anaerobic hydrolysis of fermentation of fats and proteins. In: Biology of Anaerobic Microorganisms, ed. A.J.B. Zehnder, pp. 373–415. New York: Wiley.

Nedwell, D.B. 1984. The input and mineralization of organic carbon in anaerobic aquatic sediments. *Adv. Microb. Ecol.* 7:93–131.

Oremland, R.S. 1988. Biogeochemistry of methanogenic bacteria. In: Biology of Anaerobic Microorganisms, ed. A.J.B. Zehnder, pp. 641–705. New York: Wiley.

Parkes, R.J. 1987. Analysis of microbial communities within sediments using biomarkers. In: Ecology of Microbial Communities, ed. M. Fletcher, T.R. Gray, and J.G. Jones, pp. 148–177. Cambridge: Cambridge Univ. Press.

Phelps, T.J. 1985. Ph.D. diss., University of Wisconsin, Madison, WI.

Phelps, T.J., R. Conrad, and J.G. Zeikus. 1985. Sulfate-dependent interspecies H_2 transfer between *Methanosarcina barkeri* and *Desulfovibrio vulgaris* during coculture metabolism of acetate or methanol. *Appl. Env. Microbiol.* 50:589–594.

Phelps, T.J., and J.G. Zeikus. 1984. Influence of pH on terminal carbon metabolism in anoxic sediments from a mildly acidic lake. *Appl. Env. Microbiol.* 48:1088–1095.

Platen, H., and B. Schink. 1987. Methanogenic degradation of acetone by an enrichment culture. *Arch. Microbiol.* 149:136–141.

Ponnamperuma, F.N. 1981. Some aspects of the physical chemistry of paddy soils. In: Proceedings of Symposium on Paddy Soils, pp. 59–94. Beijing: Science.

Reddy, K.R., and W.H. Patrick. 1984. Nitrogen transformations and loss in flooded soils and sediments. *CRC Crit. Rev. Env. Cont.* 13:273–309.

Rudd, J.W.M., and C.D. Taylor. 1980. Methane cycling in aquatic environments. *Adv. Aq. Microbiol.* 2:77–150.

Schink, B. 1984. Microbial degradation of pectin in plants and aquatic environments. In: Current Perspectives in Microbial Ecology, ed. M.J. Klug and C.A. Reddy, pp. 580–587. Washington, D.C.: Am. Soc. Microbiol.

Schink, B. 1988. Principles and limits of anaerobic degradation: environmental and technological aspects. In: Biology of Anaerobic Microorganisms, ed. A.J.B. Zehnder, pp. 771–846. New York: Wiley.

Schütz, H., W. Seiler, and R. Conrad. 1989. Processes involved in formation and emission of methane in rice paddies. *Biogeochem.* 7:33–53.

Sebacher, D.I., R.C. Harriss, and K.B. Bartlett. 1985. Methane emissions to the atmosphere through aquatic plants. *J. Env. Qual.* 14:40–46.

Seiler, W., and R. Conrad. 1987. Contribution of tropical ecosystems to the global budgets of trace gases, especially CH_4, H_2, CO, and N_2O. In: The Geophysiology of Amazonia: Vegetation and Climate Interactions, ed. R.E. Dickinson, pp. 133–162. New York: Wiley.

Svensson, B.H. 1984. Different temperature optima for methane formation when enrichments from acid peat are supplemented with acetate or hydrogen. *Appl. Env. Microbiol.* 48:389–394.

Thiele, J.G., and J.G. Zeikus. 1988. Control of interspecies electron flow during anaerobic digestion: significance of formate transfer versus hydrogen transfer during syntrophic methanogenesis in flocs. *Appl. Env. Microbiol.* 54:20–29.

Tiedje, J.M. 1988. Ecology of denitrification and dissimilatory nitrate reduction to ammonium. In: Biology of Anaerobic Microorganisms, ed. A.J.B. Zehnder, pp. 179–244. New York: Wiley.

Vogels, G.D., J.T. Keltjens, and C. Van der Drift. 1988. Biochemistry of methane production. In: Biology of Anaerobic Microorganisms, ed. A.J.B. Zehnder, pp. 707–770. New York: Wiley.

Ward, D.M., and M.R. Winfrey. 1985. Interactions between methanogenic and sulfate-reducing bacteria in sediments. *Adv. Aq. Microbiol.* 3:141–179.

Watanabe, I., and C. Furusaka. 1980. Microbial ecology of flooded rice soils. *Adv. Microb. Ecol.* 4:125–168.

Whitman, W.B. 1985. Methanogenic bacteria. In: The Bacteria, ed. C.R. Woese and R.S. Wolfe, vol. 8, pp. 3–84. Orlando: Academic.

Widdel, F. 1986. Growth of methanogenic bacteria in pure culture with 2-propanol and other alcohols as hydrogen donor. *Appl. Env. Microbiol.* 51:1056–1062.

Widdel, F. 1988. Microbiology and ecology of sulfate- and sulfur-reducing bacteria. In: Biology of Anaerobic Microorganisms, ed. A.J.B. Zehnder, pp. 469–585. New York: Wiley.

Williams, R.T., and R.L. Crawford. 1985. Methanogenic bacteria, including an acid-tolerant strain, from peatlands. *Appl. Env. Microbiol.* 50:1542–1544.

Zehnder, A.J.B. 1978. Ecology of methane formation. In: Water Pollution Microbiology, ed. R. Mitchell, vol. 2, pp. 349–376. New York: Wiley.

Zehnder, A.J.B., ed. 1988. Biology of Anaerobic Microorganisms. New York: Wiley.

Zehnder, A.J.B., and T.D. Brock. 1980. Anaerobic methane oxidation: occurrence and ecology. *Appl. Env. Microbiol.* 39:194–204.

Zehnder, A.J.B., and W. Stumm. 1988. Geochemistry and biogeochemistry of anaerobic habitats. In: Biology of Anaerobic Microorganisms, ed. A.J.B. Zehnder, pp. 1–38. New York: Wiley.

Zeikus, J.G. 1981. Lignin metabolism and the carbon cycle. Polymer biosynthesis, biodegradation, and environmental recalcitrance. *Adv. Microb. Ecol.* 5:211–243.

Zeikus, J.G. 1983. Metabolic communication between biodegradative populations in nature. In: Microbes in Their Natural Environments, ed. J.H. Slater, R. Whittenbury, and J.W.T. Wimpenny, pp. 423–462. Cambridge: Cambridge Univ. Press.

Exchange of Trace Gases between Terrestrial Ecosystems and the Atmosphere
eds. M.O. Andreae and D.S. Schimel, pp. 59–71
John Wiley & Sons Ltd

Biological Sinks of Methane

V.F. Galchenko[1], A. Lein[2], and M. Ivanov[1]

[1] *Institute of Microbiology*
U.S.S.R. Academy of Sciences
Moscow 7/2, U.S.S.R.

[2] *Institute of Geochemistry and Analytical Chemistry*
U.S.S.R. Academy of Sciences
Moscow B-334, U.S.S.R.

Abstract. The atmospheric methane concentration has been shown to increase globally with time. The reasons for this increase are not yet understood. In any case, human activities have contributed to the increase over the last 100–150 years. Obviously, the increase of atmospheric methane concentration results from an imbalance between methane production and oxidation. The data on methane production and oxidation rates in various ecosystems are not yet sufficient to describe the budget of this gas satisfactorily. Methane is oxidized photochemically in the atmosphere; however, we know almost nothing about biological sinks of atmospheric methane. It is believed that the major biological sink of atmospheric methane is microbial activity in soils. Aerobic and anaerobic methanotrophs consume methane in various ecosystems, but the nature of the microorganisms responsible for anaerobic methane oxidation is not certain. Nevertheless, both types of bacteria significantly limit the flux of methane from various ecosystems to the atmosphere and serve as a "bacterial filter" for methane.

INTRODUCTION

The concentration of methane in the atmosphere is incommensurably small when compared to that found in soils and sediments where methane is generated. Application of radioactive isotope methods enables one to estimate the quantities of newly generated methane in various ecosystems and the emission of CH_4 to the hydrosphere and atmosphere. It is believed (Seiler 1984; Bolle et al. 1986) that CH_4 emission to the atmosphere ranges from 225–1210 Tg CH_4 per year. Cicerone and Oremland (1988) estimated that the annual CH_4 emission is 400–640 (540) Tg.

Methane is the most abundant organic gas in the atmosphere; the average atmospheric methane concentration (AMC) ranges between 1.58 ppm in the Southern Hemisphere and 1.68 ppm in the Northern Hemisphere. The data clearly demonstrate an average temporal increase in atmospheric methane during the last decade (Rinsland et al. 1985; Blake and Rowland 1988). Analysis of gases trapped in ice cores confirms that AMC increased over the last 100 years (Craig and Chou 1982; Rasmussen and Khalil 1984; Pearman et al. 1986). This suggests that human activities are one of the causes for the AMC increase. The reader is referred to further details in Craig and Chou (1982), Raynaud et al. (1988), and Stauffer et al. (1988).

Atmospheric methane reacts with various oxidants in the troposphere, hydroxyl radical being the most important reactant. Stratospheric CH_4 reacts with Cl atoms to form HCl. These aspects are reviewed in detail in Cicerone and Oremland (1988).

It is believed that since methane is an essential component of the atmosphere it must influence global climate directly and indirectly. It is generally accepted to be one of the so-called "greenhouse" gases (Bolle et al. 1986). Blake and Rowland (1987) confirm that the increase of CH_4 and CO in the atmosphere leads to an OH decrease. According to Blake and Rowland, a permanent AMC increase may considerably influence the tropospheric ozone budget and lead to a decrease of stratospheric ozone. Also, the oxidation extent of many trace organic compounds may decrease and the quantities of stratospheric clouds increase. This, in turn, may lead to further changes in humidity, temperature, and air mass distribution (Blake and Rowland 1987).

It is necessary to consider the reasons for AMC increase from two perspectives: is AMC connected with (*a*) a decrease of the atmospheric methane sink itself or (*b*) an increase of CH_4 emission to the atmosphere from sites of CH_4 production or storage? It is not clear whether the observed AMC increase leads to decreasing atmospheric OH or whether an OH decrease is the reason for AMC growth. The OH decrease, in turn, may be related to the emissions of halocarbons and CO to the atmosphere. Obviously the increase of atmospheric methane results from an imbalance between methane production and oxidation; however, we do not know whether the production (on the global scale) increases or consumption decreases, or whether there are changes in both processes.

A considerable part of the newly generated methane is oxidized microbiologically and does not go into the atmosphere. The studies of geochemists and microbiologists point to the existence of an anaerobic methane oxidation process occurring at the same time as aerobic oxidation. More and more data are being accumulated indicating that methanotrophic bacteria represent an effective "filter" of methane on its path to the atmosphere. This chapter provides an assessment of the contribution of microbial methane oxidation to the global methane sink.

Below we list the major questions, the solutions of which would enable one to estimate the microorganisms' contribution to the global methane emission more accurately:

- Has CH_4 emission to the atmosphere increased or has its sink decreased?
- What are the natural methane sources and sinks?
- What is the contribution of individual ecosystems to the global methane emission into the atmosphere?
- What is the nature of the atmospheric methane sink: biological or photochemical? Can atmospheric methane be oxidized in soils?
- Is the process of anaerobic microbial methane oxidation real? What is the nature of methane-oxidizing microorganisms in anaerobic environments? What is the dominant process: aerobic or anaerobic?
- What sources are responsible for the increase of atmospheric CH_4?
- What is the structure and spatial extent of the methane "biofilter"? Are terrestrial and seawater bodies methane "traps"?
- What is the role of plants in methane emission to the atmosphere?
- What is methane's contribution to marine food chains?

SOURCES OF ATMOSPHERIC METHANE

The methane quantities attributed to individual sources vary in the assessments of many authors. However, the majority of researchers agree that the basic sources of methane emission to the atmosphere are ruminant animals; wetlands; swamps; marshes; lakes; shelf, sea, and ocean sediments; biomass burning; and gas and coal deposits. Methane oxidation is one of the factors which substantially influences methane flux to the atmosphere. Methanotrophic bacteria are a powerful "filter" of methane on its path from the sites of its generation to its escape to the atmosphere.

Ruminants

Ruminant animals are considered to be the major source of atmospheric CH_4. Insofar as bacterial methane utilization may occur in such a peculiar ecosystem as the gut of a ruminant, it must be addressed in this paper. We isolated a pure culture of the aerobic methane-oxidizing bacterium, *Methylobacter bovis*, from the rumen. However, the role of methanotrophs in methane oxidation and the quantitative aspects of this process in the rumen are not clear. We therefore believe that ruminants are a "pure" source of CH_4, i.e., they are without any limitations from methane oxidation (Table 1).

TABLE 1. Annual methane release rates for identified sources (Cicerone and Oremland 1988).

Identity	Annual release Tg CH_4	Range Tg CH_4
Enteric fermentation (animals)	80	65–100
Natural wetlands	115	100–100
Rice paddies	110	60–170
Biomass burning	55	50–100
Termites	40	10–100
Landfills	40	30–70
Oceans	10	5–20
Freshwaters	5	1–25
Methane hydrate destabilization	5?	0–100
Coal mining	35	25–45
Gas drilling, venting, transmission	45	25–50
Total	540	400–640

Insects

Obviously, insects also represent the case of a "pure" methane-generating ecosystem. However, the possibility of methanotrophic bacteria being present in the animals (as methanotrophic symbionts?) is not to be excluded (Table 1).

Paddy Fields

According to the calculations of various authors, this source possibly exceeds the ruminants in CH_4 generation or is at least of comparable magnitude It is believed (Ehhalt 1974) that methane oxidation changes CH_4 emission from this source insignificantly, due to the reduced environments in paddy soils. According to Ehhalt, the water layer over the soil (5 cm) is insufficient for a significant oxidation of the generated methane. However, recent studies of other authors contradict this statement. Thus, according to the data of Conrad (1987), more than 80% of methane produced in paddy fields does not reach the atmosphere due to aerobic methane oxidation in the upper centimeters of the soil and in the rhizosphere. Huang and Klug (1987) give a considerable range of values (39–92% of newly generated methane) for the fraction oxidized by aerobic methanotrophs.

Swamps, Marshes, and Wetlands

It is obvious that these ecosystems are no less powerful as sources of methane than those discussed above (130–260 Tg CH_4/yr [Ehhalt 1974];

110 Tg CH_4/yr [Matthews and Fung 1987]). The ecological situation in these ecosystems is rather complicated and diverse, which considerably hampers an evaluation of the balance of methane production and utilization. Nevertheless, these environments are rather favorable for aerobic and anaerobic (marshes) methane oxidation processes. According to the data of Yavitt and Lang (1987), 78% of newly generated methane in *Sphagnum* moss-derived peat is oxidized aerobically, and up to 95% is oxidized aerobically in *Polytricum* moss-derived peat.

Gas and Coal Deposits

It is extremely difficult to evaluate emission from coal and gas deposits since a variety of natural and anthropogenic factors are involved. The gas content in coal varies within a wide range—on the order of hundreds of cubic meters of CH_4 per ton of coal, increasing with depth. Methane concentrations in coal mine atmospheres depend both on natural gas content and the method of coal seam mining.

Methane fluxes from gas and coal deposits are evaluated at 35 and 45 Tg/yr, respectively (Table 1). A tendency towards their increase is being observed. The emission of natural gas has risen especially sharply (12 times higher in 1970 than in 1940, or by 12–15% annually [Bolle et al. 1986]). More recently this tendency has decreased somewhat, but it still remains significant (~ 4% per year). The CH_4 flux increase from coal mines is not that sharp; its annual growth averaged 2.1% from 1970 to 1980.

Aerobic methane oxidation processes have been determined in gas field soils, and methanotrophic bacteria have been isolated. However, the role of bacteria as a "filter" of methane can be significant only in the case of natural infiltration of natural gas through the soil. When methane goes directly to the atmosphere it cannot be utilized microbiologically.

An analogous picture is observed in the case of exploited coal deposits. The major part of methane released from mined coal is carried away with the ventilation flow to the surface and the atmosphere. In spite of the fact that we have discovered a considerable number of aerobic methane-oxidizing bacteria and have determined the intensity of methane oxidation in coal mines (Ivanov et al. 1978), only a very insignificant part of this methane can be oxidized directly in coal mines.

Sea- and Freshwater Bodies

The data summarized in Table 2 demonstrate that the ranges of methane content and of CH_4 generation and oxidation rates are rather wide, even for the same ecosystem. Nevertheless, we have attempted to analyze the available data and estimate the contributions of methanogenesis and methane

TABLE 2. CH_4 content and rate of CH_4 generation and oxidation in fresh- and seawater bodies*.

Sources	CH_4 content ($1m^{-3}$)	Rate ($1m^{-3}yr^{-1}$)	
		CH_4 generation	CH_4 oxidation
Lakes:			
Water	0.2–4.3	–	0.2–22.5
Sediments	4–138	0.1–616	6–130
Seas:			
Water	$6 \cdot 10^{-6}$–190	–	$16 \cdot 10^{-5}$–$12 \cdot 10^{-3}$
Sediments	10^{-2}–4.35	10^{-2}–25	10^{-2}–12.8

* Data from: Laurinavichus and Beljaev (1978); Reeburgh (1983); Lein et al. (1982); Oremland and DesMarais (1983); Ivanov et al. (1984); Adams and Fendinger (1986); Galchenko et al. (1986); Gorlatov et al. (1986); Abramochkina et al. (1987).

oxidation to the CH_4 flux from fresh- and seawater bodies to the atmosphere. CH_4 emission from lakes amounts to 2–70 Tg/yr (Table 3).

The calculations of CH_4 emission from seas (Table 4) result in a narrower range of estimates. In spite of the considerable difference in the rates of CH_4 production and oxidation from one basin to another, the net production differs insignificantly between water bodies and the total does not exceed 10–11 Tg CH_4/yr. This value is rather similar to Ehhalt's calculations (Ehhalt 1974).

ANAEROBIC AND AEROBIC METHANE OXIDATION

The problem of anaerobic methane oxidation has not yet been solved. Nevertheless, the anaerobic process has been demonstrated conclusively in many experiments with radioactive isotopes (Reeburgh 1976, 1980, 1983;

TABLE 3. CH_4 emission rates from lakes to the atmosphere*. (A sediment column of 50 cm depth has been used for the calculation of CH_4 generation and oxidation rates.)

Lake	Rate ($g\ m^{-2}yr^{-1}$) Generation	Oxidation		CH_4 emission rate**	
	Sediments	Sediments	Water	$g\ m^{-2}yr^{-1}$	$Tg\ yr^{-1}$
Kuznetchika	440	96	64	280	7–70
Dolgoë	230	96	42	92	2.3–23

* Total lakes' area: $2.3 \cdot 10^6$ km²

** Only 1–10% of the area productive

TABLE 4. CH_4 emission rate from seas to the atmosphere (for a sediment column of 2 m depth).

Sources	Rate (mg $m^{-2}yr^{-1}$) Generation		Oxidation		CH_4 emission rate	
	Sediments	Water	Sediments	Water	mg $m^{-2}yr^{-1}$	Tg yr^{-1}
Black Sea						
shelf (200 m)	14	1430	4	1422	18	2.3
deep-water zone (2000 m)	28	20018	7	20015	24	8.7
Bering Sea						
shelf (100 m)	20	–	4	19	−3	−0.4
d.–w.z. (3600 m)	208	–	16	161	31	10.8
Okhotsk Sea						
d.–w.z. (750 m)	165	–	81	79	5	1.8
Caspian Sea						
shelf (100 m)	400	–	10	?	?	?
Guaymas Basin						
d.–w.z. (2000 m)	1350	–	880	443	27	9.8

* Total area:
Shelf—$127 \cdot 10^6$ km^2
d.–w.z.—$361 \cdot 10^6$ km^2

Average:
Shelf—1.0 Tg yr^{-1}
d.–w.z.—7.8 Tg yr^{-1}

Reeburgh and Heggie 1977; Devol 1983; Oremland and DesMarais 1983; Ivanov et al. 1984; Alperin and Reeburgh 1984, 1985; Whiticar and Faber 1986; Ward et al. 1987; Alperin et al. 1988). The chemical nature of the methane oxidizer in anaerobic conditions and the identity of the dominant group of microorganisms remain unclear. Some researchers, mainly concluding from geochemical data on sulfate and methane distribution in marine sediments, believe that sulfate-reducing bacteria play the major role in anaerobic methane oxidation. We have tried to outline a range of microorganisms responsible for this process in experiments with inhibitors. Anaerobic CH_4 oxidation was not changed significantly by bromoethanesulfonic acid and molybdate, inhibitors of methanogenic and sulfate-reducing bacteria, respectively, but was fully inhibited by glutaraldehyde and autoclaving. Obviously, neither methanogenic nor sulfate-reducing bacteria are responsible for anaerobic CH_4 oxidation (Alperin and Reeburgh 1985; Iversen et al. 1987).

Application of immunology methods enabled us to determine the quantity of species and group composition of anaerobic methanotrophic microflora in water and sediments of the Black Sea and the Lake Dolgoë sediments more precisely. Aerobic methanotrophs were detected (and isolated into pure cultures) in lake and sea shelf sediments and in the oxygenated seawater only. Aerobic methanotrophs were not detected in the anaerobic zone of the Black Sea, neither in the sediments nor in the water column. Nevertheless, the rates of methane oxidation were measurable in all samples tested. Thus, in sea sediments methane oxidation is performed either by aerobic and anaerobic processes or by purely anaerobic processes. In the case of Lake Dolgoë, only the aerobic process has been observed. This has been proven by a correlation between the methane oxidation rate and the number of aerobic methanotrophs determined by the immunofluorescence method (Abramochkina et al. 1987; Galchenko, Abramochkina et al. 1988).

When estimating CH_4 emission from the sea it is necessary to mention the contribution of gas hydrates to the methane flux in marine sediments and the recently discovered deep-sea hydrothermal systems. Gas hydrates are considered to be widely distributed in marine sediments and their contribution to the total methane emission into the atmosphere may amount to 0.12–8 Tg/yr (Kvenvolden 1987). Our studies of gas hydrates in sediments of the Okhotsk Sea showed that in the direct proximity of the methane plume, considerable CH_4 oxidation rates are found which are three times higher than the CH_4 generation rate. An extremely light ($^{13}C = -48\%$) carbon isotope composition of the sediment carbonates in gas hydrate sediments is additional evidence for a considerable activity of methanotrophs in this ecosystem. The origin of these carbonates from methane is not in doubt, since the isotopic composition of the methane carbon here is $-54‰$.

A similar situation can be observed in areas of deep-sea hydrothermal activity. Noticeable quantities of methane (50–300 mM) are carried into the water with the hydrothermal fluids. In the vicinity of the hydrothermal outlets on the Juan de Fuca Ridge and in the Guaymas Basin, we determined considerable quantities of methanotrophs and a high CH_4 oxidation potential: 2–3 orders of magnitude higher than in waters not affected by hydrothermal solutions. The maxima of CH_4 oxidation and autotrophic CO_2 assimilation are found in a rather narrow layer of water (up to 200 m). The differences in the positions of the maxima of these processes can be explained by the differentiation of the hydrothermal plume in gas and mineral flows (Galchenko, Lein et al. 1988). This consequently leads to a differential distribution of methanotrophic and chemolithotrophic bacteria in various layers of water.

In hydrothermal areas, methane escapes to the ocean not only through the "black smokers"; it also infiltrates through small orifices or the bottom sediments with the geothermal fluids. Bacterial films (mats) up to 3–5 cm thick are seen at the fluids' outlets. These mats are formed mainly by hydrogen sulfide-oxidizing and methanotrophic bacteria. The fraction of methanotrophs sometimes reaches 15% of the total number of bacteria in these mats (Jannasch 1985). According to our data, the methanotroph fraction in the mats can range between 6–45%. A similar situation is also observed in areas of gas hydrate deposits (Sea of Okhotsk) and hydrocarbon seeps (Gulf of Mexico). The rich microflora of these ecosystems provides for the existence of a bottom fauna with considerable biomass. Moreover, novel symbiotrophs (symbiotic bacteria in animal tissues) have been found here: thionic bacteria in trophosomes of the vestimentifera and in gills of bivalved molluscs, and methanotrophic bacteria in molluscs (Galchenko, Abramochkina et al. 1988).

The results obtained testify to the important role of the methane oxidation process in the regulation of methane flux to the atmosphere from water ecosystems. According to our calculations, up to 60–90% of newly generated methane is oxidized in lakes and seas, respectively. Only 1.2% of newly generated methane escapes to the atmosphere from Big Soda Lake (Iversen et al. 1987), with 99% being oxidized anaerobically. Reeburgh and Alperin (1987) proposed that 70–120 Tg CH_4/yr^{-1} (10–20% of the global CH_4 emission) is oxidized anaerobically in estuary and shelf sediments. According to our data, 8–65% of methane is oxidized in anaerobic marine sediments and 50–95% in aerobic waters. Thus, the "bacterial filter" in water ecosystems retains most of the methane arriving from the sediments.

CONTRIBUTIONS TO THE INCREASE OF METHANE IN THE ATMOSPHERE

The actual contribution of various natural sources to the global CH_4 flux is not well known at this time. Cicerone and Oremland (1988) believe that the major contributions come from natural wetlands, rice paddies, and ruminants (Table 1).

The annual increase of the total methane flux for the decade (1970–1980) amounts to 1% (Bolle et al. 1986), which approximately corresponds to the increase of emission from individual sources. The exceptions are methane fluxes from gas and coal deposits which are 4.1 and 2.1% per year, respectively. If this tendency persists, methane contributions from these sources will be 160 and 70 Tg CH_4/yr by the year 2020 (Table 5). Thus, the major sources of atmospheric CH_4 will then be rice paddies, natural wetlands, and gas deposits.

In summary, we would like to emphasize again that at present we do not know which natural sources are responsible for the increase in atmospheric CH_4. With an atmospheric content of about 5000 Tg CH_4, a 1% increase corresponds to an annual growth of 50 Tg CH_4. The average increase of the total CH_4 emission flux also amounts to 1%, which corresponds to only 3–5 Tg CH_4/yr. However, this figure does not account for the annual CH_4

TABLE 5. Estimated trend of CH_4 emission rates from individual ecosystems (Tg yr^{-1}, according to Bolle et al. 1986).

Sources	Cicerone and Oremland (1988) 1988	Our calculations 2000	2020
Natural wetlands	115	130	160
Rice paddies	110	125	160
Ruminants	80	90	110
Biomass burning	55	63	80
Natural gas	45	73	160
Termites	40	40	40
Landfills	40	45	55
Coal mining	35	45	70
Oceans	10	10	10
Freshwaters	5	6	7
Total	535	627	852

increase in the atmosphere. Consequently, we can only speculate about the reasons for the AMC increase, and more research will be required to verify our hypotheses:

1. Global rates of CH_4 oxidation are decreasing in ecosystems with coupled CH_4 generation and oxidation (not less than 30–50% of the observed rates).

2. Due to the growing area of artificial water bodies and the increasing addition of mineral fertilizers and organic contamination to artificial and natural water bodies, an imbalance of CH_4 oxidation and generation processes in favor of the latter is taking place.

3. Due to contamination of the atmosphere (mostly by carbon compounds) the photochemical oxidation of atmospheric methane is decreasing.

4. Due to unknown reasons the rate of bacterial oxidation of atmospheric methane in soils is decreasing (assuming this process is essential). We believe it is necessary to separate conceptually the different types of bacterial sinks: an indigenous sink (the sink of methane inside the ecosystem: the biofilter) and a bacterial sink of atmospheric CH_4 (Fig. 1). While we have some knowledge about the indigenous sink (biofilter), we know next to nothing about the bacterial sink of atmospheric CH_4.

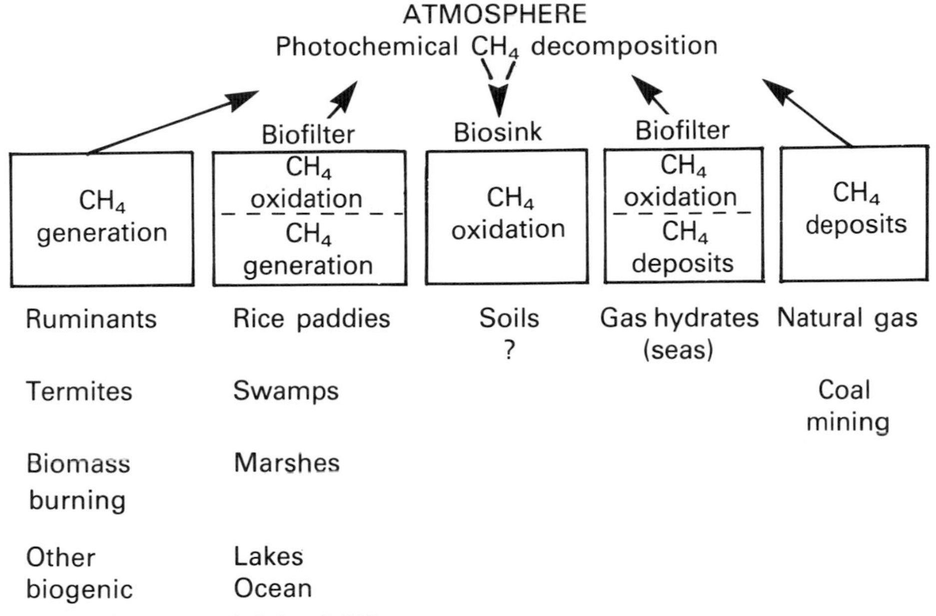

Fig. 1—Sources and sinks of CH_4.

REFERENCES

Abramochkina, F.N., L.V. Bezrukova, A.V. Koshelev, V.F. Galchenko, and M.V. Ivanov. 1987. Microbial methane oxidation in a fresh-water reservoir. *Mikrobiologija* 56:464–471.

Adams, D.D., and N.J. Fendinger. 1986. Early diagenesis of organic matter in the recent sediments of Lake Erie and Hamilton Harbor. In: Sediments and Water Interaction, ed. P.G. Sly, pp. 305–318. New York: Springer.

Alperin, M.J., and W.S. Reeburgh. 1984. Geochemical observations supporting anaerobic methane oxidation. In: Microbial Growth on C-1 Compounds, ed. R.L. Crawford and R.S. Hanson, pp. 282–289. Washington, D.C.: Am. Soc. Microbiol.

Alperin, M.J., and W.S. Reeburgh. 1985. Inhibition experiments on anaerobic methane oxidation. *Appl. Env. Microbiol.* 50:940–945.

Alperin, M.J., W.S. Reeburgh, and M.J. Whiticar. 1988. Carbon and hydrogen isotope fractionation resulting from anaerobic methane oxidation. *Glob. Biogeochem. Cyc.* 2:279–288.

Blake, D.R., and F.S. Rowland. 1987. Increasing global concentrations of tropospheric methane, 1978–1986. Abstr. Reports, 193rd ACS National Meeting, Symp. on Atmospheric Methane. Denver, CO, April 5–10, 1987.

Blake, D.R., and F.S. Rowland. 1988. Continuing worldwide increase in tropospheric methane, 1978–1987. *Science* 239:1129–1131.

Bolle, H.-J., W. Seiler, and B. Bolin. 1986. Other greenhouse gases and aerosols. In: The Greenhouse Effect, Climatic Change, and Ecosystems, ed. B. Bolin et al., pp. 157–203. SCOPE 29. Chichester: Wiley.

Cicerone, R.J., and R.S. Oremland. 1988. Biogeochemical aspects of atmospheric methane. *Glob. Biogeochem. Cyc.* 2:299–327.

Conrad, R., S. Goodwin, F.S. Lupton, T.J. Phelps, and J.G. Zeikus. 1987. H_2 turnover and interspecies H_2 transfer in methanogenic environments. Abstr. Reports, 193rd ACS National Meeting, Symp. on Atmospheric Methane. Denver, CO, April 5–10, 1987.

Craig, H., and C.C. Chou. 1982. Methane: the record in polar ice cores. *Geophys. Res. Lett.* 9:1221–1224.

Devol, A.H. 1983. Methane oxidation rates in the anaerobic sediments of Saanich Inlet. *Limnol. Ocean.* 28:738–742.

Ehhalt, D.H. 1974. The atmospheric cycle of methane. *Tellus* 24:58–70.

Galchenko, V.F., S.N. Gorlatov, and V.G. Tokarev. 1986. Microbial methane oxidation in the Bering Sea sediments. *Mikrobiologija* 55:669–673.

Galchenko, V.F., F.N. Abramochkina, L.V. Bezrukova, E.N. Sokolova, and M.V. Ivanov. 1988. The species structure of aerobic methanotrophic microflora in the Black Sea. *Mikrobiologija* 57:305–311.

Galchenko, V.F., A.Y. Lein, and M.V. Ivanov. 1988. Microbial and biogeochemical processes in ocean water as the evidences of the hydrothermal activity. *Geochimija* 12:1760–1770.

Gorlatov, S.N., V.F. Galchenko, and V.G. Tokarev. 1986. Microbial methane production in the Bering Sea sediments. *Mikrobiologija* 55:490–495.

Huang, S.N., and M.J. Klug. 1987. Methane production in and emission from paddy soils. Abstr. Reports, 193rd ACS National Meeting, Symp. on Atmospheric Methane. Denver, CO, April 5–10, 1987.

Ivanov, M.V., A.J. Nesterov, B.B. Namsaraev, V.F. Galchenko, and A.V. Nazarenko. 1978. Distribution and geochemical activity of methanotrophic bacteria in the water of coal mines. *Mikrobiologija* 47:489–493.

Ivanov, M.V., M.B. Vainshtein, V.F. Galchenko, S.N. Gorlatov, and A.Y. Lein. 1984. Distribution and geochemical activity of the bacteria in the sediments. In: Oil and Gas: Genetic Studies of the Bulgarian Sector of the Black Sea, ed. A. Geodekyan et al., pp. 150–180. Sofia: Bulgarian Acad. Sci.

Iversen, N., R.S. Oremland, and M.J. Klug. 1987. Big Soda Lake (Nevada). III. Pelagic methanogenesis and anaerobic methane oxidation. *Limnol. Ocean.* 32:804–814.

Jannasch, H.W. 1985. The chemosynthetic support of life and the microbial diversity at deep-sea hydrothermal vents. *Proc. R. Soc. Lon. A* 225:277–297.

Kvenholden, K.A. 1987. Worldwide occurrences of marine gas: implications for future atmospheric methane concentrations. Abstr. Reports, 193rd ACS National Meeting, Symp. on Atmospheric Methane. Denver, CO, April 5–10, 1987.

Laurinavichus, K.S., and S.S. Beljaev. 1978. The determination of the intensity of the microbial methane generation by the radioisotope methods. *Mikrobiologija* 47:1115–1117.

Lein, A.Y., B.B. Namsaraev, V.Y. Trotsyuk, and M.V. Ivanov. 1982. Bacterial methane generation in upper holocoenic sediments of the Baltic Sea. *Geochimija* 2:277–285.

Matthews, E., and I. Fung. 1987. Methane emission from natural wetlands: global distribution, area, and environmental characteristics of sources. *Glob. Biogeochem. Cyc.* 1:61–86.

Oremland, R.S., and D.J. DesMarais. 1983. Distribution, abundance and carbon isotopic composition of gaseous hydrocarbons in Big Soda Lake. *Geochim. Cosmo. Acta* 47:2107–2114.

Pearman, G.I., D. Etheridge, F. de Silva, and P.J. Fraser. 1986. Evidence of changing concentrations of atmospheric CO_2, N_2O and CH_4 from air bubbles in Antarctic ice. *Nature* 320:248–250.

Rasmussen, R.A., and M.A.K. Khalil. 1984. Atmospheric methane in the recent and ancient atmospheres: concentrations, trends and the interhemispheric gradient. *J. Geophys. Res.* 89(D7):11599–11605.

Raynaud, D., J. Chappellez, J.M. Barnola, Y.S. Korotkevich, and C. Lorius. 1988. Climatic and CH_4 cycle implications of glacial-interglacial CH_4 change in the Vostok ice core. *Nature* 333:655–657.

Reeburgh, W.S. 1976. Methane consumption in Cariaco Trench waters and sediments. *Earth Plan. Sci. Lett.* 28:337–344.

Reeburgh, W.S. 1980. Anaerobic methane oxidation: rate depth distributions in Skan Bay sediments. *Earth Plan. Sci. Lett.* 47:345–352.

Reeburgh, W.S. 1983. Rates of biogeochemical processes in anoxic sediments. *Ann. Rev. Earth Plan. Sci. Lett.* 11:269–294.

Reeburgh, W.S., and M.J. Alperin. 1987. Field observations of anaerobic methane oxidation. Abstr. Reports, 193rd ACS National Meeting, Symp. on Atmospheric Methane. Denver, CO, April 5–10, 1987.

Reeburgh, W.S., and D.T. Heggie. 1977. Microbial methane consumption reactions and their effect on methane distributions in freshwater and marine environment. *Limnol. Ocean.* 22:1–9.

Rinsland, C.P., J.S. Levine, and T. Miles. 1985. Concentration of methane in the troposphere deduced from 1951 infrared solar spectra. *Nature* 318:245–249.

Seiler, W. 1984. Contribution of biological processes to the global budget of CH_4 in the atmosphere. In: Current Perspectives in Microbial Ecology, ed. M.J. Klug and C.A. Reddy, pp. 468–477. New York: Am. Soc. Meteorol.

Stauffer, B., E. Lochbronner, H. Oeschger, and J. Schwander. 1988. Methane concentration in the glacial atmosphere was only half that of the preindustrial Holocene. *Nature* 332:812–814.

Ward, B.B., K.A. Kilpatrick, P.C. Novelli, and M.I. Scranton. 1987. Methane oxidation and methane fluxes in the ocean surface layer and deep anoxic waters. *Nature* 327:226–229.

Whiticar, M.J., and E. Faber. 1986. Methane oxidation in sediment and water column environments—isotope evidence. *Org. Geochim.* 10:759–786.

Yavitt, J.B., and G.E. Lang. 1987. Aerobic methane oxidation in the surface peat from a mass-dominated wetland in West Virginia. Abstr. Reports, 193rd ACS National Meeting, Symp. on Atmospheric Methane. Denver, CO, April 5–10, 1987.

Standing, left to right:
Ian Galbally, Valery Galchenko, Peter Groffman, Dieter Ehhalt, Bill Reeburgh, Ralph Cicerone, Ralf Conrad

Seated, left to right:
Friedhelm Bak, Mary Firestone, Thomas Rosswall, Dennis Baldocchi, Eugenio Sanhueza, Hans Papen

Exchange of Trace Gases between Terrestrial Ecosystems and the Atmosphere
eds. M.O. Andreae and D.S. Schimel, pp. 73–95
John Wiley & Sons Ltd

Group Report What Regulates Production and Consumption of Trace Gases in Ecosystems: Biology or Physicochemistry?

T. Rosswall, Rapporteur
F. Bak
D. Baldocchi
R.J. Cicerone
R. Conrad
D.H. Ehhalt
M.K. Firestone
I.E. Galbally
V.F. Galchenko
P.M. Groffman
H. Papen
W.S. Reeburgh
E. Sanhueza

INTRODUCTION

There are two major sources of atmospheric trace gases, including the "greenhouse" gases: microbiological processes in terrestrial and marine ecosystems, and anthropogenic emissions. We know that there are drastic changes in the composition of the atmosphere (e.g., Rowland and Isaksen 1988). Whereas direct anthropogenic and atmospheric sources and sinks are relatively easy to quantify, the magnitude of the biotic sources and sinks is less well known.

The group recognized the importance of understanding the factors that underlie the emission of nitrous oxide (N_2O), nitric oxide (NO), and methane (CH_4) and considered how such knowledge could be used to assess present and future regional and global emission rates. In developing an agenda for discussion, the group decided on topics that covered spatial scales from the organism level, to biotic and abiotic controls of emissions at the soil ecosystem level, and finally how future climate change would impact the organism/process level on a global scale.

In future discussions, the focus should not only be on the three gases selected as a topic for this conference but also on others which might interact with them or be indicators of important ecosystem processes.

The emission of trace gases from the soil to the atmosphere is regulated by the production (P) of the gas within the soil, the consumption within the soil, and the transport processes to the atmosphere. Consider the simplest model of this phenomenon, a well-mixed reservoir in steady state with exchange to the atmosphere. The model variables and their dimensions (M = moles and t = time) are as follows: E ($M\ t^{-1}$) is the rate of emission of the trace gas from the soil to the atmosphere, T (t^{-1}) is an inverse time constant that represents the rate of exchange between the reservoir and the atmosphere, C (t^{-1}) is an inverse time constant that is a measure of the rate of consumption of the trace gas within the soil, and R (M) is the number of moles of the trace gas contained in the soil reservoir. The model relationships are then:

$$E = T \times R$$

$$P = E + C \times R$$

$$E = P\,T/(T + C)$$

This illustrates clearly that production, transport, and consumption processes have major roles in trace gas emissions from soils.

Thickness of the laminar boundary layer at the soil–air interface affects the diffusion of gas from the soil to the atmosphere. Turbulence will alter the thickness of this boundary layer and facilitate the diffusion of gases into the air, but the group only considered processes in the soil system.

The group stressed the need for a basic understanding of biotic and abiotic processes at the microscopic level in order to be able to predict effects at the macroscopic level. Flux measurements alone rarely give us predictive capabilities, and process studies are needed if we are to understand present and future changes of the global environment (IGBP 1988). The group discussed only the soil processes accounting for emissions of N_2O, NO, and CH_4, but it was acknowledged that for global studies of the changing composition of the atmosphere, chemical and biological reactions in the plant canopy are essential for determining the fate of biogenic trace gases during their transport through the boundary layer.

ARE SPECIES/POPULATIONS AND MICROBIAL BIOMASS CONTROLLERS OF TRACE GAS TURNOVER?

There is some controversy over the importance of knowledge of the composition of the microbial community with respect to production and consumption of trace gases. The seriousness of the controversy as to the extent of knowledge required differs for the individual trace gases N_2O, NO, and CH_4.

There was general agreement that we need to understand trace gas exchange at the process level. Since the processes are usually due to microbial metabolism, it is necessary to know what metabolic types of microorganisms are involved and how they are regulated by environmental forcing factors. For example, it is important to know whether the N_2O flux is dominated by nitrification or denitrification since different factors control the two processes. For grassland ecosystems it was necessary to know N and C turnover rates in relation to temperature, water, and soil texture to make simple models that successfully predicted N_2O flux (Parton et al. 1988). The major question is whether this rather simple approach is adequate to be used for all relevant ecosystems or whether more detailed information must be obtained on the structure of the microbial communities involved and on the species-specific regulation of metabolism.

As an example, the knowledge of community structure seems to be essential with respect to CH_4 formation and cycling in wetland ecosystems. Degradation pathways and metabolic groups of bacteria probably differ among the various wetland ecosystems, which have different pH, temperature, and substrate regimes. Thus, the community structure would have a significant impact on CH_4 production and consumption rates, at least in vegetated wetlands.

Methane production and oxidation in the rhizosphere is influenced by the release of organic root material affecting microbial consumption of O_2. In rice paddies, root excretion of organic material also causes a direct stimulation of fermentation and therefore CH_4 production (Schütz et al. 1989). Complex patterns of diurnal and seasonal variations in CH_4 emission rates are reported in different rice paddy regions. It is likely that these complex patterns are partially due to different community structures and/or to dynamics of individual populations of fermenting, methanogenic, and methanotrophic bacteria.

The relative importance of community structure and population dynamics for the different trace gases and different ecosystems is illustrated by the following examples:

1. Production is the result of metabolic rates plus a partitioning factor; e.g., N_2O and NO production is dependent not only on rates of nitrification/denitrification, but also on how much of the metabolic flow is channeled into N_2O/NO versus other end products. The partitioning factor depends on various variables such as pH, temperature, O_2, and C/N ratios and differs between microbial species; thus, it is not easily predictable without knowledge of the microbial community structure.
2. Production may be significantly affected, if unidentified bacteria other than autotrophic nitrifiers and denitrifiers would be involved in N_2O production. For example, it has been observed that a significant

percentage of N_2O flux from soil cannot be accounted for by either autotrophic nitrification or denitrification (Robertson and Tiedje 1987). Other pathways are still very poorly known, e.g., anaerobic CH_4 oxidation and NO production and consumption.

3. Net exchange between soil and atmosphere is controlled by producing and consuming microorganisms that are sensitive to different environmental forcing factors. It is still unclear how important consumption of N_2O is in the various soil ecosystems, though it is certainly very important for NO and CH_4. In the case of CH_4, the microorganisms involved in production and consumption are frequently different. In the case of N_2O and NO, nitrifiers only act as producers, while the same denitrifier populations may both produce and consume the gases. A totally unknown population of NO consumers (perhaps N assimilatory) may also exist.
4. Production rates of trace gases are not only controlled by temperature, pH, and soil moisture, but more importantly by the quantity and quality of substrate. Substrate quality is a measure of substrate availability to the microbial community with simple sugars having a high quality and humic acids low quality. Substrate quality certainly has a large effect on the microbial community decomposing organic matter to CH_4 and, consequently, on the rate and control of CH_4 production. The same is true for heterotrophic production of N_2O and NO (i.e., denitrification and heterotrophic nitrification). It is not clear to what extent the quality of organic carbon affects N_2O/NO partitioning of denitrification. Substrate quality may also be problematic in ecosystems which are dominated by NH_4^+ supply via N_2-fixation. N_2-fixing bacteria may be very sensitive to quality and quantity of organic carbon supply, especially in the rhizosphere of plants.
5. In some systems, populations might be very low or nonexistent. Slow changes (increased atmospheric deposition of N compounds) may result in the gradual buildup of a population to such an extent that process rates become significantly different (Melillo et al., this volume). Liming of a forest soil has been shown to result in active nitrification but not until 25 weeks after liming (L. Klemedtsson, unpublished). It has also been observed that sulfate reduction started to interfere with the methane formation in a tundra mire, where the latter process 10 years ago was the dominating anaerobic electron sink. Atmospheric deposition of sulfate is most likely the reason for this change (B.H. Svensson, unpublished).

If the structure of the soil microbial community is indeed an important controller for trace gas exchange, differences in community structure between various ecosystems could be crucial in modeling exchange rates. Some participants felt that generic models should include certain aspects of population dynamics. In some instances, fluxes may be controlled by population size rather than by substrate concentration. This could play a role following changes in the ecosystem (burning, inundation, etc.) or at the onset of the rainy season after prolonged drought.

WHICH MICROBIAL PROCESSES ACCOUNT FOR N_2O, NO, AND CH_4 PRODUCTION?

Nitric and Nitrous Oxides

The biological processes which were identified as sources of N_2O in natural systems are: (*a*) chemolithotrophic (autotrophic) nitrification, (*b*) denitrification, (*c*) nonrespiratory N_2O production, (*d*) heterotrophic nitrification, and (*e*) aerobic denitrification. Of these processes, (*a*), (*b*), and (*d*) can also produce nitric oxide (NO). In addition, chemodenitrification may occur.

From the literature (e.g., Poth and Focht 1985) there is evidence that N_2O production by autotrophic nitrifiers occurs under conditions where oxygen is becoming limiting. Nitrous oxide is thus produced by a denitrification-like mechanism, where nitrite serves as an electron acceptor. However, an earlier proposed mechanism for N_2O production by the oxidative pathway from NH_4^+ to NO_2^- is still a possibility (Ritchie and Nicholas 1972).

There are few studies on the regulation of trace gas production by nitrifiers; furthermore it is difficult to use laboratory study results for extrapolation to field conditions. It is clear that under urea- or ammonium-fertilized conditions nitrification is the major source of N_2O (Bremner and Blackmer 1978). The relative importance of autotrophic nitrification for NO production in soils has not yet been established.

There is general agreement that denitrification is an important source of N_2O and possibly of NO. It is unknown why some denitrifiers stop denitrification at the N_2O step and thus waste one ATP unit, although they possess the capacity to further reduce N_2O to molecular dinitrogen. It was thought improbable that organisms lacking the ability to reduce N_2O are important in nature, even though N_2O reductase activity is known to be frequently reduced by environmental conditions. Other potential sources for N_2O that have to be taken into account are (*a*) nonrespiratory NO_3^- reduction, which could be important in acid forest soils and could be carried out by fungi, and (*b*) heterotrophic nitrification performed by a broad spectrum of bacteria and fungi. Production of trace gases by heterotrophic nitrification may be important in ecosystems from which chemolithotrophic nitrifiers cannot be isolated since NO_2^-/NO_3^- production can occur in such soils. It has to be investigated whether trace gas production observed by heterotrophically nitrifying bacteria/fungi in pure culture is of any significance in the field. Methods currently available have generally not yet been used in the field to differentiate between N_2O/NO produced by denitrifiers and nitrifiers. The use of selective inhibitors to study the role of heterotrophic nitrification should be developed. The importance of N_2O/NO production by aerobic denitrification, a reaction catalyzed by some bacteria that allows these organisms in pure culture to use NO_3^- and oxygen simultaneously as terminal electron acceptors, also remains to be assessed (Kuenen and Robertson 1988). The importance of chemodenitrification in acid soils needs to be taken into account. At the microsite level, it is possible that nitrifiers producing H^+ and NO_2^- can create suitable conditions for chemodenitrification (Firestone, this volume).

There is little information about cycling of N_2O/NO within bacterial communities. The N_2O/NO produced by one group of microorganisms can be consumed by another; who produces, who consumes, and how and when these processes are coupled is not at all understood and yet is of fundamental importance to trace gas flux.

Methane

The production of methane in natural ecosystems is as far as we know accomplished by methane-producing bacterial communities. Within these communities, different types of anaerobic bacteria cooperate in the breakdown of complex organic matter to produce CH_4 and CO_2. Methanogenic bacteria, which stand at the end of this microbial food chain, are directly responsible for CH_4 production. These bacteria can utilize only a very restricted range of energy substrates, among which H_2 and acetate are the most important. Therefore, these two compounds are the main biochemical precursors of biologically produced CH_4. The relative contribution of acetate and H_2/CO_2 to methane production can differ dramatically between different ecosystems. The reasons for these differences are not yet understood at the process or community level. Other quantitatively less important substrates are the so-called noncompetitive substrates such as methylamine or dimethylsulfide, which are believed to be converted totally to CH_4 in anoxic ecosystems.

There is still a question as to whether our list of substrates metabolized by methanogens is complete for all ecosystems or whether there are other, as yet unidentified compounds serving as precursors of biologically produced CH_4.

It is also not clear if other types of bacteria exist besides the known methanogens which can produce CH_4. The observed CH_4 formation from parathion, an insecticide, by aerobic organisms (Daughton et al. 1979) is an indication that there might exist unknown CH_4 precursors. This question is of special interest for the CH_4 emissions observed in dry aerobic soils

(Hao et al. 1988), although this may be due to seepage from natural gas reservoirs. Methane is also produced in oxic water bodies (Sieburth 1987), where strict anaerobic methanogens, if they occur, are not expected to be active.

Are There Critical Interactions between CH_4 and N-oxide Processes?

Several lines of evidence suggest that there are such interactions in various systems. The best understood interaction deals with nitrification and methane oxidation, where it has been suggested that ammonia oxygenase and methane oxygenase are able to oxidize methane and ammonia, respectively. Ammonia oxygenase in marine nitrifiers could therefore provide an important pathway for the aerobic oxidation of methane dissolved in the ocean (Ward 1987). The potential role of this process in terrestrial systems is unclear.

Studies involving nitrogen addition to experimental plots also suggest interactions between nitrogen and methane pathways but these are not well characterized. Melillo et al. (this volume) found decreases in methane oxidation rates in N-treated forest plots while Schütz et al. (1989) observed decreased CH_4 emission from paddy rice plots after application of ammonium sulfate or urea. It is not clear whether the decreased CH_4 emissions result from sulfate competition, from enhanced O_2 in the root zone, or from denitrification consuming critical substrates for CH_4 formation. Clearly, understanding the responsible mechanisms is a priority in view of the importance of rice fields as a source of atmospheric methane, of the rapidly increasing use of nitrogen fertilizers, and of increasing atmospheric deposition of nitrogen.

WHAT ARE THE MAJOR CONTROLLING FACTORS AFFECTING EMISSION RATES IN DIFFERENT ECOSYSTEMS?

Nitric and Nitrous Oxides

Based on our understanding of mechanisms which control N_2O production, the conditions under which significant N_2O flux should occur from soil systems are:

	Net effect on N_2O Production		
Condition	*Denitrif.*	*Nitrif.*	*Other*
Medium-high soil water content such that diffusional impedance of O_2 supply is high.	+	+	?

Condition	Net effect on N_2O Production		
	Denitrif.	*Nitrif.*	*Other*
High mineral-N availability such as during periods when plant consumption is absent or reduced, or during a pulse of mineralization following wetting.	+	+	+
Sporadic high C-availability such as wet-up, root die-off, etc.	+	−	+

If unidentified processes of N_2O generation (non-nitrification/denitrification) are more important than currently recognized, then this might alter the predictions. Changing soil conditions such as freeze/thaw, wet/dry, and seasonal change should be peak times for N_2O generation.

No field studies have shown clear evidence that either nitrification, denitrification, or chemodenitrification was the major source of the observed NO emissions. However, some general environmental influences are emerging. Nitric oxide emission rates are highest at intermediate soil moisture status, decreasing both in very dry soils and in soils approaching saturation. The role of temperature is only evident in moist soils, and there the emissions increase rapidly with increasing temperature. Burning off vegetation, addition of inorganic nitrogen compounds, and rewetting of parched soils all independently stimulate NO emissions. There should be integrated studies of NO and N_2O production as there are close links between the two production processes; however, they are affected to different extents by specific controlling factors.

It is essential that the relative importance of these factors for trace gas production in specific ecosystems be elucidated in order to assess the global fluxes of N_2O and NO and their response to changes in the environment.

Methane

Some methanogens live in syntrophic association with H^+-reducing bacteria, which supply the methanogenic bacteria with the energy substrate (H_2). Other bacteria, such as homoacetogens, can compete with methanogenic bacteria for H_2 and thus affect methanogenic CH_4 production. Any change in the activity of these syntrophic partners will directly affect the activity of the methanogens. Temperature has been shown to affect homoacetogens and methanogens to different extents (Conrad et al. 1989). Although models have been formulated (Lovley and Klug 1986), the quantitative nature of these microbial interactions within a bacterial community is still mostly unknown and cannot at present be extrapolated to the level of different ecosystems on a global scale or included in present generation simulation models. The same is true for the physical and chemical factors influencing biological CH_4 production. Temperature, for example, usually has a profound effect on metabolic rates of most living organisms, but measurements in rice paddy fields in China showed that there was no significant influence of temperature on CH_4 production (Schütz and Seiler, this volume). Other unrecognized factors may override the effect of temperature at these localities. Thus the factors controlling microbial CH_4 production in rice paddies, which are important sources of atmospheric CH_4, are not yet understood.

HOW DOES SUBSTRATE QUALITY AFFECT TRACE GAS EMISSION?

A major aspect of organic substrate quality is the "availability" of a given compound to microbes, i.e., how fast a unit of the compound can be consumed. There are two distinct determinants of substrate quality: (*a*) the ability to provide microorganisms with C and N for buildup of new cell material, and (*b*) the presence of enzyme systems that can catalyze the metabolism of C structures present in the substrate to provide necessary energy. Since substrates affect all microbes, substrate quality is a key regulator of trace gas fluxes from terrestrial ecosystems to the atmosphere. Although several parameters have been used as "indices" of substrate quality, there is no direct way to measure and quantify this parameter.

The main approaches to defining and assessing substrate quality have focused on either the C:N ratio of the material and/or on the presence of complex molecules that are difficult to degrade, such as lignin. C:N and lignin:N ratios are effective indices for assessing the quality of litter available to the decomposer community, with low ratios generally indicating high substrate quality. However, soil organic matter has a low C:N ratio and low lignin content, but low substrate quality. These ratios also do not address the role of micronutrients and growth-inhibiting compounds present in certain substrate, which will also affect substrate quality.

Several ecosystem factors control substrate quality. Inherent site fertility (N and water) influences vegetation type and litter quality. McClaugherty et al. (1985) found strong patterns of lignin and litter N-content across a soil fertility gradient in a deciduous forest. Dry, low N-sites are often dominated by coniferous vegetation with high lignin:N ratios. Agricultural systems, which are generally very N rich, tend to produce litter of high substrate quality, due both to the N enrichment as well as to plant breeding. Ecosystems dominated by N_2-fixing species, such as alders and black locust, have high litter quality and trace gas fluxes. Microclimate also has a strong effect on substrate quality, especially in boreal and tundra regions, where dramatic differences in vegetation type and substrate quality occur on different slope positions and faces. Soil texture is also an important environmental variable, as "you can predict the world behavior from texture data" (P. Groffman, 21 Feb. 1989, 9.53 a.m.).

Substrate quality is a useful parameter for large-scale studies of trace gas fluxes. Since there are strong patterns of substrate quality with vegetation type, large area information on vegetation type can be used as input for ecosystem trace gas flux models. It seems possible to remote-sense lignin content of plant canopies using new high resolution, aircraft mounted sensors (Wessman et al. 1988). Lignin regulates N-mineralization rates, which in turn can be a predictor of N_2O fluxes (Matson and Vitousek 1987).

There are several areas where research is needed in relation to substrate quality. Root exudates may be a significant source of highly labile substrate to microbes, but the environmental- and ecosystem-scale factors that control the nature and extent of this source are not known. Ecosystem-scale controls on substrate quality for methanogens are also poorly characterized. While the limited range of substrate that methanogens can utilize is relatively well known, the ability of different litter types to produce these substrates as they decompose has not been well studied.

Plants, and associated mycorrhizae, also compete with microorganisms for NH_4^+ and NO_3^- thus affecting the availability of substrate for NO and N_2O production.

HOW DOES TRACE GAS TRANSPORT WITHIN SOILS REGULATE EMISSION RATES?

The pathways for exchange of trace gases between the interior of the soil and the atmosphere are:

1. diffusion/advection/convection through air-filled pores in the soil,
2. diffusion through liquid water layers,
3. bubble transport through overlying water layers,
4. vascular transport within plants, and
5. displacement by water or pumping by changes in atmospheric pressure.

It is important to quantify the rate of gas transfer by these processes. Currently, diffusion processes in water and wet porous solids, and bubble transport are comparatively well understood. However, quantitative information and regulating factors are not available for advective/convective and vascular transport. The soil energy balance leads to vertical temperature and humidity profiles with temperature affecting molecular diffusion and with evapotranspiration rates regulating soil water contents. Models have been developed for diffusive fluxes including microbial production and consumption of NO (Galbally and Johansson 1989), and such models should be developed for other gases as well.

A further question concerns the chemical/biological influences that are directly coupled to transport. These are:

1. The pathway that allows diffusive escape of trace gases allows oxygen in (this does not apply to bubble transport).
2. Wetting of soils reduces the diffusive exchange of trace gases from and oxygen between the soil and atmosphere.
3. Wetting of soils stimulates heterotrophic activity, decreasing the soil oxygen concentration which is an important regulator of trace gas flux.
4. Plant growth that provides vascular transport of trace gases profoundly changes the nutrient status of the soils, adds easily available carbon sources to the system, affects oxygen status through root respiration, and through root growth provides channels for gas exchange between the soil and the atmosphere.

It is important to know the residence time of biogenic trace gases in soils. The steady-state residence time (τ_{ss}) for a gas produced in and emitted from the soil is τ_{ss} = [X]/P where P is production and [X] is the soil concentration of the trace gas. In this expression, the effects of transport from and consumption within the soil are implicit and are reflected in the soil trace gas concentration [X]. Residence time for biogenic trace gases in soil varies with environmental conditions, but in general, $\tau[N_2O] \geqslant \tau[CH_4] \gg \tau[NO]$. The residence time of these trace gases in soils is defined by the spatial extent and depth of soil that influences the trace gas emission. For NO the short lifetime corresponds to the emission being generated within the top cm or so of the soil. For N_2O, which is much more long-lived in soils, the whole active zone may contribute to the surface flux.

The size distributions of soil aggregates and pores are critically important; mean sizes are not sufficient to characterize soil textures. Macropores are important for gas transport both to and from the soil system. The importance of gravitational water flow for the transport of highly soluble gases, such as N_2O, from the biologically active soil layers to subsoil and groundwater needs to be quantified as well as the fate of gases in the groundwater system. The process rates of microbial consumption of NO and N_2O in soil is unknown. We need to know what fraction of the trace gases produced within the soil are consumed within the soil and what fraction is lost to the atmosphere.

There is a need for measurements of gas diffusion rates and of the depth distribution of trace gases in the soil atmosphere, coupled to field studies of emission rates. Dörr and Münnich (1987) have described an approach to obtaining transport-corrected CO_2 fluxes in unsaturated soils. In principle, this method should also be applicable to trace gas emissions from soils. This approach overcomes complications resulting from differences in soil physical characteristics (porosity, tortuosity, soil moisture) by using ^{222}Rn as an internal tracer. ^{222}Rn is a radioactive ($t_{1/2}$ = 3.85 d) noble gas and is produced in soils by radioactive decay of ^{226}Rn, which is distributed rather uniformly in the soil matrix. ^{222}Rn is only removed from soils by diffusion to the atmosphere and radioactive decay. ^{222}Rn and CO_2 fluxes are related as follows:

$$J_{CO_2}/J_{Rn} = (D_0^{CO_2}/D_0^{Rn})(\Delta CO_2/\Delta Rn),$$

where the J's are fluxes, $D_0^{CO_2}/D_0^{Rn}$ is the molecular diffusivity ratio, and the Δ's are concentration differences measured in a soil profile. The flux ratio is independent of soil parameters. J_{CO_2} is calculated using J_{Rn} measured by accumulation in a soil chamber, and the concentration differences are determined from samples obtained by driving a tube with a perforated tip into the soil. Thoron (^{220}Rn; $t_{1/2}$ = 55 sec) offers the potential of much shorter-term experiments. Similar measurements could be performed through studying the penetration into the soil, or release from a subsoil injection, of such stable chemical tracers as SF_6 and halocarbons.

WHAT ARE THE IMPORTANT WEATHER EVENTS AND SEASONAL CHARACTERISTICS THAT AFFECT EMISSION?

Understanding seasonal variation in biogenic trace gas flux is necessary in order to estimate annual flux rates. Analyses of seasonal variation will also be important to elucidate the critical variables that control such fluxes.

The relative importance of seasonally changing variables will change for different bioclimatic regions. Seasonal temperature variations exert a dominating influence at high latitudes but are unimportant at low latitudes, where daily temperature variations might have the overriding influence (e.g., 20°C in Venezuelan savannas). Temperature generally acts as a regulator of biological processes, although it has been observed, for example, that there is only a small temperature response of CH_4 production in some rice paddy systems. Temperature also affects physical transport processes. Increased heterotrophic activity at high temperatures consumes oxygen, which directly affects production of N_2O, NO, and CH_4.

Seasonal variability in rainfall is important not only in high-latitude and seasonal low-latitude systems, but may also strongly influence trace gas emission in tropical rain forests (Luizão et al. 1989). Rainfall, as it affects soil moisture, has a large impact on biogenic trace gas emissions because it influences heterotrophic activity (O_2 consumption) and oxygen and trace gas diffusion. Water percolation may also be an important transport mechanism for N_2O. However, changes in soil water potentials can produce results which cannot yet be explained. For example, the few available measurements of NO emission in savannas (Johansson et al. 1988; Johansson and Sanhueza 1988) indicate that there are larger fluxes during the rainy as compared to the dry season. In a tropical forest there were indications that NO was produced by denitrification rather than nitrification (Kaplan et al. 1988). Our failure to generalize from the observed emission rates reflects our poor recognition of which process is the most important for NO production under a given set of conditions. Large interannual variabilities have been observed for CH_4 fluxes in tundra and have been related to the water table height (W.S. Whalen and C.E. Reimers, unpublished). Oxidation occurs at the water table and fluxes are low when the water table is low.

Variations of solar irradiance are highly important at high latitudes, especially as they affect root exudation influencing heterotrophic gas production. UV-effects at low latitudes may also be important. Very little is known about this topic.

Seasonal variations will affect substrate input through litterfall, root turnover, and rhizodeposition (root exudation, sloughed-off root material). Biomass burning also affects soil fluxes and has been shown to increase NO flux rates (Johansson et al. 1988), but the mechanism is unknown. Nitrous oxide emissions did not increase after burning (Hao et al. 1988).

Hypothetical seasonal variations in biogenic trace gas fluxes were represented by a discussion on the flooded savannas of Venezuela (Fig. 1). Depending on the season, CH_4, NO, or N_2O would dominate the biogenic trace gas fluxes.

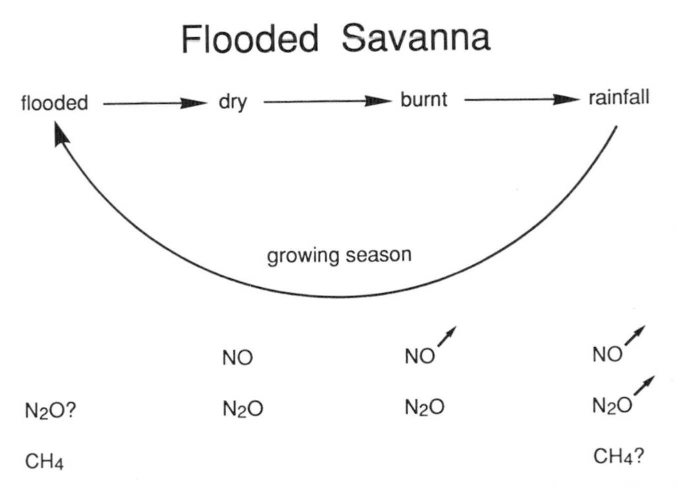

Fig. 1—Hypothetical scheme of biogenic trace gas emissions from a flooded savanna system used for agriculture.

The discussion stressed the importance of including substrate availability in any attempt to predict or model seasonal trace gas emissions. It is not adequate to simply rely on temperature and moisture dependences of the gas-producing process per se. Ecosystem carbon and nitrogen simulation models have successfully been developed for N_2O emissions in the Great Plains of the U.S. (Schimel et al. 1989). Some participants claimed, however, that events such as rainfall distribution and freeze-thaw cycles were the important controlling factors in certain systems, and present models based on simple moisture/temperature relationships and a nutrient cycling model would not be effective in describing emissions over time. The variable proportions of NO and N_2O production as a fraction of total N flow also make it uncertain whether present models are generalizable to other systems; however, this should be tested.

It is essential that simple models be developed (e.g., for NO emissions) even if they do not include the complexities of the real world. Model development and process studies must be pursued in parallel. As the need arises and as our knowledge improves, the simulation models will be refined. There is a need for a hierarchy of models ranging from detailed process models (which are useful to test hypotheses and refine our scientific understanding) up to large-scale ecosystem models in combination with Geographic Information Systems (which are needed to understand and quantify global fluxes of biogenic trace gases). Continued feedback is necessary between scientists studying organisms and process controls and those developing ecosystem simulation models. Similarly, it is necessary that field experiments and large measurement programs be developed in close cooperation between modelers, microbiologists, ecologists, and atmospheric chemists.

WHICH SOIL ECOSYSTEMS CAN ACT AS A SIGNIFICANT BIOLOGICAL SINK FOR TRACE GASES?

Nitric and Nitrous Oxides

Nitrous oxide can be consumed by soils, especially in highly anaerobic wetlands where N_2O is readily consumed as an electron acceptor by denitrifiers. While the global extent of wetlands may be small, they may have a disproportionate influence since they receive groundwater inputs from surrounding upland environments. Groundwater has been identified as a large pool in the global N_2O budget (Ronen et al. 1988), and wetlands may have a significant effect on this pool. Research addressing the levels of N_2O present in groundwater as it passes through wetlands, combined with hydrologic information on groundwater dynamics, is needed to address this question.

Studies of soil cores show that NO can be consumed in both aerobic and anaerobic soil atmospheres (Johansson and Galbally 1984). Microbial studies have shown that NO can be consumed by denitrifying bacteria but not by autotrophic nitrifying bacteria. There is also some evidence of NO uptake by chemical reactions within soil.

The outstanding problem concerning NO uptake in soils concerns the observation that in aerobic conditions there is clear evidence of NO uptake by biological processes (Johansson and Galbally 1984). It is important that the role of non-nitrifying, non-denitrifying soil bacteria in consuming NO in soil be examined. It is hypothesized that NO uptake by these bacteria could be an unrecognized form of microbial nitrogen assimiliation.

Methane

In general, methane oxidation is as important as its production in regulating emissions from ecosystems (Fig. 2). Direct and tracer measurements of CH_4 oxidation rates show that methane-oxidizing bacteria have maximum activities at interfaces separating methane sources from terminal electron acceptors. Methane-oxidizing organisms in such locations serve as an oxidizing biofilter; they have been observed to modulate methane fluxes drastically. Such zones of methane-oxidizing activity are present in marine sediments, where anaerobic methane oxidation occurs (Reeburgh 1980; Alperin and Reeburgh 1984) and at the oxycline in lakes (Rudd et al. 1974) or the water table in wetlands (W.S. Reeburgh, unpublished), where there is aerobic methane oxidation (Rudd et al. 1974; Kuivila et al. 1988). The role of anaerobic methane oxidation in soils is not known.

Methane fluxes from the atmosphere to the soil have been observed in chamber studies, indicating that soils can consume atmospheric methane. Chamber and jar experiments show oxidation of atmospheric methane down to levels of 0.1 ppm over periods ranging from one hour to several days. As above, the methane-oxidizing activity is located in a subsurface zone; rapid consumption of large additions of methane suggests that a population of nonobligatory methanotrophs may be present. We have a poor understanding of atmospheric gas transport into soils, so it is difficult to assess the importance of soils as an atmospheric methane sink.

Oxidation (biological and abiotic) of methane produces large fractionation in C and H isotopes. Fractionation factors for these reactions have been obtained experimentally from pure cultures and isolated systems, as well as by modeling environmental data. However, a concern was expressed as to how generally applicable fractionation studies are, and it was agreed that further studies are necessary. Stable isotopes provide a powerful tool for identifying atmospheric methane sources and modifications prior to its

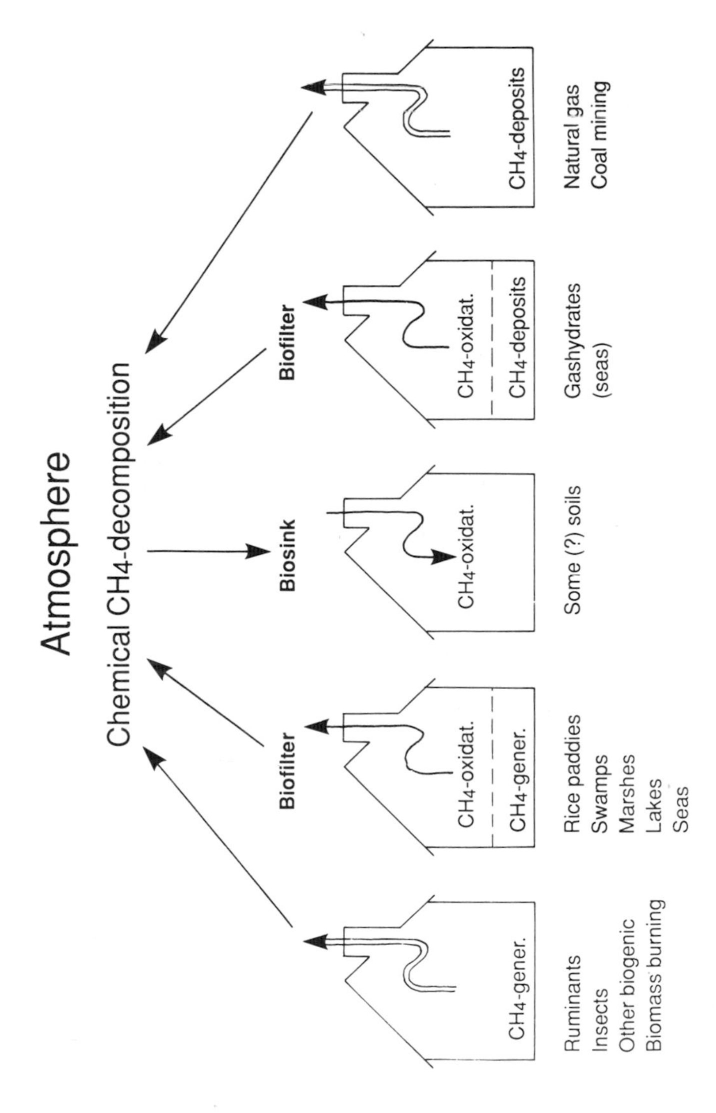

Fig. 2—Schematic diagram of different methane-producing and -consuming systems, which regulate the methane content of the atmosphere.

introduction to the atmosphere. Stable isotope information provides an additional constraint on the global methane budget.

Recent C and H isotope measurements on methane from a variety of sources suggest that methane oxidation may not be as important as the rate measurements suggested above (Wahlen, unpublished). This evidence includes:

1. CH_4 diffusing from rice paddies, a range of wetlands, and tundras, and CH_4 bubbles from the same environments grouped tightly by source on a δD vs. $\delta^{13}C$ plot. The lack of positive correlations ("mixing lines") suggests no oxidation.
2. Hydrogen isotopes from the same environments show close correspondence with parent water. This observation also suggests no methane oxidation or a balance between production and consumption.

Possible causes of the apparent conflict in the rate and the isotopic evidence for the importance of methane oxidation include: (*a*) no fractionation during oxidation, which is unlikely, (*b*) quantitative consumption (this situation appears to prevail in marine sediments [Alperin et al. 1988]), (*c*) patchy distribution of oxidation (no oxidation in isotope samples), and (*d*) balance between production and consumption producing a constant isotope effect. Careful measurements in areas where oxidation is demonstrated are needed.

HOW WILL CLIMATE CHANGE AFFECT TRACE GAS FLUXES?

By climate change we mean temporal trends in the values and quantities of the physical and chemical variables that define the environmental conditions for life on Earth. This is an overriding issue in that climate and life have co-evolved through time, and because human activities are increasingly capable of causing changes to the climate of regions and of the entire globe more rapidly than has ever been previously experienced by humanity.

We focus here on six physical and chemical climate variables: temperatures, carbon dioxide increase, precipitation changes, regional acidification, increased ultraviolet light exposures, and increased amounts of tropospheric ozone. In considering the many linkages between these changes and their effects on microbial processes that can both respond to the changes and cause further changes, we have in turn perceived and described a massive and complex system. It is not yet clear to us how we should perform research on such complex systems, e.g., how to experiment upon and model essential components of the system while trying to understand the other components and the whole system simultaneously. The strengths and time constants of many feedback processes must be determined before completely reliable research strategies can be agreed upon.

Given the tremendous complexities and linkages that are apparent between microbial process rates and climatic variables, and the rapid climate changes that are expected in the coming decades, we agreed that surprises will probably be in store for us, that researchers should perform experiments and model studies with large perturbations as well as with more modest predicted changes, and that discontinuous system responses should be watched for, and not just smooth, monotonic functions.

Temperatures

Averaged global temperatures are expected to increase due to the greenhouse effects of increasing CO_2, CH_4, CFCs, N_2O, and tropospheric O_3. Increases of even a few degrees in seasonally averaged temperatures will increase the bulk rates of many microbial processes noticeably (ignoring substrate limitations). However, larger concerns surround prolonged episodes of extreme events; more very hot and very dry summer days may be more influential than, e.g., annual increases. Such climatic extreme events are likely to become highly important in changing trace gas exchange and therefore deserve a large fraction of our scientific attention.

Climatic predictions for key regions are not yet very reliable, but the central regions of continents will probably be warmer in summer, and high-latitude regions will be warmer in winter. Likely consequences are that soil moisture will be decreased in summer at midlatitudes and that there will be more time between spring thaw and winter freezing in boreal regions. Tundra may become drier in summer and permafrost may melt at depth. As soil moisture is a key variable affecting biogenic trace gas emissions, this may have a profound effect on fluxes. The natural emissions from trees of nonmethane hydrocarbon gases, such as isoprene, may increase.

CO_2 Increases

To the extent that CO_2 concentrations in air limit plant growth, a fertilization effect can arise with increased global atmospheric CO_2 concentations. Plants whose photosynthesis employs the C_3 mechanism will respond more than C_4 plants. Profound changes could ensue in community structures of plants and microbes through, e.g., altered carbon to nitrogen ratios in plant litter and in parasite activity. Initial increases in plant growth rates (1–5 years) can lead to later changes in contents and quality of soil organic matter (5–100 years). Such changes will affect trace gas fluxes.

Other changes could occur in the use of soil water by plants through altered patterns of stomatal opening or in stomatal number.

Precipitation Changes

Of all the climatic changes that are expected to occur in the next several decades, precipitation amounts and spatial patterns are probably the least predictable, yet they are extremely important as forces of ecological change. If more rain falls in existing ecosystems, we expect more anaerobic conditions in soil, with more denitrification, more methanogenesis, and less uptake of atmospheric methane. However, if precipitation patterns change appreciably, ecosystems may fail. They may also move geographically toward more favorable locations, if the speed of the climatic transition is not too high. How to do meaningful process-based research on the effect of such changes is not obvious.

Regional Acidification

In industrially affected midlatitude areas, the principal acids in precipitation are nitric and sulfuric acid. Along with the more familiar and serious consequences of acid rain, we must be aware of effects of nutrient loading of nitrate and sulfate on trace gas processing by microbes. Recent research indicates that increased nitrate loading increases N_2O emission and decreases the ability of soils to oxidize CH_4 (Melillo et al., this volume). Such phenomena need field, laboratory, and modeling research. The possibility of increased acid and nutrient deposition in certain tropical regions should also be considered.

Ultraviolet Light (UV-B)

As stratospheric ozone is decreased globally, more UV-B will reach the Earth's surface on average. There will continue to be substantial, natural day-to-day fluctuations but high dosages will become more frequent. Larger changes in UV-B intensity are expected at high latitudes, but some shorter UV-B wavelengths will be seen closer to the equator.

UV-B may not be a major effector of soil microorganisms due to the lack of penetration in the soil proper. Plants and phytoplankton may display lower productivity and some mortality, and increased emissions of nonmethane hydrocarbons may occur. Atmospheric photochemical rates will increase in the troposphere, thus decreasing the residence times of some gases.

Tropospheric Ozone

Exposure to ozone causes several kinds of responses in plants: decreased photosynthesis rates in some, increased emissions of nonmethane hydrocarbons, and possibly increased rates of root respiration. Also, if large regions of the troposphere experience increased ozone levels, the rates of conversion of NO_x and SO_2 pollution to nitric and sulfuric acids will increase so that acid deposition will occur closer to the NO_x and SO_2 sources. Increases of tropospheric O_3 are expected to accompany increased industrialization, fossil-fuel usage, and biomass burning but the rates of increase and the eventual impact on biogenic trace gas fluxes are problematic.

CONCLUSIONS

The group decided on five priority areas for research:

1. *To determine what processes are involved in production of CH_4, N_2O, and NO in different ecosystems, and if they are constant or change with time, and why different ecosystems have evolved different production pathways.*

It is essential to understand which parameters will determine what type of process goes on in a particular ecosystem, and if there are seasonal or diurnal differences. It is necessary to identify a minimum of parameters that are necessary to predict which processes are the dominant ones (e.g., nutrients, T, pH, moisture, soil texture). Can this be extrapolated to all ecosystems? How important is it to consider differences in the soil microbial communities that catalyze processes involved in the production of CH_4, N_2O, and NO? Are differences in species diversity reflected in differences at the process level?

For this purpose it is necessary to analyze the metabolism of CH_4, N_2O, and NO with respect to metabolic processes, microbial communities, and soil environmental parameters. Metabolic processes can be studied with isotope and inhibitor techniques. Microbial community composition could be analyzed with respect to the presence of key enzymes (e.g., nitrite reductase by using antibody techniques or molecular genetic tools).

It should also be realized that many microorganisms and metabolic capabilities are simply not yet understood, e.g., H^+-reducing bacteria which are involved in the methanogenic microbial communities or heterotrophic nitrifiers that seem to be important for N_2O production in certain soils.

2. *To describe characteristics of soils that influence the areal and depth distributions of production/consumption reactions modulating trace gas emission.*

We require a much better understanding of the influence of soil physical characteristics (porosity, tortuosity, water content) on residence times, transport, and exchange of trace and atmospheric gases within soils. Future emphasis should be placed on field methods that permit determination of changes induced by precipitation events, freezing, and desiccation without modifying soil systems. Inert gas equilibrium methods (SF_6, CFC tracers) show promise in solving residence time/transport problems in seasonal studies. Methods for transport-correcting gas fluxes, such as the ^{222}Rn method, should be employed.

We also have a poor understanding of biological and chemical conditions and processes important in controlling areal and depth distributions of production/consumption reactions. Future studies should emphasize rate measurements on whole cores as well as the depth distribution of production and consumption.

3. *To develop mechanistic models that include microbiological and physical/chemical processes applicable at the scale of trace gas exchange experiments and to test these models with field and laboratory experiments.*

The roles of these models are (*a*) to identify key parameters to be measured in trace gas exchange measurements and (*b*) to aid in the design of key experiments that will verify or contradict our current understanding of these processes. Some of the key types of measurements in trace gas exchange studies include nutrient availability, microbial community activity, the physicochemical state of the environment, and the gas transfer characteristics within the system.

4. *To develop ecosystem scale models for biogenic trace gas fluxes.*

Biological/physicochemical estimates of trace gas exchanges on local to global scales rely on biogenic trace gas fluxes from the soil as a lower boundary condition. In order to provide a linkage from mechanistic soil models to the global scale, ecosystem models must be developed that are based on abiotic factors and substrate content and availability. These models must be compatible with the scale on which they will be tested, because the levels and variability of the controlling abiotic and substrate factors will vary at different scales. Ecosystem models should also be compatible for use at the plant community level for linkage to micrometeorological flux measurements and models describing trace gas fate in the boundary layer. The validity of the ecosystem models must be tested at other sites and ecosystems to evaluate their generality.

5. *To assess what quantitative changes in N_2O, NO, and CH_4 fluxes can be expected in response to physical and chemical climate changes.*

Key climate perturbations that will affect biogenic trace gas emissions are temperature, precipitation, and acidification/deposition. These perturbations will primarily be expressed through changes in soil moisture and the quality and quantity of substrate. Global climate models need to be developed that give approximate predictions of changes in soil moisture contents. Key research questions in this area include the effect of CO_2 enrichment and N deposition on the C:N ratio of plant tissue. Nitrogen deposition also directly affects substrate for nitrification and denitrification. We need to quantify changes in substrate quality in response to wide changes in climate as well as the relationships between change in substrate quantity and quality and N_2O, NO, and CH_4 flux.

REFERENCES

Alperin, M.J., and W.S. Reeburgh. 1984. Geochemical observations supporting anaerobic methane oxidation. In: Microbial Growth on C-1 Compounds, ed. R.L. Crawford and R.S. Hanson, pp. 282–289. Washington: Am. Soc. Microbiol.

Alperin, M.J., W.S. Reeburgh, and M.J. Whiticar. 1988. Carbon and hydrogen isotope fractionation resulting from anaerobic methane oxidation. *Glob. Biogeochem. Cyc.* 2:279–288.

Bremner, J.M., and A.M. Blackmer. 1978. Nitrous oxide: emission from soils during nitrification of fertilizer nitrogen. *Science* 199:295–296.

Conrad, R., F. Bak, H.J. Seitz, B. Thebrath, H.P. Mayer, and H. Schütz. 1989. Hydrogen turnover by psychrotrophic homoacetogenic and mesophilic methanogenic bacteria in anoxic paddy soil and lake sediments. *FEMS Microb. Ecol.* 62:285–294.

Daughton, C.G., A.M. Cook, and M. Alexander. 1979. Biodegradation of phosphonate toxicants yields methane or ethane on cleavage of the C-P bond. *FEMS Microbiol. Lett.* 5:91–93.

Dörr, H., and K.O. Münnich. 1987. Annual variation in the soil respiration on selected areas of the temperate zone. *Tellus* 39B:114–121.

Galbally, I.E., and C. Johansson. 1989. A model relating laboratory measurements of rates of nitric oxide production and field measurements of nitric oxide emission from soils. *J. Geophys. Res.* 94D:6473–6480.

Hao, W.M., D. Scharffe, P. Crutzen, and E. Sanhueza. 1988. Production of N_2O, CH_4 and CO_2 from soils in the tropical savanna during the dry season. *J. Atmos. Chem.* 7:93–105.

IGBP. 1988. The International Geosphere-Biosphere Programme: A Study of Global Change. A Plan for Action. IGBP Report 4. Stockholm: The Royal Swedish Academy of Sciences.

Johansson, C., and I.E. Galbally. 1984. Production of nitric oxide in a loam under aerobic and anaerobic conditions. *Appl. Env. Microbiol.* 47:1284–1289.

Johansson, C., and E. Sanhueza. 1988. Emission of NO from savanna soils during rainy season. *J. Geophys. Res.* 93D:14193–14198.

Johansson, C., H. Rodhe, and E. Sanhueza. 1988. Emission of NO in a tropical savanna and a cloud forest during the dry season. *J. Geophys. Res.* 93D:7180–7192.

Kaplan, W.A., S.C. Wofsy, M. Keller, and J.M. da Costa. 1988. Emission of NO and deposition of O_3 in a tropical forest system *J. Geophys. Res.* 93D:1389–1395.

Kuenen, J.G., and L.A. Robertson. 1988. Ecology of nitrification and denitrification. In: The Nitrogen and Sulphur Cycles, ed. Z.A. Cole and S.Z. Ferguson, pp. 161–218. Cambridge: Cambridge Univ. Press.

Kuivila, K.M., J.M. Murray, A.H. Devol, M.E. Lidstrom, and C.E. Reimers. 1988.

Methane cycling in the sediments of Lake Washington. *Limnol. Ocean.* 33:571–581.
Lovley, D.R., and M.J. Klug. 1986. Model for distribution of sulfate reduction and methanogenesis in freshwater sediments. *Geochem. Cosmochim. Acta* 50:11–18.
Luizão, F., P. Matson, G. Livingston, R. Luizão, and P. Vitousek. 1989. Nitrous oxide flux following tropical land clearing. *Glob. Biogeochem. Cyc.*, in press.
Matson, P.A., and P.M. Vitousek. 1987. Cross-system comparisons of soil nitrogen transformations on nitrous oxide flux in tropical forest ecosystems. *Glob. Biogeochem. Cyc.* 1:163–170.
McClaugherty, C.A., J. Pastor, and J.P. Aber. 1985. Forest litter decomposition in relation to soil nitrogen dynamics and litter quality. *Ecology* 66:266–275.
Parton, W.J., A.R. Mosier, and D.S. Schimel. 1988. Rates and pathways of nitrous oxide production in shortgrass steppe. *Biogeochem.* 6:45–58.
Poth, M.A., and D.D. Focht. 1985. ^{15}N kinetic analysis of N_2O production by *Nitrosomonas europea*: an examination of nitrifier denitrification. *Appl. Env. Microbiol.* 49:1134–1141.
Reeburgh, W.S. 1980. Anaerobic methane oxidation rate depth distributions in Skan Bay sediments. *Earth Plan. Sci. Lett.* 47:345–352.
Ritchie, G.A.F., and D.J.D. Nicholas. 1972. Identification of the sources of nitrous oxide produced by oxidative and reductive processes in *Nitrosomonas europea. Biochem. J.* 126:1181–1191.
Robertson, G.P., and J.M. Tiedje. 1987. Nitrous oxide sources in aerobic soils: nitrification, denitrification and other biological processes. *Soil Biol. Biochem.* 19:187–193.
Ronen, P., M. Magaritz, and E. Almon. 1988. Contaminated aquifers are a forgotten component of the global N_2O budget. *Nature* 335:57–59.
Rowland, F.S., and I.S.A. Isaksen, eds. 1988. The Changing Atmosphere. Dahlem Konferenzen. Chichester: Wiley.
Rudd, J.W.M., R.D. Hamilton, and M.E.R. Campbell. 1974. Measurement of microbial oxidation of methane in lake water. *Limnol. Ocean.* 19:519–524.
Schimel, D.S., W.J. Parton, C.V. Cole, D.S. Ojima, and T.G.F. Kittel. 1989. Grassland biogeochemistry: links to atmospheric processes. *Clim. Change*, in press.
Schütz, H., A. Holzapfel-Pschorn, R. Conrad, H. Rennenberg, and W. Seiler. 1989. A three-year continuous study on the influence of daytime, season, and fertilizer treatment on methane emission rates from an Italian rice paddy field. *J. Geophys. Res.*, in press.
Sieburth, J.M. 1987. Contrary habitats for redox-specific processes: methanogenesis in oxic waters and oxidation in anoxic waters. In: Microbes in the Sea, ed. M.A. Sleigh, pp. 11–38. Chichester: Ellis Harwood.
Ward, B.B. 1987. Kinetic studies on ammonia and methane oxidation by *Nitrosococcus oceanus. Arch. Microbiol.* 147:126–133.
Wessman, C.A., J.P. Aber, D.L. Peterson, and J.M. Melillo. 1988. Remote sensing of canopy chemistry and nitrogen cycling in temperate forest ecosystems. *Nature* 335:154–156.

Exchange of Trace Gases between Terrestrial Ecosystems and the Atmosphere
eds. M.O. Andreae and D.S. Schimel, pp. 97–108
John Wiley & Sons Ltd

Regional Extrapolation of Trace Gas Flux Based on Soils and Ecosystems

P.A. Matson[1], P.M. Vitousek[2], and D.S. Schimel[1]

[1]*Ecosystem Science and Technology Branch*
NASA–Ames Research Center
Moffett Field, CA 94035, U.S.A.

[2]*Department of Biological Sciences*
Stanford University
Stanford, CA 94305, U.S.A.

Abstract. Spatial and temporal variability in trace gas fluxes is often seen as a major impediment to the development of regional and global budgets. We suggest that much of that variability is controlled by a discrete number of factors, including soil/parent material, climate, vegetation, and topography. Examination of fluxes and the processes that control fluxes along gradients of those factors provides a basis for extrapolation strategies that use rather than attempt to average out ecosystem variability. Such an approach should not only improve extrapolations and budgets, but also provide information about ecosystem functioning and serve as the basis for predictive modeling.

INTRODUCTION

One of the major impediments to the development of regional and global budgets of trace gases is the substantial spatial and temporal variability in biogenic gas fluxes. Numerous ground-based studies have examined fluxes from a variety of ecosystems; these studies have revealed great variability with time in a given site, among ecosystems of a particular type (e.g., temperate forests), and between systems of different types (e.g., temperate forest vs. temperate wetland) (Keller et al. 1983, 1986; Goodroad and Keeney 1984; Harriss et al. 1985; Sebacher et al. 1986; Matson and Vitousek 1987; Schmidt et al. 1988; to name only a few).

Most regional and global budgets have been extrapolated from flux estimates collected at reasonably well-studied sites that are assumed to be representative of the region (Keller et al. 1983; McElroy and Wofsy 1986). This approach has provided a much needed starting point for global budgets and has served to focus further research in the more important regions. Nevertheless, the assumption of "representativeness" is often arbitrary and may lead to erroneous estimates. On the other hand, any attempt to account for variability by randomly selecting sample sites within a biome would require an overwhelming number of samples, many with daunting logistical problems. We believe that a better approach is to measure and attempt to understand the spatial and temporal variability in fluxes and the biological processes that control them.

Understanding and accounting for this variability is profitable in several ways. First, as will be discussed later, such information can be used as a basis for extrapolation to regional and global budgets. Second, it will tell us something about distribution and timing of fluxes. This is useful because the episodic co-occurrence of high levels of trace gases (especially the strongly interacting ones) can be more important in controlling atmospheric processes than are globally averaged fluxes or concentrations. Global and regional flux values (such as those obtained from atmospheric models or regional aircraft measurements) provide only relatively crude information on distribution.

Third, the timing and magnitude of biogenic gas fluxes reveal much about the nature and dynamics of source ecosystems. If the distribution of fluxes is known, it can be used to understand important elements of the functioning of ecosystems. Finally, it is only through an understanding of the control of fluxes that we can predict how anthropogenic and natural disturbances and global climate change will affect global fluxes in the future.

So the question is, "How can we take advantage of spatial and temporal variability rather than being overwhelmed by it?" Most biogenic trace gas fluxes arise from relatively well-defined biological processes; these processes are variable because they are controlled by factors that vary in space and time. In a general sense, these factors are the state factors of Jenny (1980), i.e., climate, parent material, topography, organisms, and time. Some of these factors vary discontinuously in space and thus may be classified (e.g., soil types and parent material; the mosaic of ecosystem types that results from human or natural disturbance). Others (including climate) vary more or less continuously in space. In such cases, ecosystem properties can be thought of as continua on scales from local to global (Whittaker 1975).

The examination of biogeochemical processes and trace gas fluxes as they vary along gradients of these controlling state factors can provide quantitative relationships between known or measurable variables and flux. These relationships then provide a basis for stratified sampling on regional and global scales, and they can be used directly for extrapolation using ground-based global data bases. Such relationships are also amenable both to remote sensing and to modeling approaches. In the following sections, we present several approaches that utilize knowledge about spatial and temporal variability in terrestrial ecosystems to estimate trace gas flux on regional to global scales.

STRATEGIES FOR EXTRAPOLATING FLUXES

Estimates Using Empirical Relationships between Ecosystem Classes and Fluxes

Where distinct and easily identified differences in ecosystem structure and vegetation composition coincide with differences in ecosystem function and trace gas flux, delineation of ecosystems into functionally different types is a useful basis for measurement strategies and for extrapolation. Measured fluxes for distinct ecosystem types within a region can be multiplied by areal extents of the types to yield an estimate for a region or the globe; this estimate should be an improvement over those that have assumed a "representative" flux because it acknowledges some of the variability in natural systems.

This approach has in fact been used quite extensively, but only for rather coarse scale applications. For example, most global budgets place natural ecosystems into general groupings such as temperate forests, tropical forests, grasslands, and agricultural lands; a flux that is assumed to be representative of each stratum is then multiplied by the areal extents of each stratum. The sources of error in this kind of estimate arise not only from the enormous variability in flux within the groups, but also from inaccuracies in estimates of areal extent.

On the other hand, some global and regional budgets have used finer-scale detail to improve estimates. Matthews and Fung (1987) used global data bases of vegetation, soil, and inundation characteristics to stratify all wetlands into five major groups. Global methane emission and the distribution of methane fluxes from wetlands were calculated using different methane fluxes for the five types. Similarly, Robertson and Rosswall (1986) stratified West Africa into cover classes on the basis of savanna vs. forest and disturbance or land-use type, and estimated fluxes of nitrogen in the soil and between the soil and atmosphere using that stratification.

These studies have used global data bases of various types to provide information on areal extent. Another obvious source of such data is vegetation classifications based on satellite and aircraft remote sensing data. While this approach seems quite straightforward and is widely discussed, it has not been used often. One recent study used this approach to estimate methane fluxes from the Florida Everglades. The region was stratified into major wetland types using Thematic Mapper (TM) classifications, and fluxes for representatives of each type were measured through an annual cycle (D. Bartlett et al., submitted). We have used a similar approach combining TM-based classifications of Wyoming sagebrush steppe ecosystems (Reiners et al. 1989) with ground-based measurements of nitrogen turnover in the soil and nitrous oxide flux over a 2-year period (P. Matson, unpubl. data). The products of both of these studies are region-wide estimates of annual trace gas flux and information about the distribution of flux on the landscape. These studies also incorporate information on the factors that control flux.

The combination of remote sensing and ground-based flux measurements can also be applied to examining the effects of land-use change and disturbance on trace gas fluxes. We compared fluxes in a variety of soils and in pastures near Manaus, Brazil and found wide variation in fluxes in forests which differed in soil fertility (Matson and Vitousek 1987), and greatly elevated fluxes in pastures that had been converted from terra firma forest (P. Matson, P. Vitousek, and G. Livingston, unpubl. data). After extrapolation using TM and SPOT classifications of the area, it was estimated that while pastures accounted for only 11% of the area (Swanberg, unpubl. data), they were responsible for over 40% of the nitrous oxide flux.

In general, this approach has a number of strengths. It makes use of natural variability in ecosystems instead of attempting to average out that variability, thereby accounting for at least part of a large source of error in global budgets. If the classes selected are appropriate and reliable, it reduces the number of sites which must be sampled in order to find a statistically sound average flux value for a region. It also provides information about the distribution of flux and identification of hot spots—sites that are disproportionately important in atmosphere–biosphere interactions—that may be important in terms of their response to change.

The approach also has limitations. Classifications that are not based on characteristics strongly associated with trace gas fluxes are not useful; in many areas it may be possible to group ecosystems into types that are different in structure or vegetation composition but that are not substantially different in soil or vegetation processes and trace gas flux. Ground-based measurements are essential to show that the types are indeed *functionally* different. Moreover, because the approach relies on empirical flux data collected at a given time, it cannot account for variations in flux that result from year-to-year variations in climate and inundation patterns. Finally, estimates of flux can only be as good as the ground-based or satellite data bases; for many variables, data are not georeferenced; for others, data are of uneven quality and detail depending on region and political base. (For further discussion of data bases see Stewart et al., this volume.) Because of these limitations, we believe that classification-based extrapolations are better used in combination with gradient and modeling approaches discussed below.

Extrapolations Based on Classifications Coupled with an Understanding of Controls

An alternative way to examine fluxes of trace gases is through consideration of the gradients of factors which control both trace gas flux and ecosystem properties and processes. This approach utilizes elements of the classification schemes outlined above, but incorporates spatial and temporal information on the factors that vary more or less continuously. This conceptual approach is particularly useful where the gradients which control ecosystem properties also control trace gas fluxes more or less directly. For example, in the case of methane flux from natural wetlands, climate and topography (with some contribution from other factors) interact to control the hydroperiod and rate of decomposition in wetland ecosystems. These in turn control the balance between methane sources and methane sinks in soils. Classification of wetlands can be based on slowly varying topographic and vegetative characteristics; continuously varying climatic time series can be used to modify the empirical relationships between the classified type, soil moisture, and flux.

The approach is useful even where the connections between controlling gradients and microbial mechanisms have not been fully characterized—but here it must be applied with considerable caution. For example, nitrogen generally is present in much greater supply (relative to other elements) in lowland tropical forests in comparison with temperate forests; the mean C:N:P ratio of forest litterfall is 810:23:1 in lowland tropical forest versus 630:11:1 in temperate and boreal forest (Vitousek and Sanford 1986). Nitrous oxide fluxes from tropical forests are much greater as well, to the extent that they dominate the natural source budget (Keller et al. 1983, 1986). Nevertheless, there is substantial variation among tropical forests in the amount and relative availability of nitrogen, and this variation is highly correlated with fluxes of nitrous oxide (Fig. 1 [Matson and Vitousek 1987]). Here stratification and subsequent extrapolation can be developed on the basis of soil fertility, determined substantially by climate, parent material, soil age, and topography.

Even on local scales there may be correlated variation in ecosystem processes and trace gas flux resulting from topographic, edaphic, or other state factor controls. Schimel et al. (1988) reported a consistent variation in nitrogen fluxes as a function of landscape position in short-grass steppe.

We believe that these patterns can be useful for the development of regional and global estimates of trace gas sources and sinks. Techniques for estimating regional fluxes based on continuous variation along gradients of controlling factors are not widely used, but some methods exist (e.g., Robertson 1987) and should prove rewarding here. Moreover, we believe that analysis of trace gas fluxes in terms of controlling gradients can stimulate

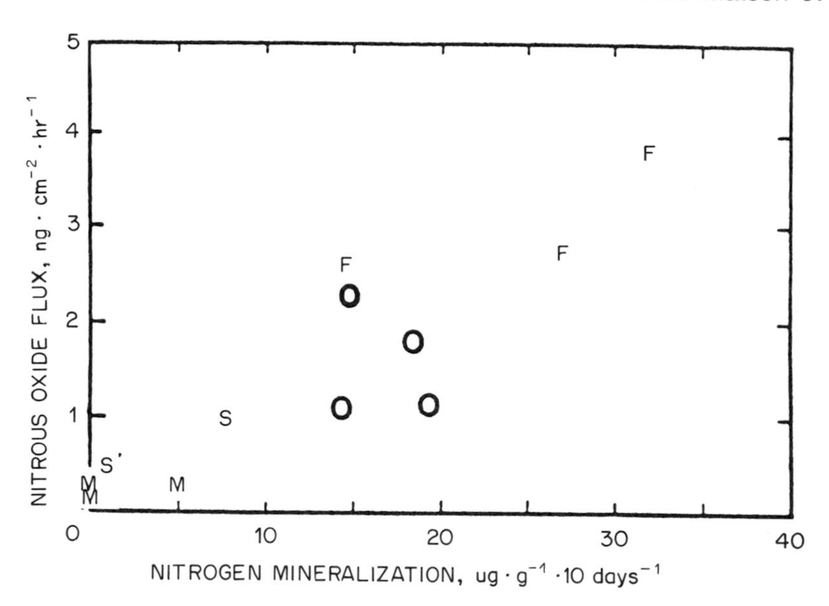

Fig. 1—The association between net nitrogen mineralization (release of inorganic N from decaying organic matter) and nitrous oxide flux in a range of tropical forests which differ in soil fertility. "O" represents forests on oxisols and ultisols; "F" represents forests on fertile soils (e.g., alfisols); "S" on sandy soils; "M" tropical montane forests. Adapted from Matson and Vitousek (1987).

more focused work on the microbial mechanisms controlling flux, as well as provide interim improvements in regional and global estimates.

The Use of Simulation Models to Estimate Gas Flux

The examination of ecosystem processes and fluxes as a function of controlling factors leads quite logically to the use of simulation models to predict and extrapolate flux. This approach is similar to the use of ecosystem or vegetation type for extrapolation, but it incorporates mechanistic components directly and is thus better able to predict the effects of novel combinations of controlling factors than are approaches based on empirical relationships. Moreover, simulations can better reflect the interactions of the controlling variables both spatially and temporally.

Simulation of regional trace gas emissions requires modeling at both ecosystem (seasons to year) and physiological (minutes to days) time scales

(Schimel et al. 1989). For example, the availability of substrate and of limiting nutrients can be simulated by an ecosystem model on subannual time steps, while physiological models can simulate consequences of diurnal physiological regulation. In this way, the effects of soils, topography, or long-term climate can be combined with the effects of weather to calculate fluxes.

Simulation of spatial variability in fluxes can be driven by the response of ecosystem processes and trace gas fluxes to spatial variation in soils and climate. The microbiological model of Parton, Mosier et al. (1988) accurately predicted differences in soil nitrogen cycling and trace gas fluxes as they varied between fine-textured lowland and coarse-textured upland soils (Fig. 2); it also captured some but not all of the temporal variation in flux. Burke et al. (1989) predicted regional nitrogen gas fluxes for northeastern Colorado using a geographic data base containing soil and climatc data as drivers for an ecosystem model (CENTURY) (Fig. 3).

Clearly, questions remain concerning the identification of climatic, topographic, soil, and other variables required to drive simulation models. Even where the key driving variables are known, more information is needed on the fundamental relationships between these variables and flux. Development of simulation models is best done through examination of ecosystem processes and trace gas fluxes as they vary along regional gradients of state factors. Implementing this information for use in global budgets requires the development of remotely sensed or ground-based geographically referenced data bases on hydrologic, edaphic, climatic, and vegetation attributes worldwide.

Validation on a Regional Scale

The most promising approach for regional-scale model validation involves the use of aircraft measurements of regional flux (Matson and Harriss 1988). Aircraft-based flux and boundary layer studies provide integrated estimates of exchange between terrestrial systems and the atmosphere. It should be possible to use this data as an independent data set against which to test results of extrapolations of flux measurements or of model simulations. The major impediment to this validation approach is that most ecosystem models predict fluxes for seasons annually, yet the aircraft data can only provide measurements for short time periods under specific environmental conditions. The solution to this problem lies in the development of models that account for geographic variability but that also simulate short time-scale ecosystem response to continua of climatic and environmental conditions. A number of such models are under development.

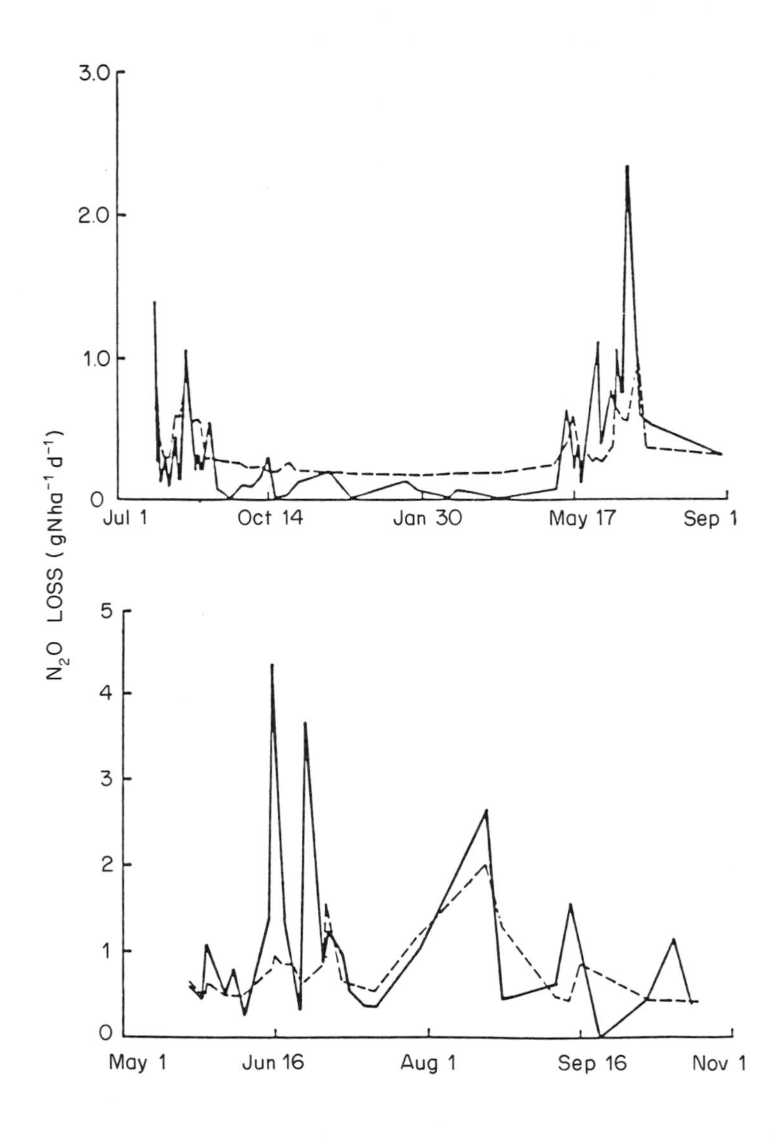

Fig. 2—Simulated and observed nitrous oxide fluxes from upland (top graph) and lowland (bottom graph) soils within the shortgrass steppe. The lowland soils are coarse in texture, with lower inorganic nitrogen turnover. Simulations (shown with the dashed lines) were made using a physiologically based model of gas production, driven by daily weather data. Detailed information on soil inorganic nitrogen is required by this model; it may be derived from field data or simulated by an ecosystem model. Adapted from Parton, Mosier et al. (1988).

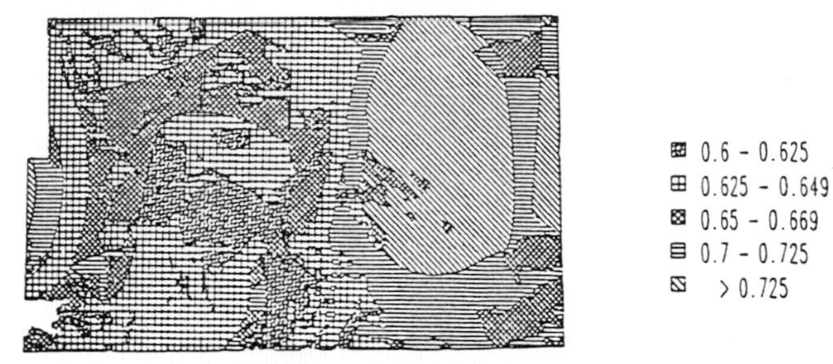

Fig. 3—Regional nitrous oxide emission from northeastern Colorado (g ha^{-1} yr^{-1}) simulated by an ecosystem model coupled to a geographic data base containing monthly weather data and soil properties. This type of model requires simple input data and simulates trace gas production with little temporal detail. Adapted from Burke et al. (1989).

CONCLUSIONS AND POINTS OF CONTROVERSY ON EXTRAPOLATION APPROACHES

Extrapolations based on ecosystem characteristics and ecosystem modeling have demonstrated the utility of this approach in:

1. The identification of biomes which dominate global fluxes of trace gases (e.g., tropical forest for nitrous oxide), so that measurements and mechanistic work can be concentrated where they are most important.
2. The identification and characterization of local "hot-spots" of flux (e.g., marginal fens within boreal wetland complexes). Moreover, in both 1 and 2 the ecosystem approach forces a systematic consideration of possible source regions.
3. The development of models which are based on factors controlling trace gas flux and which thereby focus attention on the microbial and physical mechanisms causing this flux.

Each of these three have led in particular cases to better regional and/or global estimates of trace gas flux than can be obtained from studies of well-characterized "representative" sites; their contribution will increase as the insights they yield are applied more widely.

A number of questions or points of controversy remain. First—why worry about extrapolating trace gas fluxes to large areas when they can be measured over those same areas by aircraft or by global sampling networks? We believe that extrapolation is necessary so that estimates and predictions can be made for areas and for time spans which cannot be sampled by aircraft. Moreover, information on which ecosystems are responsible for fluxes (information that is not often available on a sufficiently fine scale in aircraft-based estimates) is necessary so that predictions of responses to local or global change are possible (Harriss 1988). Of course, extrapolations are difficult or impossible to validate in the absence of integrated aircraft-based measurements of flux.

Second—isn't spatial and temporal variability within sites or types simply too large to permit any kind of classification-based stratification and extrapolation? Certainly variability in trace gas flux is extraordinary compared to many other ecosystem characteristics: single chamber measurements within most sites could yield very high or very low fluxes. However, there are many cases in which differences between sites or soil types are consistent and statistically very clear (e.g., Fig. 4, which contrasts clay and sand soils in Amazonia), and it is more useful to include than to ignore such differences in the construction of regional or global budgets.

Third—do we know enough about the mechanisms controlling trace gas fluxes to use ecosystem-level simulation models to estimate regional or global fluxes? Probably not for most gases and for most areas. Even the model of Matthews and Fung (1987), which simply classified wetland types based on vegetation, inundation, and soils (rather than simulating fluxes as a function of each of those factors separately), had to rely on incomplete

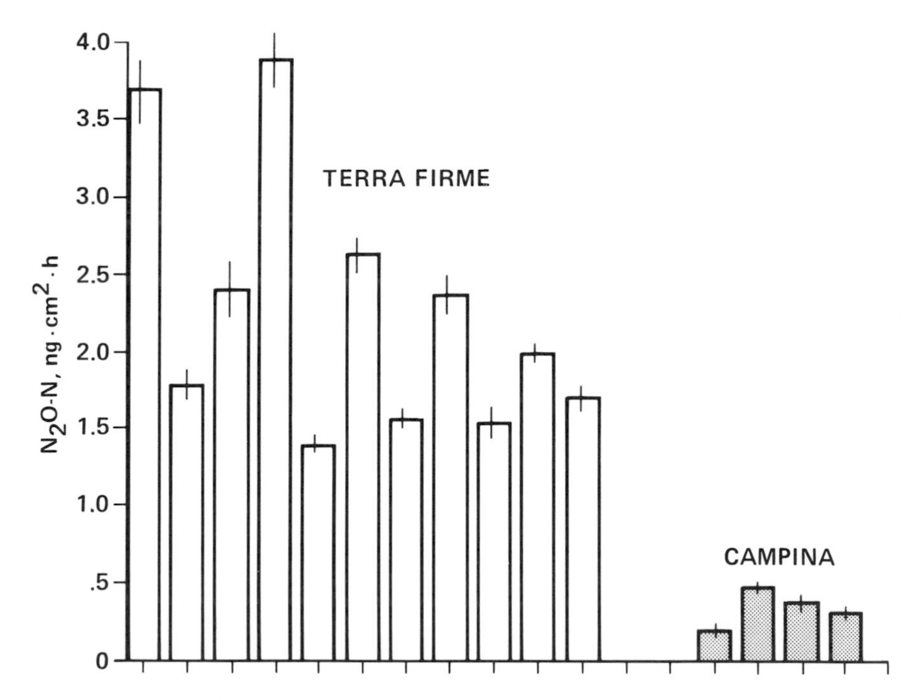

Fig. 4—Nitrous oxide fluxes from Amazonian forest ecosystems near Manaus, Brazil during April 1987. Terra firme are upland forests on clay soils; campina are forests on sand soils. Data from Matson, Vitousek, and Livingston (unpublished).

and even assumed fluxes in order to extrapolate. Most often, we do not have a clear understanding of the individual and interacting effects of fertility, moisture, temperature, substrate availability, pH, and soil texture on ecosystem processes and trace gas flux; this lack limits the accuracy of simulation models. However, ecosystem models have reached a stage of development where broad spatial and temporal trends in nutrient turnover and productivity can be predicted.

Initial efforts to link trace gas emissions to such models are promising. More importantly, however, such models can and should be used to shape the design of experiments analyzing the mechanisms controlling trace gas flux. The understanding gained thereby can feed back to improve the models. In some ways, ecosystem models for forest and grasslands are at a stage comparable to atmospheric models: they are capable of simulating large-scale patterns, but simulation of more detailed dynamics and mechanisms is still under development. This development will be stimulated by the use of existing ecosystem models and by a willingness to modify them as more information on mechanisms becomes available, in part as a consequence of their utilization.

Acknowledgements. Support for P.A. Matson and P.M. Vitousek was provided by NASA's Terrestrial Ecosystems Program and Biospheric Research Program. A National Research Council Fellowship and the National Science Foundation Ecosystem Studies Program provided support for D.S. Schimel. We thank J. Melillo for comments on an earlier draft.

REFERENCES

Burke, I.C., D.S. Schimel, C.M. Yonker, W.J. Parton, and L.A. Joyce. 1989. Regional modeling of grassland biogeochemistry using GIS. *Landsc. Ecol.*, in press.

Goodroad, L.L., and D.R. Keeney. 1984. Nitrous oxide emission from forest, marsh, and prairie ecosystems. *J. Env. Qual.* 13:448–452.

Harriss, R.C., et al. 1988. How has the atmosphere already changed? In: The Changing Atmosphere, ed. F.S. Rowland and I.S.A. Isaksen, pp. 201–216. Dahlem Konferenzen. Chichester: Wiley.

Harriss, R.C., E. Gorham, D.I. Sebacher, K.B. Bartlett, and P.A. Flebbe. 1985. Methane flux from northern peatlands. *Nature* 315:652–654.

Jenny, H. 1980. Soil Genesis with Ecological Perspectives. New York: Springer.

Keller, M., T.J. Goreau, S.C. Wofsy, W.A. Kaplan, and M.B. McElroy. 1983. Production of nitrous oxide and consumption of methane by forest soils. *Geophys. Res. Lett.* 10:1156–1159.

Keller, M., W.A. Kaplan, and S.C. Wofsy. 1986. Emission of N_2O, CH_4, and CO_2 from tropical soils. *J. Geophys. Res.* 91:11791–11802.

Matson, P.A., and R.C. Harriss. 1988. Prospects for aircraft-based exchange measurements in ecosystem studies. *Ecology* 69(5):1318–1325.

Matson, P.A., and P.M. Vitousek. 1987. Cross-system comparisons of soil nitrogen transformations and nitrous oxide flux in tropical forest ecosystems. *Glob. Biogeochem. Cyc.* 1(2):163–170.

Matthews, E., and I. Fung. 1987. Methane emission from natural wetlands: global distribution, area, and environmental characteristics of sources. *Glob. Biogeochem. Cyc.* 1(1):61–86.

McElroy, M.B., and S.C. Wofsy. 1986. Tropical forests: interactions with the atmosphere. In: Tropical Rain Forests and the World Atmosphere, ed. G.T. Prance, pp. 33–60. Boulder, CO: Westview.

Parton, W.J., A.R. Mosier, and D.S. Schimel. 1988. Rates and pathways of nitrous oxide production in a shortgrass steppe. *Biogeochem.* 6:45–58.

Reiners, W.A., L.L. Strong, P.A. Matson, I.C. Burke, and D.S. Ojima. 1989. Estimating biogeochemical fluxes across sagebrush-steppe landscapes with thematic mapper imagery. *Rem. Sens. Env.*, in press.

Robertson, G.P. 1987. Geostatistics in ecology: interpolating with known variance. *Ecology* 68:744–748.

Robertson, G.P., and T.H. Rosswall. 1986. Nitrogen in West Africa: the regional cycle. *Ecol. Monogr.* 56:43–72.

Schimel, D.S., W.J. Parton, C.V. Cole, D.S. Ojima, and T.G.F. Kittel. 1989. Grassland biogeochemistry: links to atmospheric processes. *Clim. Change*, in press.

Schimel, D.S., S. Simkins, T.H. Rosswall, A.R. Mosier, and W.J. Parton. 1988. Scale and the measurement of nitrogen gas fluxes from terrestrial ecosystems. In: Scales and Global Change, ed. T.H. Rosswall, R.G. Woodmansee, and P.G. Risser, pp. 179–193. SCOPE. Chichester: Wiley.

Schmidt, J., W. Seiler, and R. Conrad. 1988. Emission of nitrous oxide from temperate forest soils into the atmosphere. *J. Atmos. Chem.* 6:95–115.

Sebacher, D.I., R.C. Harriss, K.B. Bartlett, S.M. Sebacher, and S.S. Grice. 1986. Atmospheric methane sources: Alaskan tundra bogs, an alpine fen, and a subarctic boreal marsh. *Tellus* 38B:1–10.

Vitousek, P.M., and R.L. Sanford, Jr. 1986. Nutrient cycling in moist tropical forest. *Ann. Rev. Ecol. Syst.* 17:137–167.

Whittaker, R.H. 1975. Communities and Ecosystems. New York: MacMillan.

Exchange of Trace Gases between Terrestrial Ecosystems and the Atmosphere
eds. M.O. Andreae and D.S. Schimel, pp. 109–118
John Wiley & Sons Ltd

Regional Extrapolation: Vegetation–Atmosphere Approach

B.B. Hicks

NOAA/ATDD
456 South Illinois Avenue
Oak Ridge, TN 37831, U.S.A.

Abstract. Available data on the exchange of trace gases from the surface into the atmosphere is generated primarily by methods that yield locally representative flux information over short time intervals. On the other hand, regional-scale and global models require exchange data averaged over grid cells that are typically 100 km × 100 km, or larger. The extension from local data to grid-cell averages is far from trivial. Micrometeorological formulations of atmosphere–surface exchange are potentially well suited for use in such extrapolation, because they describe area averages at the outset (usually several hectares) and include descriptions of the terrain and vegetation properties that influence the exchange. However, these formulations lack the biological and chemical detail provided by models developed as a consequence of laboratory studies of soils and vegetation, and of field studies using cuvettes and chambers. Methods are required, therefore, to include the detail of these biological and chemical models in the micrometeorological routines, and then to use the modified micrometeorological methods to produce area averages. Here, methods are proposed for both purposes, based on the techniques developed to describe fundamental characteristics of vegetation in atmospheric models. This, then, constitutes the "Vegetation–Atmosphere Approach" of the title of this chapter. Micrometeorological relationships are presented in a format that offers an opportunity to include the results of detailed point-by-point emission models in their specification of appropriate zero plane displacements, roughness lengths, and average surface–air concentrations. Extension to larger areas is then proposed to make use of a replicated application of the modified micrometeorological relations that are derived.

INTRODUCTION

Basic information on the exchange of N_2O, NO_x, and CH_4 from the surface derives mainly from studies of emissions from samples of soils and vegetation, usually in carefully controlled conditions (e.g., see Delwiche et al. 1978). These studies relate emission rates to factors that are indicative of (or that determine) relevant biological processes. Companion papers in this volume present detailed information on the experimental methods that are employed and on emission models that are developed from the field and laboratory observations that are made. These methods rarely produce emission estimates that are representative of areas greater than square meters. In the context of regional-scale models and global balances, these methods constitute a "bottom-up" approach.

The variability of emission rates in field conditions can be addressed directly using chambers of a wide range of configurations, also described elsewhere among the companion papers. In intensive field campaigns, these small-scale flux-measuring devices produce data that are often primarily designed to develop and test emission models, over areas of special interest. These same data provide much of the currently available information on emission rates. In many studies, global estimates of emissions are then derived by straightforward, linear extrapolation based on representative or stratified sampling of fluxes and an estimate of the total amount of surface to which such observations apply. It is well known that this kind of extrapolation is highly susceptible to error and that significant bias could be introduced as a consequence. Better methods are required to estimate large area averages from limited small area measurements of fluxes and of the quantities that control them.

An additional strong need for improved confidence in constructing area averages from limited site-specific data arises in the case of numerical modeling. Global-scale models have grid cells that are typically 100 km $\times$ 100 km, if not larger. It becomes necessary to link emission estimates that are characteristic of small areas to the substantially larger areas of interest to the modeling community. Satellite data are the most frequently used basis for such extrapolation (see Salop et al. 1983, for example). The problems that arise are likely to be substantially more severe than can be resolved by satellite imagery, especially because of the wide range of scales of processes that contribute—from the scales of microbes to those of terrain and synoptic meteorology. A stepwise approach to the problem would subsume small-scale complexities in the parameterizations used in increasingly larger-scale models. This must be accomplished without loss of either accuracy or confidence in the results.

It is sometimes assumed that the results derived from application of chamber techniques or by use of models derived from chamber measurements can be averaged over time and space, without the need to account for interactions involving terrain and meteorology. However, it is also well appreciated that emission rates from soils can be affected by atmospheric turbulence (Kimball and Lemon 1971). The interaction between atmospheric properties and emissions from soils remains a subject of research.

AREA AVERAGING BY MICROMETEOROLOGY

Just as chambers and detailed biological models focus on producing point-by-point estimates of surface fluxes (F_c, yielding concentrations in air in contact with the surface C_o), so micrometeorological methods consider area averages, as viewed from aloft. In micrometeorological relationships, individual sources and sinks at the surface are not considered in detail. Instead, these relationships consider average exchange rates associated with a distribution of sources and sinks throughout the lowest layer of the atmosphere containing the vegetation canopy. This is the case even for heat, moisture, and momentum, but the same arguments apply to scalar quantities in general (see Denmead and Bradley 1987).

It is informative to consider the way in which surface fluxes can be formulated in terms of in-air gradients. The relationships between fluxes and gradients are complex, especially in and above vegetation canopies. The complexity is largely due to the fact that the sources and sinks that need to be considered are not necessarily coincident within canopies (e.g., see Smith et al. 1985). Momentum and heat fluxes are mostly derived from foliage in the upper part of a canopy, but CH_4 and N_2O are mainly from the soil itself (e.g., see Keller et al. 1986); however, gaseous nitrogen species can also be emitted and adsorbed by leaves (Farquhar et al. 1983).

Consider the relationship between the wind at some height z and the momentum flux $M = -d_o u_*^2$. Here, d_o is air density and u_* is the friction velocity. If the surface is horizontally homogeneous and if conditions are not changing with time, then the local wind speed gradient can be written as:

$$(\partial u/\partial z) = u_*/(k(z - d)) \tag{1}$$

where k (= 0.4) is the von Karman constant and d is the displacement height of the wind profile due to the presence of the canopy. Hence, by integration,

$$u = (u_*/k).1n((z - d)/z_o) \tag{2}$$

where z_o is a constant that is a measure of the roughness of the site, the so-called roughness length. In practice, d is conveniently associated with the level of action of the drag force of the wind on the plant canopy (see Thom 1971). The roughness length is a measure of the spatial variability and size of roughness elements that make up the surface. Experience has shown that d is in the range 0.7 h to 0.8 h, and z_o is typically 0.1 (h − d) to 0.2 (h − d).

It should be emphasized that Eq. 2 applies above the level at which individual roughness elements influence the wind, i.e., above a height typically taken to be more than ten times z_o above d. Thus, Eq. 2 should not be expected to apply below about 1.3 times the average height of the trees. It should also be emphasized that the present treatment is of the neutral case, in which there is no effect of buoyancy.

The derivation of the wind profile relationship given above is quite standard. The relevance of such arguments to the case of present interest arises when similar consideration is given to the local gradient $(\partial C_z/\partial z)$ of some chemical species, above the same canopy. The spatial average flux density of this chemical species is $\langle F_c \rangle$. (Angle brackets are used to indicate a spatial average.) If conditions are horizontally uniform, and not changing with time, then the flux can be considered constant with height above the canopy and can be related to the vertical concentration gradient via

$$\langle F_c \rangle = -d_o K_c \, (\partial C_z/\partial z) \qquad (3)$$

where K_c is the relevant eddy diffusivity. As before, conditions are taken to be neutral, so that eddy transfer is adequately formulated in terms of the friction velocity u_*. Rearrangement of Eq. 3 and integration yields a relationship that parallels the wind profile law:

$$\langle C_o \rangle - \langle C_z \rangle = (\langle F_c \rangle / d_o k u_*).1n((z - d)/z_{oc}) \qquad (4)$$

where z_{oc} is another constant of integration, again representing surface roughness but not necessarily the same as the momentum roughness length z_o; C_z is concentration at height z. In some literature (e.g., Hicks et al. 1979; Munro 1985) the displacement height d is also taken to be quantity- or pollutant-specific, but this remains the subject of research and debate (see Garratt 1978; Raupach and Thom 1981).

A major limitation resulting from the simplification of the surface in meteorological relations such as those developed above is that biological considerations (which influence exchange at the surface itself in a complicated and highly variable manner) cannot be included in great detail in micrometeorological relationships. Instead, the consequences of the complexity of surface behavior are generally contained within three key terms of Eq. 4: the displacement height and the roughness length appropriate for the situation under consideration, and the specification of the average surface concentration $\langle C_o \rangle$. This issue has been the subject of considerable attention in research on the deposition of trace gases (see Hicks and Matt 1988).

The quantities that are used in micrometeorological relations to describe exchange between the air and the underlying surface are necessarily averages. The size of the area over which these averages apply cannot be specified with precision, since the size will vary with the prevailing conditions.

However, it is unlikely that micrometeorological relations will be applicable to surfaces smaller than about 100 m × 100 m. Above an area of this size, it is the average flux density that is the subject of micrometeorological formulations. Likewise, the variability of the flux across the averaging area influences the "roughness" scale length, z_{oc}, in Eq. 4. The actual effective height at which the source or sink is located might well differ from that appropriate for momentum transfer, especially in the present cases for which the emission is primarily (if not entirely) from the ground surface and not from the foliage. In this instance, it seems highly likely that there will be a need to take into account that the displacement height, d_c, is different from the momentum displacement height d.

A link with the chamber and point-specific modeling predictions can now be contemplated. The micrometeorological relations require the use of average flux densities and simplified descriptions of the appropriate source and sink distributions. On the other hand, the detailed microbiological models describing emission rates from soils produce data (on both F_c and C_o) on a far smaller spatial scale than the micrometeorology can accommodate directly. An appropriate linkage might be to employ point-specific, biologically oriented techniques to quantify the average flux densities needed by the micrometeorology. Spatial variability might then also be described quantitatively, such as by using some form of probability distribution. In this way, the consequences of biological detail needed to assess surface exchange rates might be properly included in the computation of surface emissions, and the products of these computations might be handed over to micrometeorological methods. It is these micrometeorological relationships that might then provide the capacity for extrapolation to greater scales. In essence, the intent would be to subsume the consequences of local biological and subsurface chemical complexity (and the resulting variability of surface emissions) in comparatively simple descriptive variables (analogous to z_o, d, and $\langle C_o \rangle$), and to express these properties in terms of appropriate controlling variables such as land use, water availability, soil type, etc.

EXTENSION TO GRID–CELL AREAS: THE MULTIPLE RESISTANCE MODEL

Figure 1 illustrates how various processes combine to control transfer through the lower atmosphere. The scheme that is shown derives directly from studies of deposition, in which a trace gas present in the air with concentration $\langle C_z \rangle$ is transferred to an underlying medium in which the concentration is negligible. The model is a multiple resistance scheme, in which different resistances represent the effects of turbulent transfer through the atmosphere (R_a), transfer by molecular diffusion across the layer of air in immediate contact with the surface is represented by a spatial average boundary layer

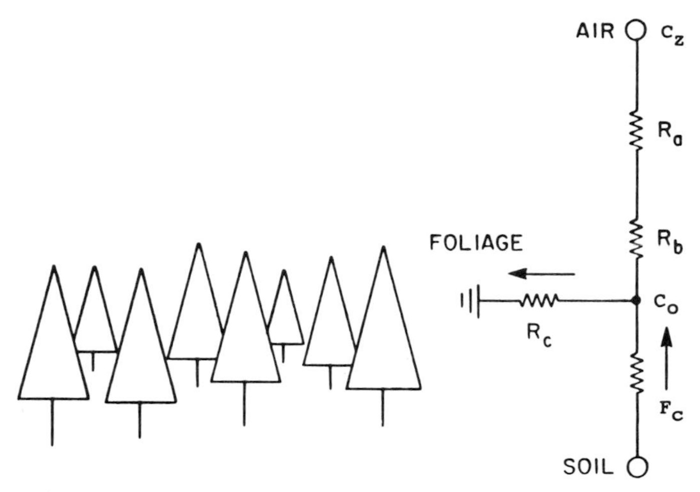

Fig. 1—A schematic drawing showing the various resistance pathways associated with transfer involving soil, foliage, and air. The case represented is that of emission of a trace gas from the soil, with a pathway involving foliage.

resistance (R_b), and the resistance associated with uptake by the receiving medium is R_c. Hence, Fig. 1 represents a more general situation than that of prime interest here.

Following an analogy with the flow of current through an electrical circuit, the spatial average atmosphere–surface exchange rate $<F_c>$ is normally expressed in terms of atmospheric concentration $<C_z>$ and the average concentration in air in contact with the surface $<C_o>$ as

$$F_c = d_o(<C_o> - <C_z>)/(R_a + R_b). \tag{5}$$

Comparison between Eqs. 4 and 5 reveals that the aerodynamic resistance can be written as

$$R_a = (1/ku_*).\ln((z - d)/z_o) \tag{6}$$

where it is now assumed that $(z - d) >> z_o$. The boundary layer resistance can be expressed as

$$R_b = (1/ku_*).\ln(z_o/z_{oc}). \tag{7}$$

In the case of deposition, a third resistance term is usually considered to represent transfer across the surface on which deposition is actually occurring. In the present case of emission from soils under a vegetation canopy, the role of this resistance is difficult to see.

Now consider some target area, say 100 km × 100 km, and divide the area with a 1 km grid. (The reasons for selecting these scales are related to model resolution at the larger end and to the areal extent of micrometeorological averaging methods at the other.) The intent, then, is to describe the area average flux $<F_c>$ when the spatial distributions of the component resistance and concentration fields are not well resolved by either measurements or theory.

Expansion of Eq. 5 and integration across the array of sampling locations reveals that the critical considerations are the spatial coefficients of variation of each of the contributing properties ($<C_z>$, $<C_o>$, R_a, and R_b and the correlations that exist among them. Separate parts of any overall program must therefore quantify each of these factors.

Coefficients of Variation

— For $<C_z>$, aircraft sampling programs could provide the necessary answers, although with limited temporal coverage. Atmospheric and canopy models of varying levels of complexity could be used to interpolate among data provided by aircraft.

— For $<C_o>$, there appears no option but to rely on a surface-sampling program yielding fine resolution data on C_o from which $<C_o>$ can then be derived arithmetically, or on the outputs (averaged similarly) of emissions models that adequately simulate the roles of controlling biological and other surface variables.

— For R_a, appropriate data could be obtained using an array of measurements of appropriate properties, "benchmarked" against field data obtained in intensive studies or employing dynamic wind field models. A relatively simple method for considering possible effects of terrain complexity would be to adapt the "bounding" methods described by Hicks and Meyers (1988).

— For R_b, calculations might be based on detailed topographic and surface cover information.

Spatial Correlation Distributions

Expansion of Eq. 5 and deriving an area average from the point-by-point application of it leads to additional terms that are not easily addressed. In particular, a number of cross-correlation terms arise.

— For the correlation terms, there appears no option but to rely on an experimental program, involving flux measurements at locations selected to represent the range of conditions encountered in the target area.

In concept, a selected area could be investigated by independently studying the spatial fields of each of the various resistance and cross-correlation components, and then by combining these fields as in a series of "overlays."

Once a spatial average flux is derived, then it is necessary first to verify the results and second to explore ways by which the result might be extended to other areas that are not so intensively studied.

Verification

Suppose that a suitable test area is selected. It is then required to measure the average rates of exchange between the surface and the atmosphere as air moves across this area. The methods that are available are considered in the companion papers. Three methods offer promise:

1. Budget methods, in which horizontal fluxes across the upwind and downwind edges of the area are measured and intermediate surface exchange is deduced from the difference.

   In practice, budget methods have been tried and have been shown to be extremely limited; the average surface flux is small in comparison to other error terms in the mass budget equation (such as "leakage" between the lower and upper troposphere and the consequences of wind field variability) which cannot be measured well.
2. Aircraft methods and towers, in which eddy correlation measurements are made along paths across the area in question.

   Aircraft methods are available and can be applied for trace species such as ozone. Application to other trace gases is presently problematic; we are at the mercy of the limitations of chemical detection technology.
3. Tracer techniques in which the air or the surface is enriched in some chemical species selected to mimic the behavior of the chemical of interest.

   Tracer techniques have been explored in concept, but have not yet been exploited because of the cost of the tracers. However, these methods are worthy of continuing attention.

Table 1 shows some characteristic time and space scales of the various available techniques for assessing wide area averaged exchange rates.

Extrapolation Using Probability Density Functions

Local measurements (or model predictions) of localized emission rates might be used to specify the input for micrometeorological models of area averaged air–surface exchange. In this way, predictions of localized emissions over small areas could be combined to drive micrometeorological relations, by specifying the appropriate surface roughness parameter. For some trace gases, the extension from localized and detailed flux specification to areal micrometeorological averaging could be tested using either tower gradient studies or eddy correlation measurements.

TABLE 1.

	Mass Budget	Aircraft Fluxes	Tracer Methods
Chemical Species Tested so far	SO_2	O_3, CO_2 H_2O CH_4	
Sample Integration Time:	Many hours	Fraction of an hour	Not yet determined
Spatial Sampling Capability: (km × km)	10 × 10 to 100 × 100	5 × 20 to 5 × 100	1 × 1 to 10 × 10

It is not feasible to conduct such detailed studies everywhere, nor to perform enough to "inventory" every kind of surface that would be confronted in constructing global averages. Instead, some method is needed to interpolate among the data points that are obtained and to extrapolate to situations outside the existing range of field experience. It seems likely that probabilistic methods might be appropriate, in which each of the surface emission fields, as well as resistance and correlation fields addressed above, is described in terms of probability distributions (or probability density functions) rather than gridded point values.

It is not clear what form would be best for such a probabilistic approach. Indeed, several options are immediately apparent without much indication of how to select one rather than another. Among the techniques that might prove worth investigating are variations on structure function analysis and fractal analysis.

CONCLUSIONS

Outputs of detailed evaluations of surface fluxes, either by chamber measurements or by the application of detailed biophysical models, do not mesh readily with micrometeorological models, since the latter require spatial averages as inputs. To combine the products of emission investigations conducted across a nonuniform area, a probabilistic approach is suggested in which the effects of surface flux heterogeneity would be described in parameters analogous to the roughness length, displacement height, etc. of micrometeorological relationships. This approach offers promise of a method by which detailed microbiological models of trace gas emission from soils could be used to drive micrometeorological relations, which would then be suitable for spatial extrapolation.

Extension from areas where microbiological and micrometeorological relations are available to the larger areas that constitute grid cells of numerical models would require extensive development. The research methods could be based on a point-by-point application of the micrometeorological techniques, but alternative methods would need to be considered. Among the possible alternative approaches are methods based on combining individual probabilistic descriptions of separate contributing factors across the areas of interest.

Acknowledgements. This work was conducted under the sponsorship of the National Oceanic and Atmospheric Administration, as part of work for the U.S. National Acid Precipitation Assessment Program, Task Group on Atmospheric Chemistry.

REFERENCES

Delwiche, C.C., S. Bissell, and R. Virginia. 1978. Soil and other sources of nitrous oxide. In: Nitrogen Behavior in Field Soil, vol. 1 of Nitrogen in the Environment, ed. D.R. Nielsen and J.G. MacDonald, pp. 459–476. New York: Academic.

Denmead, O.T., and E.F. Bradley. 1987. On scalar transport in plant canopies. *Irrig. Sci.* 8:131–149.

Farquhar, G.D., R. Wetselaar, and B. Weir. 1983. Gaseous nitrogen losses from plants. In: Gaseous Loss of Nitrogen from Plant-soil Systems, ed. J.R. Freney and J.R. Simpson, pp. 161–180. The Hague: M. Nijhoff.

Garratt, J.R. 1978. Flux profile relations above tall vegetation. *Q. J. Roy. Meteor. Soc.* 104:199–211.

Hicks, B.B., G.D. Hess, and M.L. Wesely. 1979. Analysis of flux-profile relations above tall vegetation—an alternative view. *Q. J. Roy. Meteor. Soc.* 105:1074–1077.

Hicks, B.B., and D.R. Matt. 1988. Combining biology, chemistry, and meteorology in modeling and measuring dry deposition. *J. Atmos. Chem.* 6:117–131.

Hicks, B.B., and T.P. Meyers. 1988. Measuring and modeling dry deposition in mountainous areas. In: Acid Deposition at High Elevation Sites, ed. M.H. Unsworth and D. Fowler, pp. 541–552. Dordrecht: Kluwer.

Keller, M.K., W.A. Kaplan, and S.C. Wofsy. 1986. Emissions of N_2O, CH_4, and CO_2 from tropical forest soils. *J. Geophys. Res.* 91:11791–11802.

Kimball, B.A., and E.R. Lemon. 1971. Air turbulence effects upon soil gas exchange. *Soil Sci. Soc. Am. Proc.* 35:16–21.

Munro, D.S. 1985. Internal consistency of the Bowen ratio approach to flux estimation over forested wetland. In: The Forest Atmosphere Interaction, ed. B.A. Hutchison and B.B. Hicks, pp. 395–404. Dordrecht: Reidel.

Raupach, M.R., and A.S. Thom. 1981. Turbulence in and above plant canopies. *Ann. Rev. Fluid Mech.* 13:97–129.

Salop, J., N.T. Wakelyn, G.F. Levy, E.M. Middleton, and J.C. Gervin. 1983. The application of forest classification from Landsat data as a basis for natural hydrocarbon emission estimation and photochemical oxidant model simulations in southeastern Virginia. *J. Air Poll. Cont. Assn.* 33:17–22.

Smith, M.O., J.R. Simpson, and L.J. Fritschen. 1985. Spatial and temporal variation of eddy flux measures in the roughness sublayer above 30 m Douglas Fir forest. In: The Forest Atmosphere Interaction, ed. B.A. Hutchison and B.B. Hicks, pp. 563–590. Dordrecht: Reidel.

Thom, A.S. 1971. Momentum absorption by vegetation. *Q. J. Roy. Meteor. Soc.* 97:414–428.

Exchange of Trace Gases between Terrestrial Ecosystems and the Atmosphere
eds. M.O. Andreae and D.S. Schimel, pp. 119–133
John Wiley & Sons Ltd

Global-scale Extrapolation: A Critical Assessment

I. Aselmann

Max-Planck-Institute for Chemistry
Division of Atmospheric Chemistry
P.O. Box 3060
6500 Mainz, F.R. Germany

Abstract. Extrapolation of gas flux measurements on a global scale usually lacks statistical significance due to the paucity of data and sampling and is therefore often speculative. This is illustrated in examples for radon, carbon dioxide, and methane. Speculation may be scientifically helpful, however, if discussion and research are stimulated. Limited financial funding, among other constraints which will certainly accompany future measurement programs, makes it likely that such programs will also suffer from unrepresentativeness according to strict statistical requirements. However, data, theory, and models mutually stimulate each other. Improved data bases of environmental parameters and increased model capability together with isotope data will constitute a tool of increasing potential for checking extrapolations.

INTRODUCTION

Over the past several decades environmental sciences have increasingly changed from being more qualitative to more quantitative disciplines. The characterization of ecological systems in terms of stocks (e.g., biomass), their temporal variations, and the associated fluxes became particularly prominent after World War II. This shift towards quantification was supported not only by the scientific community but was further promoted by political groups as public awareness of anthropogenically induced environmental changes grew. The issue of a global atmospheric CO_2 increase, for example, was qualitatively addressed by scientists several decades ago. Extensive effort to quantify the CO_2 budget, however, was not supported until the 1970s, when the global implications of the CO_2 increase first penetrated public recognition. As the publicity of the CO_2 problem grew, large research programs in atmospheric, oceanic, and biospheric sciences were initiated to study the global carbon cycle. Around 1980 a considerable number of scientific papers were published on the "missing carbon sink," which was required to balance the global budget. However, all lacked numerical significance since the general problem of balancing large numbers associated with high degrees of uncertainty remained unresolved. Since then the interest in global matter cycling has broadened, particularly due to the quick development of atmospheric science, and today includes most chemically reactive trace gases. The problems of quantifying the relevant sources and sinks for these other gases are in principle similar to those for CO_2, although each trace gas has unique characteristics.

VARIATION AND HETEROGENEITY

The key problem of quantifying stocks and fluxes is the continuously changing nature of this world. The degree of constancy depends on the entity involved and on the time scale considered. We are facing areal heterogeneity and temporal variation on every scale, from seconds to millenia and from millimeters to thousands of kilometers, in the atmosphere, oceans, soils, and vegetation (Rosswall et al. 1988). In respect to vegetation, long-term changes over decades or centuries occur in the structure of communities driven by changing competitiveness of individuals and external impacts such as storms, lightning, fires, or pest outbreaks. These long-term dynamics of vegetated areas are expressed in the concept of successional phases. The various phases of succession have distinct features of species composition and of matter and energy turnover. They further influence soil properties and microclimate, which in turn may change the competitiveness of individuals, driving the system to a theoretical final or climax phase. Relevant changes on smaller time scales comprise seasonal, daily, and diurnal changes that become apparent in buildup and senescence or consumption of plant matter. More subtle changes occur, for example, through translocation of carbohydrates within the plants, different photosynthetic production and gas exchange rates of leaves, turnover rates of fine roots, or seasonally and daily varying decay and soil respiration rates. All these may influence flux rates of biogenically released trace gases, such as CO_2, CO, CH_4, N_2O, isoprene, and terpene, just to name a few.

Furthermore, heterogeneity in space is introduced by the innumerable variations of environmental factors that define the habitat of any plant community and which may be summarized in the two broad categories of soil properties and climate. The anthropogenic factor has been shown to become increasingly important as humanity introduces changes within decades and centuries, changes which otherwise would take place in geological time scales.

As a consequence of this complex nature, direct determination of any mean global flux or stock is the determination of its heterogeneity and variability. This requires "fair sampling," sampling that is representative, objective, and unbiased. Suitable techniques comprise (*a*) systematic sampling with repetitive measurements to evaluate the variability of the data and (*b*) random sampling which yields unbiased samples and which allows appraisal of heterogeneity and precision of the data (Jeffers 1988). Ideally, this first implies that all units with similar characteristics in respect to the flux or stock studied have been quantified, and second, that a mean value for each unit, based on sound sampling, has been derived.

The following sections focus on the extent to which these prerequisites have been met for three different species: radon, carbon dioxide, and methane. These gases have primarily terrestrial sources but are quite different in their chemical and radiative behavior and may thus be taken as examples for the whole range of trace gases relevant to atmospheric and marine science.

^{222}Radon

^{222}Rn is a noble gas and the first gaseous product of the uranium decay chain. Its sources and sinks are well defined because it is almost exclusively released from soils (oceans are insignificant) and the only sink is radioactive decay. Its half-life time is 3.82 days. Emanation from soils has been found to be quite variable from place to place. The main influencing factors are: ^{226}Radium content of the soil, structure of soils (mainly porosity and grain size), and turbulent exchange of air in the boundary layer. There is not much seasonal variation unless the ground is snow or ice covered, which would inhibit the radon flux from soil to atmosphere. Vegetation may play a role in the gas exchange, when under waterlogged conditions radon-enriched soil water is transported through the plants and transpired via the stomata.

In principle, two methods have been used to measure flux rates from the ground. The first is the widespread direct flux determination which employs a box covering a certain area in which the concentration increase is measured over a known period. This method has been used by a number of investigators in various regions of the world and the observed fluxes range between 0.004–2.5 atoms $cm^{-2}s^{-1}$ (Turekian et al. 1977). The second approach integrates radon profiles in the atmosphere, usually taken from aircraft, and estimates the flux by balancing between emanation and radioactive decay within the sampled domain. It is assumed that the measured profiles result from soil emanation of the sampled domain and from vertical air motion which is calculated from temperature, humidity, and pressure gradients concurrently measured with radon. This method has the advantage that it levels out much of the site-to-site variability encountered in direct flux measurements at the ground. The range of radon fluxes based on air profiles taken over various regions is thus smaller, i.e., 0.18–0.91 atoms $cm^{-2}s^{-1}$ (Turekian et al. 1977). Since the profile integration method encounters much less variability, a global measuring campaign requires less sampling than direct flux determinations on the ground to derive a representative mean global flux. However, apart from funding and staff constraints, such a program would suffer from methodological problems inherent to the assumption of steady state between emanation and decay. In the case of radon, for example, no exchange with air masses from outside the sample domain must have taken place within 3–4 days. As this time period is sufficiently long to bring marine air masses depleted in radon to any continental location, this method is prone to underestimation. The subject of aircraft-based measurements is dealt with in more detail by Desjardins and MacPherson (this volume).

Aircraft-based and direct flux measurements have been used to derive global mean flux estimates. Extrapolation has to account for the different geography of influencing factors which are not available on the global scale. Up to now, all calculations of a mean radon flux have been based on insufficient data augmented with skilled assumptions or models needed for extrapolation. Hence, estimates range between 0.6 and 1.2 atoms $cm^{-2}s^{-1}$ on a global average, which appears to be a rather wide range considering the relatively simple nature of radon as a tracer.

Carbon Dioxide

The study of carbon cycling has contributed significantly to the current understanding of source–sink dynamics and of interface exchanges at soil–atmosphere, vegetation–atmosphere, and ocean–atmosphere boundaries. However, open questions on phasing of fluxes and on trends of carbon sinks and sources at regional to global scales remain. Two key parameters in the global carbon cycle are net primary productivity (NPP) of vegetation and soil respiration, as these define the yearly in- and outflux between the biosphere and the atmosphere. Estimates of global NPP all stem from a limited number of studies with various vegetation types. The studies employ a variety of techniques to estimate the increase of plant matter within a growing season. Since thorough studies on all components in an ecosystem, particularly complex forest sites, are very time-consuming and costly, such detailed investigations are few and are lacking from such important systems as tropical rain forests. The bulk of NPP data are estimates based on regression analyses on single plant species, which have been applied to easy-to-measure variables in different sample plots (Whittaker and Marks 1975). In view of the complexity of vegetation, this method involves to a large extent expertise and intuition. The same is true for the extrapolation of NPP data to regional or global scales, as this involves the choice of a suitable vegetation map and the designation of a mean (and hopefully representative) productivity value to each of the distinguished vegetation types (e.g., Whittaker and Likens 1975).

Alternatively, fluxes may be correlated with environmental parameters, which preferentially should have a better regional or global documentation. This has been done, e.g., for soil respiration and NPP (e.g., Lieth 1975), although in a simplistic manner since all variation in the data is attributed to one or two single parameters, i.e., evapotranspiration or air temperature and precipitation. This leads to large deviations between measured data and modeled prediction at a certain location. For global estimates this approach may be suitable when one assumes that variations between sites and regions compensate on the global scale. Comparison of various NPP estimates agrees within about 15%, which may give some confidence in the global carbon flux of 45–55 $\times$ 10^{15}g C annually from the atmosphere into the biosphere. However, part of this agreement may come from the fact that all approaches principally rely on the same data base.

In recent years, remote sensing from satellites has been useful for vegetation monitoring on various scales. Since the spectral reflectance characteristics of green leaves in the visible and near infrared radiation are unique, they can be combined into indices which determine the greenness or photosynthetic activity of plant canopies. A commonly used index is the NDVI (normalized difference vegetation index) derived from the Advanced Very High-Resolution Radiometer (AVHRR) aboard NOAA 7, a polar orbiting satellite providing global radiation data at a resolution of 1 and 4 km. Tucker et al. (1985) have used this index to estimate the NPP of grassland in the Senegalese Sahel during the period 1981–1984. As the index is only a relative number, it cannot directly be converted to production values. This requires sampling of herbaceous biomass over the range of vegetation types for which the NDVI values are to be calibrated. The relationship between the stock of herbaceous tissue (taken as NPP) and the integrated NDVIs of a series of satellite data over the growing season is then used to estimate the entire NPP of the region. This approach has the great advantage that it can monitor changes over subsequent years and may be used for forage prediction. Application of this method for global NPP appraisals, up to now, suffers from surface data for calibration. Data are needed from more complex systems like forests, in which the relationship of NPP and NDVIs is not as straightforward since a large fraction of the NPP is used for woody tissue. However, seasonally integrated NDVIs show a strong correlation with estimated mean biome NPP and also with atmospheric CO_2 data, emphasizing the potential of AVHRR observations to improve the understanding of macroscale vegetation dynamics.

TABLE 1. Methane source characteristics.

Sources: / Flux Characteristics	Domestic Animals	Wild Mammals	Termites	Natural Wetlands	Rice Paddies	Waste Dumps	Biomass Burning	Natural Gas Leaks	Coal Mining	Others ?
anthropogenic	+				+	+	+	+	+	
natural		+	+	+			+			
Area				+	+		+			
Population	+	+	+							
Bulk						+	+	+	+	
~constant	+	(+)?	(+)?			(+)?		+	+	
seasonal				+	+					
diurnal				+	+					
Range[1]										
– less than one order of magnitude	+	+					(+)	(+)?	(+)?	
– several orders of magnitude			+	+	+	+				
critical factors	feed levels, management schemes	population statistics, methane yield per animal	flux per colony, population statistics	wetland statistics, emission rates, seasonality	emission rates	composition of wastes, mode of disposal, emission rates	area and biomass burned, CH_4/CO_2 ratio	suitable statistics	suitable statistics	

[1] Ranges of observed fluxes within samples or for species.

Methane

Unlike CO_2 and ^{222}Rn, methane is a chemically reactive species in the atmosphere. It is broken down by OH radicals which constitute the main sink of CH_4. Less important, and only tentatively quantified, is the destruction of CH_4 at the ground. The total sink of CH_4 is estimated to be around 500 Tg with an uncertainty range of ca. 35% (WMO 1985). This constrains the range of possible sources. The major CH_4 emittors have probably been identified but quantification of individual sources is still a major problem. Most important is CH_4 from biogenic sources produced in anaerobic decay of organic matter and enteric fermentation. Anaerobic environments with significant CH_4 production potential are natural wetlands, rice paddies, waste dumps, and the rumen of livestock and other animals. Abiotic sources which may contribute substantial amounts of CH_4 to the global methane cycle are biomass burning and CH_4 escaping from natural gas leaks and coal mining. These sources are listed in Table 1 together with their flux characteristics.

Methane produced from livestock is primarily a function of animal species and quantity and quality of feed stuff, as well as energy expenditure of the animal. This has been documented in investigations on CH_4 yields from the major domestic animals, mainly from the developed world. These CH_4 yields, combined with population statistics and information about the dominant scheme of animal raising and management in the various regions of the world, have been used to estimate the global CH_4 flux from domestic livestock (Crutzen et al. 1986). Since the data from developed countries are comparably well documented, the CH_4 contribution from livestock in these regions may be considered one of the safer estimates in the global CH_4 budget. Nonetheless, CH_4 emissions from livestock in developing countries are poorly documented and new studies, preferentially on Third World cattle and buffalos, may lead to revised figures. Large uncertainties are associated with the estimated contribution from wildlife and, in particular, from termites. Individual CH_4 yields for wildlife species may be deduced from livestock, but global extrapolation suffers from the lack of population statistics and generates at best a very tentative range based on crude population estimates. The global CH_4 flux from termites has been estimated by various groups, but all have based their analysis on very limited samples. Crucial problems in these estimates are ingested carbon to CH_4 ratios that may vary within two factors of magnitude between different species, quantity of food intake per termite or colony, and the subsequent extrapolation, which in view of lacking population statistics remains highly uncertain. Better estimates can only be achieved on the basis of more flux measurements, preferentially from colonies in the field, and improved knowledge of the world termite population.

Source strength estimates of wetlands are likewise tentative. Wetlands have been poorly documented in statistics and land maps, as they usually cover only small fractions of total land area and since they represent land of low economic value. This has led to low values of estimated global wetland area of 2–3.6 $\times$ 10^6 km^2, a range that has commonly been used in carbon and methane studies. New compilations show that the actual area covered by wetlands is more likely between 5–6 $\times$ $10^6 km^2$ (Matthews and Fung 1987; Aselmann and Crutzen 1989). However, regional disagreement between the two compilations suggests that the knowledge of present wetland distributions and the characteristics important for CH_4 production is still imperfect. Particularly in many cases, knowledge on the seasonal status, which defines the period of potential CH_4 production and which often shows large year-to-year variations, can at present be only tentatively drawn from climatological data. Moreover, CH_4 flux measurements mirror the complex nature of CH_4 production as found in a variety of wetland types in all major climate regions. A set of influencing factors has been identified, but variability within and between sites has, so far, limited the understanding of flux rates from wetlands. Table 2 (Aselmann and Crutzen 1989) comprises data on methane emissions from wetlands published until 1988. All measurements were taken by direct flux determination, but data are integrated over different periods, i.e., short term, seasonal, or yearly; thus, they are not strictly comparable. The data are mean values, either cited unaltered or calculated as the geometric mean in cases where a set of individual data was published. Note that the ranges given in parentheses represent one standard deviation about the geometric mean, assuming log-normal distribution, and not the range of measured values which is much larger. Table 2 documents that fluxes for the seven broad wetland categories overlap widely and that differences, although apparent in the overall mean for each category, are statistically not significant. It should also be noted that the assignment of sample sites to the various wetland categories depends to a large extent on the interpretation of the sometimes meager information about sample sites in the publication. In conclusion, present wetland statistics, their seasonal status, and the extrapolation of flux data can only yield rather large ranges. According to the data in Table 2, 30–170 Tg and 60–140 Tg may be annually released from natural wetlands and rice paddies, respectively.

Large amounts of biodegradable materials in wastes are produced each year, which are potentially a large source of CH_4 when anaerobic conditions in landfills develop. An upper limit of this source may be deduced from the statistics of the U.S., the world leader in per capita trash production (Sheppard et al. 1982). A more elaborate estimate, however, requires data from various countries or regions with different economies on the amount of waste, its composition, and the dominant mode of dumping and cover.

Such data are only available for some industrial countries and a few large cities of the developing world. Furthermore, studies on landfills show that production, transport, and consumption of CH_4 in dumpsites are complex, impeding a simple relationship between biodegradable carbon and methane flux to the atmosphere. Extrapolation of the available data has, therefore, to rely on many assumptions and yields tentative ranges open to future revision (Bingemer and Crutzen 1987).

Methane emission from abiotic sources is primarily man induced and comprises methane escaping from fossil fuel exploitation and the burning of biomass. Here I concentrate on CH_4 from biomass burning, since appraisal of these sources involves factors which challenge global extrapolation. Methane has been measured in fire plumes, in both laboratory-controlled and natural fires. It is commonly expressed as the CH_4/CO_2 ratio, which allows convenient conversion of burnt biomass to CH_4 yield. Measured ratios under controlled conditions differ within a factor of 5 and depend primarily on the material burnt, the combustion temperature, and the smoldering stage of the fire. Measurements in fire plumes of prescribed burning and wildfires show much larger ranges with averages of about $0.8–2.2 \times 10^{-2}$ for the ratio of CH_4/CO_2. Another crucial problem in estimating the magnitude of CH_4 produced in biomass burning is the amount of biomass which is burnt annually. Biomass burning takes place due to natural fires, slash and burn practices in shifting agriculture, burning of savannas for forage improvement for human livestock, combustion of agricultural residues, and fuelwood combustion. Quantification of these processes depends to a large degree on population statistics, in particular from tropical and subtropical countries, distinguished into groups with different agricultural practices and fuelwood consumption. Highly uncertain factors in this respect are the annually burned area of savannas and the conversion of tropical forests, either permanent or temporal, to agricultural land. Estimates on tropical forest conversion have also been based on FAO land-use statistics and on individual information about forest transformation for certain regions or countries, which show rather conflicting data (Woodwell et al. 1983). Mean biomass estimates of tropical forests are either based on data derived from a number of investigations, in which destructive sampling was applied, or are deduced from data of standing stock of trunk volumes by using mean wood density and volume to total biomass ratios. The ensuing estimates differ significantly (Brown and Lugo 1984). Finally, burning efficiency and the number of successive burns introduce further uncertainties into global appraisals. For example, logs felled in shifting agriculture commonly remain unburned on the field and are subject to subsequent fires, so that the burning efficiency is low, whereas fires in pure grass savannas may burn almost 100% of the aboveground biomass. It is easy to conceive that the burning efficiency in the field depends on many factors which makes

TABLE 2. Measured methane emissions from wetlands (mg CH_4 $m^{-2}d^{-1}$).

Bogs	Fens	Swamps	Marshes	Floodplains	Lakes	Rice Paddies	Month	Location	Reference
			157 (4?) (68–246)				autumn/winter	Delaware, U.S.A.	Swain (1973)
	122 (n=?) (61–183)						winter/spring	Minnesota, U.S.A.	ibid.
4.0 (14) (0.7–23)	**59** (56) (18–195)		**175** (13) (106–289)				August	Alaska, U.S.A.	Sebacher et al. (1986)
4.7 (95) (1.1–21)	**95** (45) (26–350)						June–Sept.	Stordalen, Sweden	Svensson & Rosswall (1984)
		72 (7) (40–104)	**67** (49) (59–75)				Jan/Febr.	Florida, U.S.A.	Bartlett et al. (1985)
113 (18) (60–211)	**65** (23) (9–485)		**572** (5) (493–664)				August	Minnesota, U.S.A.	Harriss et al. (1985)
		4 (9?) (1.9–9)					wet period (Sept.–May)	Virginia, U.S.A.	Harriss et al. (1982)
		23.3 (15) (7–73)					summer	S. Carolina, Georgia, Florida, U.S.A.	Harriss & Sebacher (1981)
			304 (7) (168–550)				June–Sept.	Michigan, U.S.A.	Baker-Blocker et al. (1977)
23.2 (120) (< 1–62)							yearly mean	Moore House, England	Clymo & Reddaway (1971)
					49 (63) (24–74)		summer mean	Manitoba, Canada	Rudd & Hamilton (1978)

			587 (39) (223–951)		**49** (n=?) (24–74)		yearly mean	Barataria basin, Louisiana, U.S.A.	DeLaune et al. (1983)
					18.4 (10) (1.9–180)	**180** (6) (75–300)	Aug.–Nov.	S. California, U.S.A.	Cicerone & Shetter (1981)
						32 (9) (13–68)	Aug.–Nov.	S. California, U.S.A.	ibid.
						250 (46) (125–375)	entire growth season	Davis, California, U.S.A.	Cicerone et al. (1983)
						96 (100) (79–113)	entire growth season	Andalusia, Spain	Seiler et al. (1984)
						384 (227) (257–511)	entire growth season	Vercelli, Italy	Holzapfel-Pschorn & Seiler (1986)
		192 (90) (162–219)	**230** (55) (158–302)		**27** (41) (22–32)		July/Aug.	Central Amazon, Brazil	Bartlett et al. (1988)
		108 (31?) (54–162)	**590** (31?) (295–885)		**120** (31?) (60–180)		July/Aug.	Amazon, Brazil	Devol et al. (1988)
	31 (87) (1.3–45)						entire growth season	Smith Lake, Alaska	Whalen & Reeburgh (1989)
13 **(1–42)**	**52** **(6–137)**	**105** **(68–148)**	**238** **(136–328)**	**(100)** **(50–200)**	**47** **(24–75)**	**357** **(228–485)**			

Values are either cited as given by the authors or calculated as the geometric mean, when individual data were given. Bold-faced numbers denote the emission rate followed by the number of measurements in parentheses and the minimum–maximum values derived from the standard deviation. The mean emission rate for each wetland category shown in the last row was calculated as the geometric mean weighted with the number of measurements. Ranges in the last row were likewise calculated using the minimum and maximum values in each column.

quantification for global estimates rather arbitrary. Again, as stated before with other methane sources, estimation of CH_4 flux from burning biomass is confronted with large uncertainties in the relevant factors.

CONCLUSIONS

It is clear from these examples that global extrapolation commonly relies on insufficient data that are compensated for by supplementary assumptions. In a statistical sense they are not valid, since in most cases random sampling or systematic sampling requirements by mathematical statistics have not been met. Does this imply that global extrapolation, based on insufficient data, is speculative and not scientific? Indeed, measurements of CH_4 fluxes from rice paddy soils in a flower pot and subsequent extrapolation to the global area of irrigated rice are purely speculative, taking for granted (after a few assumptions about soil type and temperature effects) that the sampled pots represent global means (Koyama 1963). As a first step in quantifying fluxes or stocks, which have not been determined before, these bold extrapolations may be justified and helpful, as they indicate a first order of magnitude. This is even more helpful when from the samples taken it is known that they represent maximum values, which may then be used to set an upper limit for a global figure. However, between Koyama's first "guesstimate" 25 years ago and today, only a few more (although *in situ*) measurements of CH_4 production from paddies have been made and most of the authors who made these measurements took the chance to derive their own global flux estimates, which remain open to revision until new data become available. However, the tendency to speculate is only one side of the coin. The other side is time constraints, staff constraints, and financial bottlenecks, which commonly impede thorough statistical settings and which finally determine which and how often a locality may be sampled. Even if sound and representative data are available, they are usually "historic," since they were taken in the past and will thus not include changes that meanwhile may have taken place. Furthermore, scientifically challenging research areas are those which are not exploited and in which data are few and provoke speculation. This has always been a stimulating part of science and will be so in future. What is going to change in the future is the ability to prove data and their extrapolation on the basis of increased modeling capability. Simple box models of the oceans in the 1970s revealed inconsistencies in the global carbon cycle, in particular the role of the vegetation as a net sink or source for atmospheric carbon. The seasonal behavior of soil respiration and NPP has been simulated with three-dimensional atmospheric circulation models using different functions of biospheric carbon uptake and release (e.g., Fung et al. 1987). Comparison of CO_2 time series at various stations with model results showed reasonable results in respect to the general phasing and magnitude of carbon uptake and release, but also revealed some deviations that are not well understood. This may be due not only to inconsistencies in the exchange function and underlying data sets, but also to shortcomings of the models. Feichter and Crutzen (1989) have tested a 3-D atmospheric transport model with radon, including a subgrid parameterization scheme for deep convection. The best results between observations and model results were obtained when using a mean radon emission rate of 1.2 atoms $cm^{-2}s^{-1}$, i.e., higher than most previous estimates based on flux samples. Vertical radon profiles appropriate for model comparison are, however, only available for some temperate regions. They are lacking from the tropics, where convection is most pronounced, which weakens the model evidence. In other words: models need data and vice versa. This can only be an iterative process of mutual stimulation between theory, measurement, and modeling (Junge 1987). Improved direct extrapolation as well as modeling capability will strongly depend on global geographic data bases of environmental parameters, such as soil properties, vegetation cover and associated characteristics, land use, climate, etc. Digital data bases suitable for global modeling are available but need to be supplemented and improved. As discussed for NPP, remote sensing from satellites holds great potential for data acquisition, in both compiling data sets and monitoring (cf. Stewart et al., this volume). Finally, isotopic composition of trace gases such as CO_2 or CH_4 in the atmosphere already provides powerful constraints on global extrapolations. The ^{13}C and ^{14}C content of methane has been shown to be different for various sources, which together with atmospheric carbon isotope composition of methane sets the possible range of individual sources (e.g., Wahlen et al., submitted). According to Wahlen et al. (submitted), present estimated sources of dead carbon, i.e., carbon devoid of ^{14}C, are too small to account for the global $\delta^{14}C$ signal in the atmosphere. Extended isotope measurements including deuterium will yield strong constraints on the other biogenic and nonbiogenic sources in the methane budget. Furthermore, in combination with atmospheric transport models, isotope data constitute a tool to check regional flux estimates based on direct measurements.

In conclusion the discussion may be summarized as follows: In view of the complexity and rapid transformation of nature, together with logistic and financial constraints which will certainly accompany future measurement programs, it is hard to conceive that they will ever meet strict statistical requirements. Isotope data, global geographic data bases, and remote sensing together with improved models, however, may reveal inconsistencies in global extrapolation and point to areas that need further sampling.

Acknowledgements. I thank Hans Feichter and Wei Min Hao for helpful discussions, as well as Karen Valentin for careful proofreading and upgrading of the linguistics.

REFERENCES

Aselmann, I., and P.J. Crutzen. 1989. Freshwater wetlands: global distribution of natural wetlands and rice paddies, their net primary productivity, seasonality and possible methane emissions. *J. Atmos. Chem.*, in press.

Baker-Blocker, H., T.M. Donahue, and K.H. Mancy. 1977. Methane flux from wetland areas. *Tellus* 29:245–250.

Bartlett, K.B., D.S. Bartlett, D.I. Sebacher, R.C. Harriss, and D.P. Brannon. 1985. Sources of atmospheric methane from wetlands. 36th Congress International Astronautical Federation, Stockholm. Oxford: Pergamon.

Bartlett, K.B., P.M. Crill, D.I. Sebacher, R.C. Harriss, J.O. Wilson, and J.M. Melack. 1988. Methane flux from the Central Amazonian floodplain. *J. Geophys. Res.* 93 (D2):1571–1582.

Bingemer, H.G., and P.J. Crutzen. 1987. The production of methane from solid wastes. *J. Geophys. Res.* 92(D2):2181–2187.

Brown, S., and A.E. Lugo. 1984. Biomass of tropical forests: a new estimate based on forest volume. *Science* 223:1290–1293.

Cicerone, R.J., and J.D. Shetter. 1981. Sources of atmospheric methane: measurements in rice paddies and a discussion. *J. Geophys. Res.* 86(C8):7203–7209.

Cicerone, R.J., J.D. Shetter, and C.C. Delwiche. 1983. Seasonal variation of methane from a California rice paddy. *J. Geophys. Res.* 88(15):11022–11024.

Clymo, R.S., and E.J.F. Reddaway. 1971. Productivity of *Sphagnum* (bog-moss) and peat accumulation. *Hydrobiol.* 12:181–192.

Crutzen, P.J., I. Aselmann, and W. Seiler. 1986. Methane production by domestic animals, wild ruminants, other herbivorous fauna, and humans. *Tellus* 38B:271–284.

DeLaune, R.D., C.J. Smith, and W.H. Patrick. 1983. Methane release from Gulf Coast wetlands. *Tellus* 35B:8–15.

Devol, A.H., J.E. Richey, W.A. Clark, S.L. King, and L.A. Martinelli. 1988. Methane emissions to the atmosphere from the Amazon floodplain. *J. Geophys. Res.* 93 (D2):1583–1592.

Feichter, J., and P.J. Crutzen. 1989. Parameterization of deep cumulus convection in a global tracer transport model and its verification with ^{222}Radon. *Tellus*, in press.

Fung, I.Y., C.J. Tucker, and K.C. Prentice. 1987. Application of advanced very high resolution radiometer vegetation index to study atmosphere–biosphere exchange of CO_2. *J. Geophys. Res.* 92(D3):2999–3015.

Harriss, R.C., C. Gorham, D.I. Sebacher, K.B. Bartlett, and P.A. Flebbe. 1985. Methane flux from northern peatlands. *Nature* 315(20):652–653.

Harriss, R.C., and D.I. Sebacher. 1981. Methane flux in forested freshwater swamps of the southern United States. *Geophys. Res. Lett.* 8(9):1002–1004.

Harriss, R.C., D.I. Sebacher, and F.P. Day. 1982. Methane flux in the Great Dismal Swamp. *Nature* 297:673–674.

Holzapfel-Pschorn, A., and W. Seiler. 1986. Methane emission during a cultivation period from an Italian rice paddy. *J. Geophys. Res.* 91(D11):11803–11814.

Jeffers, J.N.R. 1988. Statistical and mathematical approaches to issues of scales in ecology. In: Scales and Global Change, ed. T. Rosswall et al., pp. 47–56. SCOPE 35. Chichester: Wiley.

Junge, C. 1987. Kreisläufe von Spurengasen in der Atmosphäre. In: Atmosphärische Spurenstoffe, ed. R. Jaenicke, pp. 19–30. Weinheim, F.R. Germany: VCH.

Koyama, T. 1963. Gaseous metabolism in lake sediments and paddy soils and the production of hydrogen and methane. *J. Geophys. Res.* 68:3971–3973.

Lieth, H. 1975. Modeling the primary productivity of the world. In: Primary Productivity of the Biosphere, ed. H. Lieth and R.H. Whittaker, pp. 55–116. New York: Springer.

Matthews, E., and I. Fung. 1987. Methane emission from natural wetlands: global distribution, area, and environmental characteristics of sources. *Glob. Biogeochem. Cyc.* 1(1):61–86.

Rosswall, T., P.G. Risser, and R.G. Woodmansee, eds. 1988. Scales and Global Change. SCOPE 35. Chichester: Wiley.

Rudd, J.W.M., and R.D. Hamilton. 1978. Methane cycling in an eutrophic shield lake and its effect on whole lake metabolism. *Limnol. Ocean.* 23(3):337–348.

Sebacher, D.I., R.C. Harriss, K.B. Bartlett, S.M. Sebacher, and S.S. Grice. 1986. Atmospheric methane sources: Alaskan tundra bogs, an Alpine fen, and a subarctic boreal marsh. *Tellus* 38:1–10.

Seiler, W., A. Holzapfel-Pschorn, R. Conrad, and D. Scharffe. 1984. Methane emission from rice paddies. *J. Atmos. Chem.* 1:241–268.

Sheppard, J.C., H. Westberg, J.F. Hopper, and K. Ganesan. 1982. Inventory of global methane sources and their production rates. *J. Geophys. Res.* 87(C2):1305–1312.

Svensson, B.H., and T. Rosswall. 1984. *In situ* methane production from acid peat in plant communities with different moisture regimes in a subarctic mire. *Oikos* 43:341–350.

Swain, F.M. 1973. Marsh gas from the Atlantic coastal plain, United States. In: Advances in Organic Geochemistry, ed. B. Tissot and F. Bremer, pp. 673–687. Paris: Technip.

Tucker, C.J., C.L. Van Praet, M.J. Sharman, and G. van Ittersum. 1985. Satellite remote sensing of total herbaceous biomass production in the Senegalese Sahel: 1980–1984. *Rem. Sens. Env.* 17:233–249.

Turekian, K.K., Y. Nozaki, and L.K. Benninger. 1977. Geochemistry of atmospheric radon and radon products. *Ann. Rev. Earth Plan. Sci.* 5:227–255.

Whalen, S.C., and W.S. Reeburgh. 1989. A methane flux time-series for tundra environments. *Glob. Biogeochem. Cyc.*, in press.

Whittaker, R.H., and G.E. Likens. 1975. The biosphere and man. In: Primary Productivity of the Biosphere, ed. H. Lieth and R.H. Whittaker, pp. 306–328. New York: Springer.

Whittaker, R.H., and P.L. Marks. 1975. Methods of assessing terrestrial productivity. In: Primary Productivity of the Biosphere, ed. H. Lieth and R.H. Whittaker, pp. 55–116. New York: Springer.

WMO (World Meteorological Organization). 1985. Atmospheric Ozone 1985. Global Ozone Research and Monitoring Project No. 16, vol. 1–3. Geneva: WMO.

Woodwell, M., J.E. Hobbie, R.A. Houghton, J.M. Melillo, B. Moore, B.J. Peterson, and G.R. Shaver. 1983. Global deforestation: contribution to global carbon dioxide. *Science* 222 (4628):1081–1086.

Exchange of Trace Gases between Terrestrial Ecosystems and the Atmosphere
eds. M.O. Andreae and D.S. Schimel, pp. 135–152
John Wiley & Sons Ltd

Aircraft-based Measurements of Trace Gas Fluxes

R.L. Desjardins[1] and J.I. MacPherson[2]

[1]*Land Resource Research Centre, Agrometeorology*
Central Experimental Farm, Bldg. 74
Ottawa K1A 0C6, Canada

[2]*Flight Research Laboratory, National Aeronautical Establishment*
National Research Council
Ottawa K1A 0R6, Canada

Abstract. Information on trace gas exchange between the biosphere and the atmosphere is important to predict future atmospheric concentration. Because of measurement difficulties, very little quantitative information is available on the exchange rates for major ecosystems which could affect the global atmospheric chemistry. A review of techniques to measure gaseous exchange is presented. These involve direct and indirect methods frequently used with ground-based systems and which can be adapted to use on aircraft. Airborne flux measurements of CO_2, H_2O, and O_3 are presented as examples of results obtained on a regional scale by taking into account factors such as time lag adjustment, frequency response of sensors, air density corrections, flow distortion, and averaging requirements.

INTRODUCTION

Carbon, nitrogen, and sulfur compounds are some of the most important trace gases on Earth. There is well-substantiated evidence that the atmospheric concentrations of these trace gases are increasing. This may change the energy balance of the Earth and the vigor of the biosphere, leading to further imbalance in the chemical composition of the atmosphere (Mooney et al. 1987). Land-use change and industrial activities are important causes of this increase. Because of the scale involved, the net effect is difficult to predict but it is potentially so significant that it is widely thought that actual measurements of trace gas fluxes are essential to address the problem better.

Of the carbon elements, carbon dioxide (CO_2), carbon monoxide (CO), methane (CH_4), and chlorofluorocarbons (CFCs) are the most important greenhouse gases. The contribution of CO_2 approximately equals that of all other trace gases combined. Its annual increase of 0.4% is primarily due to the combustion of fossil fuels and to the reduction of the global carbon storage in the terrestrial biota, which has decreased by approximately 50% from pre-agricultural times. The absorption of CO_2 by vegetation and transfer into the oceans are the main factors contributing to its removal from the atmosphere. A decrease in photosynthetic vigor of the biosphere due to pollution would lead to an even more rapid increase in atmospheric CO_2.

CH_4 concentration is also increasing rapidly, approximately 1% per year. Even though its concentration is 200 times lower than CO_2, its effectiveness in trapping long wave radiation is approximately 20 times that of CO_2, making it an important greenhouse gas. The reasons for its increase are not fully understood. Freshwater systems and oceans are considered important sources. It is biological in origin and its rate of emission is strongly affected by climatic conditions. Emission rates over wetlands, which have been estimated to produce 25% of the global methane, are expected to increase greatly with an increase in temperature.

Carbon monoxide, which originates mainly at the Earth's surface, is increasing at a rate of 2 to 6% per year. It reacts with oxygen to produce CO_2 and ozone (O_3) and can react with the hydroxyl radical (OH), or oxidize CH_4. It has a relatively short atmospheric lifetime of a few months.

CFCs, which are synthetic industrial chemicals, are considered the major culprits in the massive destruction of O_3 over the Antarctic. Even though far less abundant than CO_2, one molecule has 10,000 times more greenhouse effect than a CO_2 molecule and it is estimated CFCs now account for approximately 15% of the unnatural greenhouse effect.

The most important trace nitrogen gases are nitrous oxide (N_2O), nitric oxide (NO), and ammonia (NH_3), which result from biomass and fossil fuel burning and agricultural and biological activity. NO and NH_3 are short-lived gases while N_2O has a long atmospheric lifetime. It is increasing at a rate of 0.2% per year (Mooney et al. 1987).

O_3 is another important atmospheric constituent. It is produced and destroyed in the stratosphere and the troposphere. Its destruction in the stratosphere, which permits the sun's biologically damaging ultraviolet radiation to reach the Earth's surface, is linked to CFCs, CH_4, and N_2O. At ground level, O_3 is a threat to human health. It also affects vegetation directly by destroying chlorophyll molecules and indirectly by leading to nutrient deficiencies. Since it can be formed from NO, which has a highly inhomogeneous distribution, some regions act as O_3 sources while others act as O_3 sinks. Flux divergence is then important in determining the budget and distribution of O_3 (Lenschow et al. 1981, 1982). Because it reacts on contact with surfaces, dry deposition is also important.

Sulfur compounds are present in the atmosphere in large quantities. Both natural and anthropogenic sources are thought to contribute in equal proportions to the global atmospheric sulfur budget. The main sulfur compounds are sulfur dioxide (SO_2), hydrogen sulfide (H_2S), and dimethyldisulfide (DMS). Anthropogenic sulfur emissions are dominated by the release of SO_2 from fossil-fuel burning, while natural emissions come from biomass burning, volatilization of sulfur gases from soil, plants, oceans, and volcanic activities.

Sources and sinks for most of these gases are not well known (Mooney et al. 1987). Sometimes it is not clear if an ecosystem is a net source or sink of a particular chemical. Because of the diversity of ecosystems, flux measurement is the only realistic approach to quantify exchange rates of trace gases over large areas. Transfer processes of highly soluble and chemically reactive trace gases such as NO_2, SO_2, and SO_3 at a water surface are analogous to those of H_2O. When the liquid phase resistance is important, as is the case for gases such as CO_2, N_2O, and CO, then the deposition velocities are decreased to a limiting value specific for each gas (Hicks and Liss 1976). In the mixed layer, away from the surface, most trace gases are transported by the same physical mechanisms.

The objective of this paper is to review the present capabilities for measuring gas exchange with aircraft flux systems. Because of a lack of fast response sensors for most trace gases, aircraft flux measurements have been limited to CO_2, H_2O, and O_3. Using examples from such observations, we will discuss the accuracy of flux measurements and the extrapolation of these measurements to the ground surface. We will also discuss the possibilities for the development of techniques for quantifying the biological regulation of trace gas exchanges on a regional scale via momentum, sensible heat, and latent heat flux measurements.

REVIEW OF TECHNIQUES TO MEASURE GASEOUS EXCHANGE

A review of current micrometeorological techniques for the measurement of dry depositions of atmospheric gases using tower-based systems was recently prepared by Businger (1986). These techniques include gradient, eddy correlation, budget studies, variance, Bowen ratio, and eddy accumulation. Most of these have potential applications using aircraft-mounted sensors. We will briefly review what has been done with aircraft and suggest what could be done.

The use of aircraft for flux measurements offers considerable advantages over tower-based measurements. It provides averages over large areas and is suitable for remote studies. However, aircraft can be used for only short periods and are costly to operate, and the frequency response of sensors for many important species is not fast enough for use with aircraft systems. Qualitative and quantitative information can, however, be obtained from mean concentration measurements. Aircraft measurements of mean atmospheric concentrations at different heights have been used to calculate the flux of NH_3 in Germany (Lenhard and Gravenhorst 1980). During the ABLE-2A experiment, Harriss et al. (1988) demonstrated that isoprene, CH_4, CO_2, NO, dimethylsulfate, and organic aerosol emissions from soils and vegetation play a major role in determining the chemical composition of the mixed layer over undisturbed forest and wetlands. Emissions from biomass burning were also found to be important sources of CO, NO, and O_3.

GRADIENT OR PROFILE METHOD

Many researchers have estimated the vertical flux of trace gases using profile methods. The flux is usually related to the gradient by the turbulent diffusivity, K. Recently, both the conceptual foundation of the K theory and the practical consequences of using it near sources and sinks have been widely questioned (Finnigan and Raupach 1987). It is considered not valid for heights less than 50 to 100 z_0, where z_0 is the roughness length. This puts in question many of the flux measurements of trace gases reported in the literature with ground-based systems (Businger 1986). At higher elevations, the K theory is satisfied but gradients are very small. Lenhard and Gravenhorst (1980) estimated ammonia fluxes by aircraft measurements of concentration values at 100 m and 700 m above ground. However, because of the difficulty of accurately determining exchange coefficients on this scale, the eddy correlation technique is preferred with aircraft.

EDDY CORRELATION TECHNIQUE

Eddy correlation methods for measuring sensible and latent heat, momentum, CO_2, and O_3 fluxes using aircraft-based systems are well developed (Lenschow et al. 1981, 1982; Desjardins et al. 1989b). It is generally agreed that with *in situ* calibration of boom-mounted velocity sensors, most errors due to flow distortion can be minimized (Lenschow 1986). The availability of suitable trace gas sensors has been the main limiting factor in using this technique for measuring the flux of trace gases. Significant progress has been made in recent years in developing fast response sensors for O_3, SO_2, NO, NO_2, and CO_2. Sensors are under development for CH_4 and CO, but no fast response sensors are yet available for N_2O and NH_3. A frequency response of approximately 10 Hz and a sensitivity of approximately 0.1% of the absolute value are needed. A relaxation of the sampling rate is also possible with this technique. For example, with tower data, it has been shown that the 1 h variances and covariances computed from fast response sensors recorded at only one observation per 10 s were within 5% and 10% respectively of their true value (Haugen 1978).

BUDGET STUDIES

CO_2 exchange has been estimated by measuring the rate at which air mass modification occurs while passing over a surface. Similar studies have been carried out to estimate regional evaporation. Such estimates require accurate measurements of both concentration and velocities through the depth of the mixed layer. It also requires the assumption that the observed concentration modification is not a result of processes other than the surface flux. Lenschow et al. (1981, 1982) carried out ozone budget studies over cropland and over the Gulf of Mexico. The flight patterns over cropland consisted of a series of L-shaped patterns at four levels in the first 300 m. The main terms that govern the time rate of change of the concentration of a gas, $\partial\bar{s}/\partial t$, consist of advection, convection, and vertical flux divergence:

$$\frac{\partial\bar{s}}{\partial t} - \bar{u}\frac{\partial\bar{s}}{\partial x} - \bar{v}\frac{\partial\bar{s}}{\partial y} - \bar{w}\frac{\partial\bar{s}}{\partial z} - \frac{\partial\overline{w's'}}{\partial z} = \text{sources or sinks.} \qquad (1)$$

Values found by Lenschow et al. (1981) were

$$\frac{\partial\bar{s}}{\partial t} = 2.62 \text{ ng } O_3 \text{ m}^{-3}\text{s}^{-1}, \bar{u}\frac{\partial\bar{s}}{\partial x} = 0.13 \text{ ng } O_3 \text{ m}^{-3}\text{s}^{-1}, \text{ and}$$

$$\frac{\partial\overline{w's'}}{\partial z} = 0.07 \text{ ng } 0_3 \text{ m}^{-3}\text{s}^{-1}.$$

They then concluded that 2.4 ng O_3 $m^{-3}s^{-1}$ were due to photochemical production of ozone. The validity of their conclusion was primarily based on the fact that sensible heat and humidity budgets determined by the same method were shown to balance reasonably well.

In order to minimize the effect of advection, flight patterns can also be flown in a coordinate system that is advected with the mean wind. Lenschow et al. (1982) flew four squares each about 40 km on a side at 15, 160, and 325 m on two successive days to estimate the various budget terms. Since the flight tracks were advected along the mean wind, the advection term $\bar{u}(\partial\bar{s}/\partial x) = 0$. The remaining terms, that is, $\partial\bar{s}/\partial t$ and $\partial\overline{w's'}/\partial z$, were found to be 0.2 to 0.3 ng O_3 $m^{-3}s^{-1}$, which was assumed to be due to photochemical production of ozone.

VARIANCE TECHNIQUE

Use of the variance technique permits preliminary answers to be obtained when sensors for true covariance are not available. Once the flux of some quantity such as heat or H_2O is determined, it is possible to estimate fluxes of trace gases using concentration sensors of less perfect capabilities. Wesely (1988) has proposed several variations of this technique. For example, one might start from the fact that the flux of a trace gas, c, can be obtained from the following relationship:

$$F_c = r_{wc}\,\sigma_w\,\sigma_c, \tag{2}$$

where r_{wc} is the correlation between vertical wind and the concentration of the gas of interest, and σ_w and σ_c are the standard deviations of vertical wind and of the gas of interest. r_{wc} is a function of stability and is usually assumed to be equal to the correlation of vertical wind with temperature $r_{w\theta}$. This value is considered to be equal to 0.55 under unstable conditions and 0.40 under slightly unstable to slightly stable conditions. However, as shown in Table 1, data confirms this approximation for sensible and latent heat, but not always as clearly for a gas such as CO_2.

Another form is based on the ratio of the standard deviations, where the flux of a scalar quantity, c, is given by

$$F_c = F_s\,\sigma_c/\sigma_s, \tag{3}$$

where σ_c and σ_s represent the standard deviations of the gas of interest and that of a substance for which the flux, F_s, is known. F_s could be sensible heat. A third approach is based on

TABLE 1. Sensible heat, F_θ, latent heat, F_q, and CO_2 fluxes estimated from an empirical relationship (F_{c1}), and using mixing ratio of CO_2 with respect to dry air (F_{c2}), and corresponding correlation coefficients, r, for data collected at 100 m above a grassland ecosystem on 15 August 1987.

Time CDT	F_θ W m^{-2}	$r_{w\theta}$	F_q W m^{-2}	r_{wq}	F_{c1} mg m^{-2}s^{-1}	F_{c2} mg m^{-2}s^{-1}	r_{wc}
1102	48	0.53	230	0.51	−0.45	−0.43	−0.47
1108	52	0.51	191	0.53	−0.27	−0.26	−0.35
1115	46	0.52	195	0.50	−0.31	−0.30	−0.40
1121	55	0.53	220	0.55	−0.34	−0.31	−0.36
1127	46	0.49	205	0.51	−0.27	−0.27	−0.37

$$F_c = \sigma_c \frac{u_*}{f(z/L)} \tag{4}$$

where u_* is the friction velocity which can be estimated from the mean windspeed for various surfaces (Wesely 1988) and

$$f(z/L) = 1.85 \qquad \text{for } z/L > -.31$$
$$= 1.25\,(-z/L)^{-1/3} \qquad \text{for } z/L < -.31,$$

where z is the height of measurement and L is the Monin-Obukhov length.

These techniques have not been tested on aircraft yet but are most attractive for cases where advection and flux divergence can be neglected and fluxes of heat and water vapor can be determined through energy balance considerations. One can then bypass the need for direct flux measurements and avoid the need of costly vertical wind sensors.

Sensors with high signal-to-noise ratios are required, otherwise considerable overestimation of the flux will result. Wesely (1988) has reported an overestimation of 200% for SO_2 flux using a sensor with a signal-to-noise ratio of 0.5. The main argument for using such a variance technique is that considerably less time is required to obtain representative variances as compared with covariances (Lenschow and Stankov 1986). Although not fully tested, this technique is considered to be less demanding on both sensors and terrain conditions (Hicks and McMillen 1988).

BOWEN RATIO

Another possible approach is a modified Bowen ratio technique. The ratio of potential temperature, θ, versus mixing ratio of water vapor with respect to dry air, r, and potential temperature versus the mixing ratio of a trace gas, c, can provide valuable information to estimate the flux of these scalars. Micrometeorologists have frequently used Bowen ratio values to estimate sensible and latent heat fluxes from energy balance considerations (Hicks and McMillen 1988). They have also extended the technique to estimate the fluxes of CO_2. Similar estimates are possible from aircraft by finding the relationship between θ, r, and c along a run at constant pressure altitude.

Figure 1 presents the relationship between 0.5 s averages of θ and r for the data collected during a flight over a distance of 15 km at approximately 100 m above a grassland ecosystem in Kansas. Using these data, one can determine a so-called "Bowen ratio" either through the ratio of F_θ/F_q or

$$B = (\rho/\rho_a)\,(C_p/L_v)\,(\Delta\theta/\Delta r) \tag{5}$$

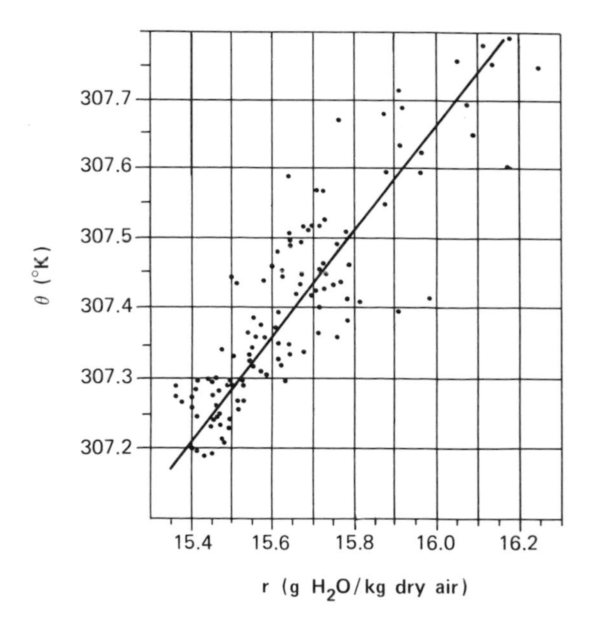

Fig. 1—Relationship between potential temperature, θ, versus mixing ratio of water vapor, r, for 0.5 second averages along a 15 km flight at 100 m above a grassland ecosystem on 15 August 1987.

where $\Delta\theta/\Delta r$ represents the slope of θ versus r in Fig. 1, C_ρ is the specific heat of air at constant pressure, L_v is the latent heat of vaporization, and ρ and ρ_a are the density of air and dry air, respectively. Assuming the coefficients of turbulent diffusion to be all equal, it is then possible to estimate the flux of a trace gas "F_c" from K theory and energy balance considerations as:

$$F_c = (Rn-S)/(\rho\ C_\rho\ \Delta\theta/\Delta c + \rho_a L_v\ \Delta r/\Delta c) \tag{6}$$

where $\Delta\theta/\Delta c$ and $\Delta r/\Delta c$ are the corresponding slopes as obtained in Fig. 1. The net radiation, Rn, can be estimated from radiative measurements and S, the storage term, from the soil heat flux and the kind of vegetation cover involved (Hicks and McMillen 1988). This again avoids the need for vertical wind measurements, which are difficult and expensive to do on an aircraft. Since mean values are involved rather than covariances, shorter time averages are needed to obtain results of acceptable accuracies. Figure 2 presents modified Bowen ratios, B, obtained from Eq. 5 versus the ratios of the eddy flux measurements of sensible (F_θ) and latent (F_q) heat for 3.8 km runs during a grid flight on August 15, 1987 over a grassland ecosystem. The flux values were high-pass filtered at 0.012 Hz in order to minimize low frequency contributions associated with measuring conditions. B was found to be $1.02\ F_\theta/F_q - 0.05$ with a correlation coefficient of 0.93 and an rms error of 0.03. The agreement between the two estimates indicates that Bowen ratios obtained using these two approaches provide similar results and that Eq. 6 might be useful to estimate the flux of trace gases.

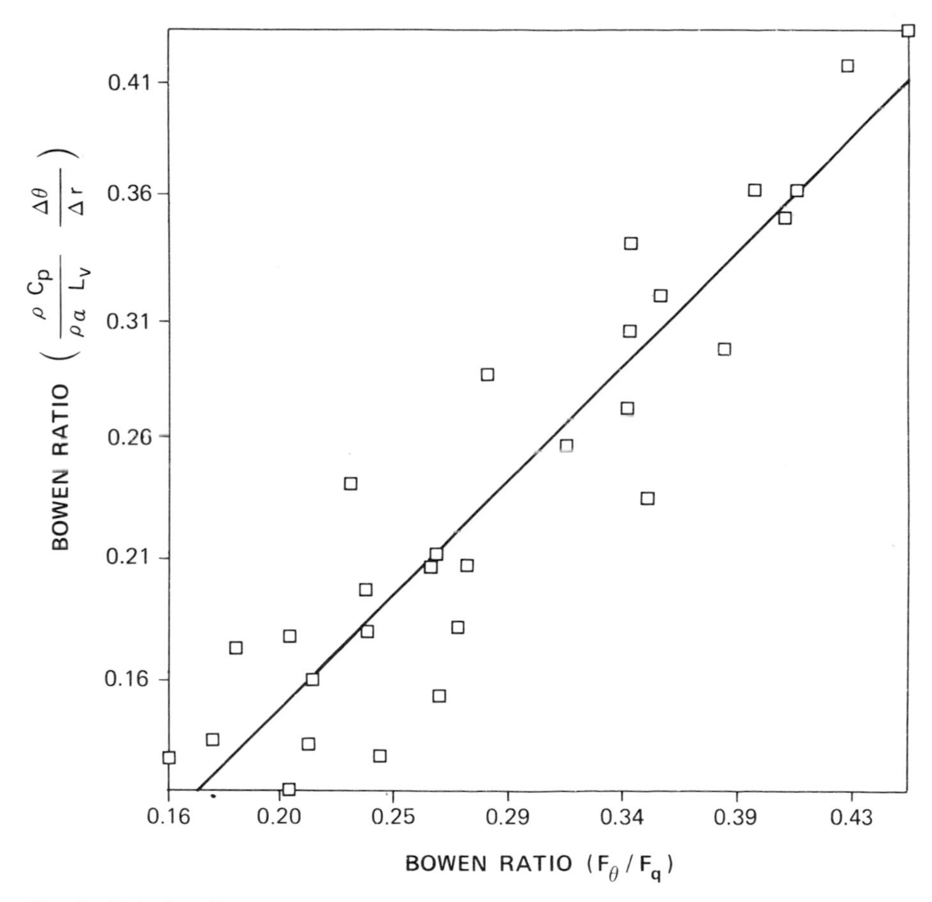

Fig. 2—Relationship between Bowen ratio estimates based on potential temperature, θ, and mixing ratio of water vapor, r, versus Bowen ratio estimates based on the filtered sensible heat, F_θ, and latent heat, F_q, fluxes all measured 100 m above a grassland ecosystem. All values are based on 3.8 km average.

EDDY ACCUMULATION

This technique bypasses the need for fast response gas sensors and still provides a direct measurement of the eddy flux. It requires the sampling of air associated with updrafts and downdrafts in two systems at a rate proportional to the vertical velocity. Desjardins (1972) has shown that the compositional differences of updrafts and downdrafts, obtained from mass spectrometer traces of air samples, could be used to estimate the flux of most trace gases simultaneously. While discussing some of the difficulties in using this technique, Hicks and McMillen (1984) have emphasized the need for further development due to its great potential. Although it has not been attempted on aircraft, with high-pass filtering of the vertical wind and fast response valves, it could provide the opportunity to obtain eddy flux for a whole series of gases for which there is a lack of fast response chemical detectors.

ASSESSMENT OF AIRCRAFT-BASED EDDY FLUX MEASUREMENTS

In this section, experimental requirements for eddy flux systems will be described, based on the experience to date with H_2O, CO_2, and O_3 measurements. Sensor response, flight patterns, and data analysis procedures will be discussed.

Time Lag Adjustment

In order to minimize flow distortion, turbulence sensors are usually placed on a noseboom far away from other sensors. A lag occurs between the measurement of the wind component and the concentration of the gas. Time lag adjustment of the signal can correct most of this effect. Errors of 24 to 33% can result in water vapor fluxes at an altitude of 50 and 25 m if a lag of $\frac{1}{4}$ s between the vertical wind measurement and the water vapor concentration is not considered (MacPherson et al. 1987).

Cospectral Contribution

The inadequacy of sensor response is the primary reason for delaying the use of this technique for measuring the flux of trace gases. It can best be evaluated using spectral and cospectral analysis (Lenschow et al. 1982; Desjardins et al. 1989b). Cospectral analysis provides information about the range of time scales contributing to the transfer of mass and energy. Average cospectral estimates for a wheat crop in Manitoba are shown in Fig. 3. They are based on 25 runs at 25 and 50 m, using two infrared gas analyzers: one with a frequency response of 1 Hz mounted outside the aircraft and the other with a frequency response of 15 Hz mounted in a duct within the aircraft. The cospectra from both sensors are similar at low frequencies but the slower sensor incurs approximately a 20% loss at high frequencies.

The limiting effect of the frequency response can also be quantified by making repeated passes at various altitudes at the same location. Figure 4 presents cospectral estimates of vertical wind and CO_2, as a function of the nondimensional wave number Kxz. The cospectral estimates are based on the average of 25 runs, 4 km in length, made at 10, 25, and 50 m. These runs were carried out in Manitoba around midday on six days between

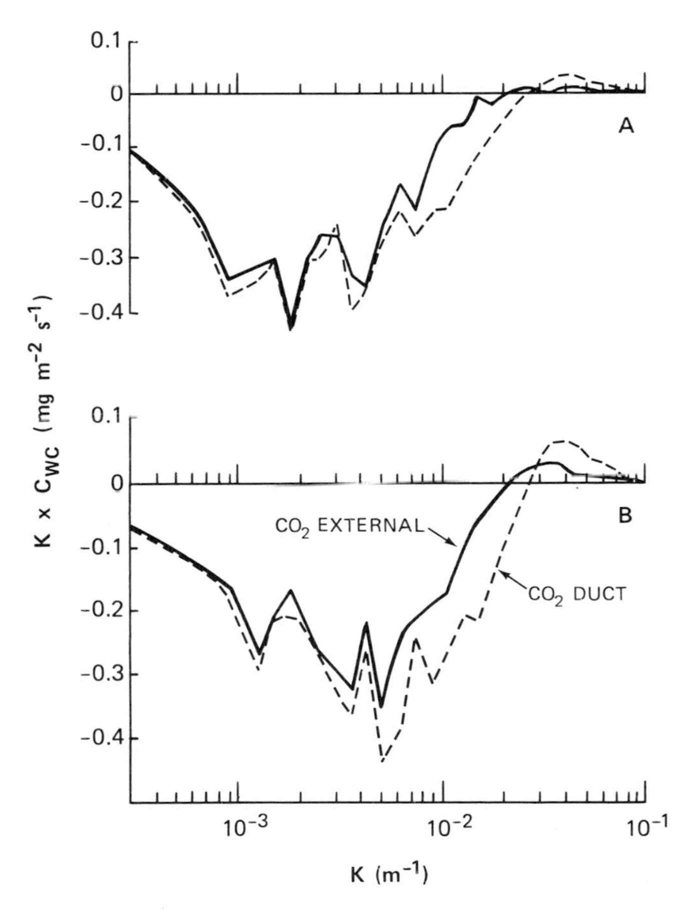

Fig. 3—Average cospectra of vertical wind and CO_2 from slow (solid) and fast responding (dashed) analyzers for 25(B) and 50(A) m above wheat fields in Manitoba (July 1985).

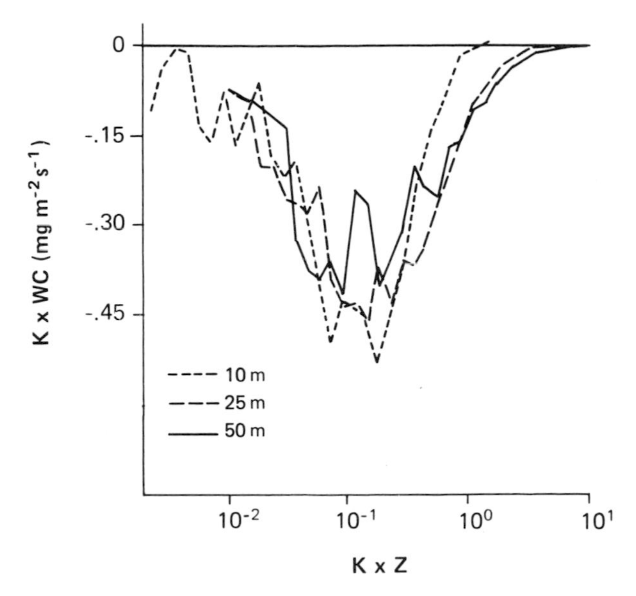

Fig. 4—Average cospectra of vertical wind and CO_2 as a function of dimensionless wave number for three altitudes above a wheat growing area in Manitoba, Canada.

July 4 and July 11, 1985. The results show that flux values scale well with height and that considerable information is lost at high frequencies at the 10 m level under these circumstances. Discrepancies between aircraft- and tower-based measurements, particularly for sensible heat fluxes, remain to be resolved (Desjardins et al. 1989b).

The cumulative contributions to CO_2 fluxes based on the cospectral estimates at 25 and 50 m are presented as a function of wavelength (Fig. 5). Such information can be used to estimate the loss due to high-pass filtering on short flight trajectories. This is most important for estimating flux loss at long wavelengths. Repeated passes over short distances can be used to eliminate the variability in the flux estimates but the long wavelength contribution can only be taken into account with long runs. Based on 45 km runs over a homogeneous area in Manitoba, at 50 m above ground during unstable conditions, wavelengths longer than 5 km contributed less than 5% to the flux. For less homogeneous areas, involving mixed farmland in Kansas, contributions of up to 40% sometimes occurred for wavelengths between 5 and 15 km at an altitude of 150 m (Desjardins et al. 1989a).

Air Density Corrections

The need for density corrections for air temperature fluctuations has long been recognized for flux measurements of CO_2 using infrared gas analyzers,

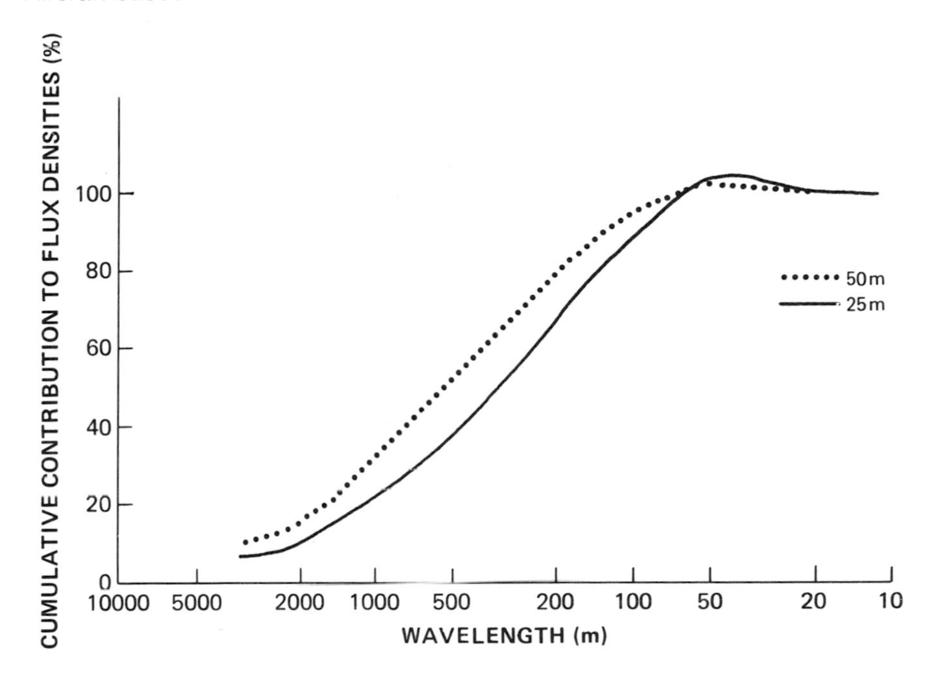

Fig. 5—Average cumulative contribution to CO_2 flux as a function wavelength for flights at 25 and 50 m altitude.

because such sensors measure density that is not a conservative quantity (Desjardins 1972). Heat and water vapor transfers induce a mean vertical mass flux which is not taken into account in most eddy covariance calculations. This correction can be done using either Eq. 7, which is expressed in terms of these fluxes, or by converting species density signals to mixing ratio with respect to dry air (Webb et al. 1980).

$$F_c = F_c \text{ raw} + (ma/mv)\,(\bar{\rho}_c/\bar{\rho}_a)\,F_q + (1 + \bar{\rho}_w/\bar{\rho}_a\,(ma/mv))\,(1/\bar{T})\,F_\theta\bar{\rho}_c \quad (7)$$

where ma and mv are the molecular weight of air and water vapor; $\bar{\rho}_w$, $\bar{\rho}_a$, and $\bar{\rho}_c$ are the mean density of water vapor, dry air, and the scalar of interest "c"; and $\bar{T}$ is the mean air temperature. The mixing ratio can be calculated by measuring the pressure, temperature, and water vapor simultaneously with the density of the species of interest. Table 1 shows CO_2 flux, F_{c2}, calculated using mixing ratio of CO_2 with respect to dry air for five runs with substantially different sensible heat, F_θ, and latent heat, F_q, fluxes as well as the CO_2 flux, F_{c1}, estimated using Eq. 7.

Averaging Requirements

Because gas transfer involves turbulent exchange, a certain averaging distance is required in order to obtain a reliable flux estimate in the atmosphere. We need to determine ensemble averages but can only calculate a time average over short records. Lumley and Panofsky (1964) demonstrated that the averaging time, T, required to determine a mean flux, $\overline{w'c'}$, to an accuracy "a" compared to its ensemble average is given by

$$T \simeq \frac{2\,\lambda\,\overline{(w'c')^2}}{a^2(\overline{w'c'})^2} \tag{8}$$

where λ is the integral scale of covariance, which is the time interval for a ground-based system, or the distance for an aircraft-based system, over which a flux maintains some degree of correlation with itself, and $\overline{(w'c')^2}$ is the variance of the flux. The integral scale can be estimated from the frequency of the peak of the spectra of the flux (Panofsky and Dutton 1984). Lenschow and Stankov (1986) have evaluated the accuracy of boundary layer turbulence estimates due to random sampling errors as a function of measuring distance. For example, they have estimated that sampling distances 10 to 100 times the sampling height are needed to obtain 20% accuracy of scalar fluxes. This factor is affected by the kind of sensor used, sampling level, surface properties, flight direction with respect to wind direction, data handling, and many other factors. For example, the higher the flight level, the greater the contribution of low frequency fluctuations and the longer the run has to be. On the other hand, the lower the flight level, the greater the contribution of high frequency fluctuations and the more rapid the sensor response has to be.

PARAMETERIZATION OF GASEOUS EXCHANGE

Parameterization of exchange processes has gone a long way towards describing biosphere–atmosphere interactions (Sellers et al. 1986). Gas exchange between vegetation and the atmosphere involves transport through the leaf boundary layer and diffusion through the stomata. In order to describe quantitatively the transport of gases, F_c, the concept of deposition velocity, V_{dc}, defined by

$$V_{dc} = -F_c/\bar{c} \tag{9}$$

has been widely used (Businger 1986). This assumes that the concentration, c, at the surface is equal to zero. Then, by definition,

$$V_{dc} = \frac{1}{r_{ac} + r_s} \qquad (10)$$

where the aerodynamic resistance, r_{ac}, can be determined from the momentum flux and stability conditions. The surface resistance, r_s, which is very much a function of environmental conditions, is often described as being due to molecular diffusivity plus stomatal resistance.

A complicating factor in parameterizing gaseous exchange is that there exists for many trace gases a concentration within the stomata which is not always zero. For example, in the case of a gas like NH_3, it can be given up or absorbed depending on the relative value of the concentration within the stomata as compared with the concentration in the atmosphere. This is thought to be the case also for nitrogen- and sulfur-containing gases (Farquhar et al. 1983). For cases for which the internal concentration is not known, these parameterization schemes do not provide useful information.

Aircraft flux measurements, which integrate gaseous exchange over large areas, can be used to relate ground-based measurements to a regional scale. Figure 6 shows the diurnal cycle of the CO_2 and sensible heat flux over a wheat crop with the corresponding regional fluxes obtained with an aircraft for four short periods. This combination of tower- and aircraft-based measurements can be used to generate regional estimates of the diurnal flux patterns.

SUMMARY

A good qualitative understanding of trace gas exchange processes is now available. However, more specific information is needed to identify the factors controlling uptake and release of trace gases by ecosystems at the global scale. Considerable effort is being made to quantify exchange rates over important ecosystems such as tropical forests, grasslands, and tundra (Harriss et al. 1988; Desjardins et al. 1989a). Other ecosystems such as boreal forests, wetlands, oceans, savanna, peatlands, and agricultural regions need to be studied.

Much research is still required to improve aircraft flux measuring procedures, particularly under less than ideal conditions. As fast response gas sensors become available for various trace gases, aircraft flux systems can be used to evaluate their budgets and understand their chemical transformation rates. This is important in order to quantify the role of the biosphere in influencing global atmospheric changes of greenhouse gases. Feedback processes that may affect future concentration of trace gases, such as the response of large ecosystems to climate change, need to be studied. New observation programs are required to supply the data necessary to

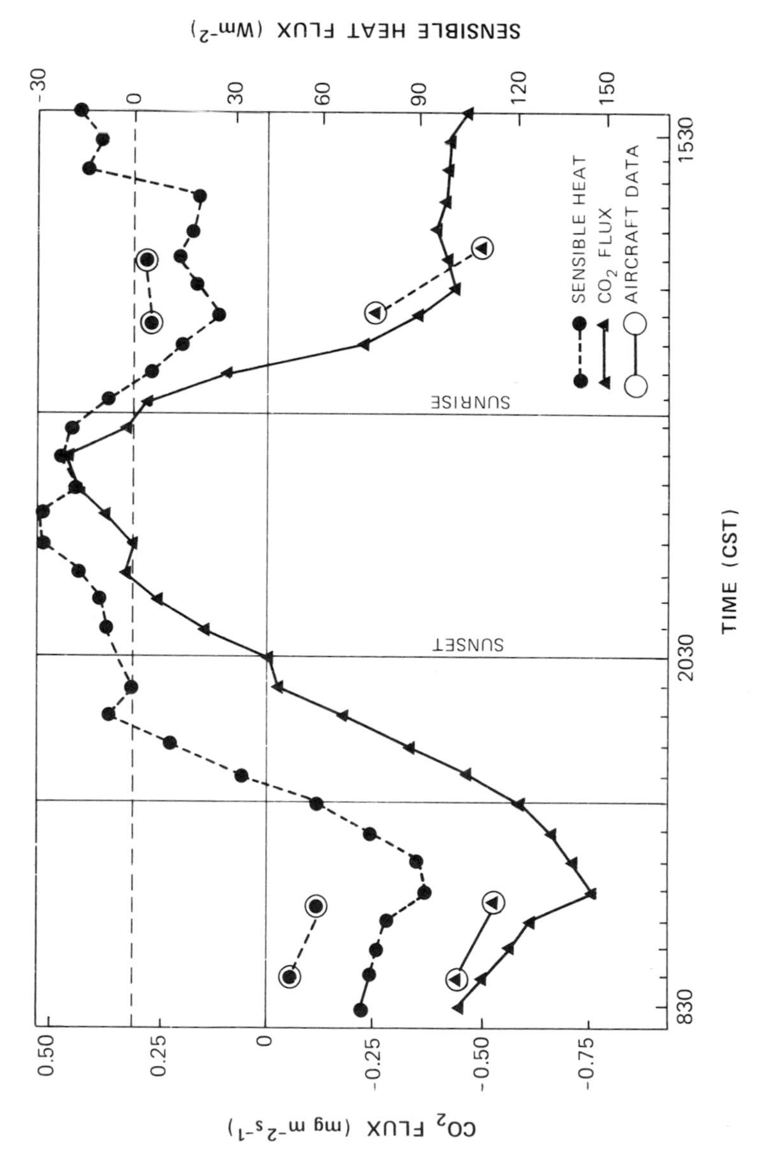

Fig. 6—Diurnal cycle of CO_2 and sensible heat flux using a ground-based flux system combined with four regional estimates using an aircraft-based system.

develop a better understanding of the regulation of gaseous exchange between the atmosphere and biosphere. Case studies using aircraft-mounted sensors should provide necessary information for quantifying the sources and sinks of important trace gases. Aircraft flux measurements should also provide independent data for testing regional and global extrapolations of trace gas fluxes.

Acknowledgements. We are grateful to the organizers of the Dahlem Konferenzen for inviting us to participate in this workshop. We would also like to acknowledge the assistance of the personnel of the Land Resource Research and the Engineering and Statistical Research Centres of Agriculture Canada and of the Flight Research Laboratory of the National Aeronautical Establishment of NRC in collecting and analyzing some of the data presented. Thanks are also due to P.H. Schuepp of McGill University and P. Rochette for their helpful discussion in reviewing this manuscript. The financial support of NASA as part of the FIFE program is also gratefully acknowledged. This manuscript is LRRC contribution no. 88–75.

REFERENCES

Businger, J.A. 1986. Evaluation of the accuracy with current micrometeorological techniques. *J. Clim. Appl. Meteor.* 25:1100–1124.

Desjardins, R.L. 1972. A study of carbon dioxide and sensible heat fluxes using the eddy correlation technique. Ph.D. diss., Cornell Univ. #73–34 Univ. microfilms, Ann Arbor, MI.

Desjardins, R.L., J.I. MacPherson, and P.H. Schuepp. 1989a. Long wavelength contributions to aircraft-based flux estimates of CO_2, H_2O and sensible heat. In: 19th Conference of Agriculture and Forest Meteorol., reprint vol., pp. 125–128. Boston: Am. Meteorol. Soc.

Desjardins, R.L., J.I. MacPherson, P.H. Schuepp, and F. Karanja. 1989b. An evaluation of aircraft flux measurements of CO_2, water vapor and sensible heat. *Bound.-Lay. Meteor.* 47:55–70.

Farquhar, G.D., R. Wetselaar, and B. Weir. 1983. Gaseous nitrogen losses from plants. In: Gaseous Loss of Nitrogen from Plant-Soil Systems, ed. J.R. Freney and J.R. Simpson, pp. 161–180. The Hague: M. Nijhoff.

Finnigan, J.J., and M.R. Raupach. 1987. Transfer processes in plant canopies in relation to stomatal characteristics. In: Stomatal Function, ed. E. Zeiger, G. Farquhar, and I. Cowan, pp. 376–429.

Harriss, R.C., S.C. Wofsy, M. Garstang, L.C.B. Molion, R.S. McNeal, J.M. Hoell, R.J. Bendura, S.M. Beck, R.L. Navarro, J.T. Riley, and R.C. Shell. 1988. The Amazon Boundary Layer Experiment (ABLE 2A): dry season 1985. *J. Geophys. Res.* 93:1351–1360.

Haugen, D.A. 1978. Effects of sampling rates and averaging periods on meteorological measurements. In: Fourth Symposium on Meteorological Observations and Instrumentation, reprint vol., pp. 15–18. Boston: Am. Meteorol. Soc.

Hicks, B.B., and P.S. Liss. 1976. Transfer of SO_2 and other reactive gases across the air–sea interface. *Tellus* 28:348–354.

Hicks, B.B., and R.T. McMillen. 1984. A simulation of eddy accumulation method for measuring pollutant fluxes. *J. Clim. Appl. Meteor.* 23:637–643.

Hicks, B.B., and R.T. McMillen. 1988. On the measurement of dry deposition using imperfect sensors and in non-ideal terrain. *Bound.-Lay. Meteor.* 42:79–94.

Lenhard, U., and G. Gravenhorst. 1980. Evaluation of ammonia fluxes into the free atmosphere over Western Germany. *Tellus* 32:48–55.

Lenschow, D.H. 1986. Probing the Atmospheric Boundary-Layer. Boston: Am. Meteorol. Soc.

Lenschow, D.H., R. Pearson, Jr., and B.B. Stankov. 1981. Estimating the ozone budget in the boundary layer by use of aircraft measurements of ozone eddy flux and mean concentration. *J. Geophys. Res.* 86:7291–7297.

Lenschow, D.H., R. Pearson, Jr., and B.B. Stankov. 1982. Measurements of ozone vertical flux to ocean and forest. *J. Geophys. Res.* 87:8833–8837.

Lenschow, D.H., and B.B. Stankov. 1986. Length scales in the convective boundary layer. *J. Atmos. Sci.* 43:1198–1209.

Lumley, J.L., and H.A. Panofsky. 1964. The Structure of Atmospheric Turbulence. New York: Interscience.

MacPherson, J.I., R.L. Desjardins, and P.H. Schuepp. 1987. Gaseous exchange measurements using aircraft-mounted sensors. In: Sixth Symposium on Meteorological Observations and Instrumentation, reprint vol., pp. 128–131. Boston: Am. Meteorol. Soc.

Mooney, H.A., P.M. Vitousek, and P.A. Matson. 1987. Exchange of materials between terrestrial ecosystems and the atmosphere. *Science* 238:926–932.

Panofsky, H.A., and J.A. Dutton. 1984. Atmospheric Turbulence. New York: Wiley.

Sellers, P.J., Y. Mintz, Y.C. Sud, and A. Dalcher. 1986. A simple biosphere model (SIB) for use within general circulation models. *J. Atmos. Sci.* 43:305–331.

Webb, E.K., G.I. Pearman, and R. Leuning. 1980. Correction of flux measurements for density effects due to heat and vapor transfer. *Q. J. Roy. Meteor. Soc.* 106:85–100.

Wesely, M.L. 1988. Use of variance techniques to measure dry air-surface exchange rates. *Bound.-Lay. Meteor.* 44:13–31.

Standing, left to right:
Ingo Aselmann, David Schimel, Bruce Hicks, Raymond Desjardins, Pamela Matson, Wen-Xiang Yang, Henning Rodhe
Seated, left to right:
Bo Svensson, Reiner Wassmann, Alexander Bouwman, Michael Whiticar, John Stewart

Exchange of Trace Gases between Terrestrial Ecosystems and the Atmosphere
eds. M.O. Andreae and D.S. Schimel, pp. 155–174
John Wiley & Sons Ltd

Group Report
Extrapolation of Flux Measurements to Regional and Global Scales

J.W.B. Stewart, Rapporteur
I. Aselmann
A.F. Bouwman
R.L. Desjardins
B.B. Hicks
P.A. Matson
H. Rodhe
D.S. Schimel
B.H. Svensson
R. Wassmann
M.J. Whiticar
W.-X. Yang

INTRODUCTION

Current estimates of trace gas fluxes confirm that the accuracy of such predictions in most cases is very low. Our group discussions centered on the problem of improving these estimates and addressed the question, "How should we extrapolate flux measurements to regional and global scales?" After a brief overview of current flux estimates and a short discussion on the possibility of important unidentified sources, we spent most of the time discussing the problems encountered in accurate extrapolation from a few point measurements to a whole region. Several approaches that may eventually provide regional and global estimates were critically evaluated.

HOW GOOD ARE GLOBAL FLUX ESTIMATES? ARE WE MISSING IMPORTANT SOURCES AND FLUXES?

These questions were mainly discussed with respect to fluxes of two gases: methane and nitrous oxide. Some consideration was also given to more reactive gases such as the oxides of nitrogen.

Methane (CH_4)

The methodology to detect CH_4 accurately in the atmosphere has been available for less than fifty years, and it is only in the last two decades that a significant scientific effort has focused on estimating sources and fluxes. It is therefore not surprising that considerable uncertainty exists with regard to the sources and amounts as shown in Table 1. Of particular concern is the increase in atmospheric methane recorded from direct measurements during the last 15 years and inferred from ice core data extending back several hundred years. The rate of increase in atmospheric methane concentration since 1965 is about 1% per year. Whether this is due to a constant increase in several fluxes or from one or more specific sources is unknown. It is of fundamental importance to determine the causes of this increase so that rational decisions about emission reductions can be made.

The total flux into the atmosphere has to be approximately balanced by the photochemical decomposition in the atmosphere, estimated to be 550 ± 20% Tg CH_4 per year, and by a biological sink at the Earth's surface (Cicerone and Oremland 1988). The main sources of methane are included in Table 1 but considerable uncertainty exists in the emission rates. For instance, although the geographic extent of flooded cultivated soils can be estimated with reasonable accuracy on a global basis, the proportion of the year that they are anaerobic fluctuates greatly (20–365 days) with location and management practices. An equally great uncertainty surrounds CH_4 fluxes from wetlands (Schütz and Seiler, this volume). These data are based

TABLE 1. Current estimates of uncertainty: CH_4 emission rates from various sources (adapted from Schütz and Seiler, this volume).

Source	Range of reported CH_4 emission ($\times 10^{12}$ g yr^{-1})
Waterlogged soils (paddy rice)	70 – 170
Natural wetlands	40 – 160[a]
Landfill sites	30 – 70
Oceans/lakes/other biogenic	15 – 35
Ruminant digestive tract	66 – 90
Termites	2 – 5
Exploration of natural gas	30 – 40
Coal mining	35
Biomass burning	55 – 100
Other nonbiogenic	1 – 2
TOTAL	374 – 717

[a] Aselmann (pers. comm.)

on relatively few measurements and do not take into account the diversity in types of wetland (swamp, marshes), the seasonality and duration of flooding, or their chemical status (pH, N, S, and P content, salinity, etc.). Factors controlling CH_4 emission have been researched under controlled conditions (Conrad, this volume). However, too few field measurements are available to extrapolate with any degree of certainty using global data bases of vegetation, soils, climate, etc. Future model development depends on expansion of these measurements.

Based on the comparison of the isotopic composition ($\delta^{13}C$, $\delta^{2}H$, and ^{14}C) of atmospheric methane sources, some overall constraints can be derived. Measurements of ^{14}C in atmospheric methane indicate a relative contribution of fossil CH_4 of 21 ± 3% at the present time (Wahlen et al. 1987, 1989). Based on $\delta^{13}C$ and $\delta^{2}H$ isotopic evidence, the relative contributions by the remaining sources of atmospheric methane have been estimated to be: wetlands (including tundra) 25%, ruminants 20%, rice production 24%, and biomass burning 10% (Wahlen et al. 1987, 1989). This is in general agreement with the listing in Table 1. Biomass burning is difficult to estimate, production being highly dependent on fire intensity, temperature of the fire, and the material being burnt. Estimates of CH_4 production from ruminants were thought to be reasonably accurate but earlier estimates of the contribution from termites (Zimmerman et al. 1982) need to be refined (Cicerone and Oremland 1988). The volcanic emission of methane (sporadic), emission from water columns, asphalt, and seepage of hydrocarbons are poorly known but are not thought to be major sources. A few data points are available on methane uptake and oxidation by soils, although the nature of this reaction needs to be clarified. For instance, the CH_4-oxidizing capacity of the soil covering landfills leads to a major uncertainty in estimating net CH_4 emissions from such sources. For other sources of CH_4, estimates of the current budget are calculated by difference as no direct measurements are available from coal mining and natural gas production (Wahlen, pers. comm.).

The ranges given in Table 1 should not be taken to represent the uncertainty in any strict scientific sense. In most cases the estimates of the ranges are based on the observed variability of a very limited number of real measurements, many of them of unknown representativity, as well as on rough estimates of the frequency of occurrence of the conditions under which the measurements were made. The true fluxes may thus lie well outside those ranges.

Nitrous Oxide (N_2O)

The data on N_2O fluxes also exhibit considerable variability due to the fact that the sources are poorly defined and heterogeneous. The largest fluxes are believed to occur from tropical forest soils. Source flux values have mainly been derived from a few chamber flux measurements. Little attention has been paid to variability due to soil, vegetation, land use, and climate.

Considerable uncertainty is also associated with N_2O contribution by fossil fuel combustion. The highly variable, seasonal nature of N_2O loss with fertilizer and waste application also lends uncertainty to anthropogenic source values. Other missing sources may include flux from offshore marine systems and aquifers (Luizão et al., pers. comm.). At the present time, no satisfactory balance has been established between the estimated loss of N_2O through photochemical processes in the stratosphere (14 ± 3.5 Tg N_2O–nitrogen per year) and the various emission fluxes. The estimated anthropogenic sources (combustion, fertilization) also do not seem to be large enough to match the observed increase in atmospheric concentration of approximately 3.5 Tg N_2O–nitrogen per year.

Nitrogen Oxides (NO_x)

For species such as NO_x, with a short atmospheric residence time, global flux estimates are less useful while regional/continental-scale budgets are much more relevant. Our current understanding of surface exchange of NO_x and other nitrogen-containing species is very limited. In the highly polluted regions of Europe and North America, the anthropogenic NO_x emissions dominate the input. In those regions, an approximate balance has been established with deposition of nitrate in precipitation and export of NO_x. However, in other less polluted regions, biological sources and fluxes become much more important and the total atmospheric budgets are less well established.

Very high emissions of NO_x from natural soils have recently been reported from some tropical ecosystems (savanna and tropical forests) and tentative budgets have been estimated.

WHAT IS A REASONABLE APPROACH FOR STRATIFYING THE WORLD TO PERMIT DEVELOPMENT OF TRACE GAS MODELS AND BUDGETS, AND FOR STUDIES OF PRODUCTION AND CONSUMPTION?

Trace gas fluxes vary widely on temporal and spatial scales; because of this, extrapolating fluxes from point measurements to regional and global scales can at times seem a monumental undertaking. If the basis for extrapolation required statistically appropriate random sampling over the entire surface of Earth, the task would be impossible.

The terrestrial organization of soils, vegetation, and climate lends itself to stratification. It is possible, for example, to divide the surface of Earth into biomes based on vegetation–climate characteristics (e.g., tropical forest, temperate forest, boreal forest, etc). Sampling for trace gas fluxes can then be carried out for each stratum. Most budgets are based on such stratifications: limited amounts of flux data are acquired from "representative" sites for a given biome, and then extrapolated across the entire area of the biome. Thus, site selection is not random but is based on the assumption that one particular ecosystem can represent all systems within a biome. This assumption ignores the huge amount of variability in flux within the broad class resulting from variability in environmental factors within the biome.

One alternative to this approach is to use some of the variability in ecosystems to improve extrapolation approaches. We know, for example, that ecosystem processes vary with many environmental factors, including soil fertility, climate, flooding regimes, and natural or anthropogenic disturbance. Selection of sites that span a range in variability, and measurement of fluxes in association with these factors, provide a basis for direct extrapolation using global data bases and for the development of process models. In the group's discussion, two examples of this gradient approach and its limitations were detailed. Tropical forests have been recognized in several budgets as the single most important source of N_2O (estimated flux ca. 7 TgN/yr). However this flux is based on multiplication of the entire area of tropical forest (including rain forest, dry forest, and savanna) by an average flux derived from a few rain forest measurements. We know that one of the factors that controls N_2O flux across a range of forests is soil nitrogen turnover, which will vary with soil nitrogen status. Therefore, we expect that stratification of tropical forests into groups based on soil fertility (aligned along a soil fertility "gradient") should give information on distribution of fluxes and of hot spots. Such information can then be used in conjunction with global data bases to refine budgets. Using this approach for a range of rain forests differing in fertility, Matson and Vitousek (pers. comm.) calculated a total humid tropical forest source of 2–3 Tg/yr.

In the example outlined above, forests were actually stratified into groups on the basis of soil type, which provided a range in soil fertility. Within each soil type, a number of forests were sampled, with the assumption that they were representative of others in the same soil type. With this direct extrapolation approach, regional and global estimates of flux can be improved to the extent that the selected gradient incorporates the important controlling factors. However, this approach cannot easily incorporate dynamic controls such as daily to seasonal moisture changes and so has limitations in predicting changes in flux under novel climatic conditions.

Use of gradients for ecosystem studies, and the development of relationships between fluxes and controlling factors changing along those gradients, lends itself to the development of simulation models to estimate and extrapolate flux. Several ecosystem models have been developed to predict ecosystem processes and response based on information from soils and climate data bases (see Matson et al., this volume). These models utilize relationships developed at sites representing a range in soil and climatic variability for a region. Because the range in variability is incorporated in model functions, the models can estimate fluxes at unstudied sites and under varying climatic regimes for a given region. The data requirement is for information (including remote sensing data) on spatial and temporal patterns in soil, climate, and management that can drive the model on the appropriate time and space scales. The utility of such models in estimating trace gas flux is obvious; what is not clear is the potential for development of "generic" trace gas models that can be applied beyond given regions. Global models based on mechanistic understanding of controlling processes have not yet been developed. It is probable that application of modeling approaches will be done within the confines of ecosystem stratification (i.e., different models will be developed for different regions or biomes).

There are some limitations to the so-called "gradient" approach. First, the approach assumes that enough is known about the processes controlling fluxes to identify the key factors. For many trace gases, this is probably true. For example, it is reasonable to look at N_2O flux as a function of soil fertility and soil moisture. For those gases where such information is lacking, initial coarser-scale stratification (by latitudinal belts, biomes, etc.) may be required.

A second limitation may be a lack of protocol for extrapolating flux data in statistically meaningful ways. We can deal with non-normal distributions on a site-by-site basis, but the approach is less clear when site data must be combined, averaged, and stratified for use in direct extrapolations. There are several assumptions that must go into model development when flux data fit several different distributions, and when distributions vary as a function of site characteristics and position on the gradient. Is statistical variability subsumed within each higher scale, or does it propagate to the next higher scale? Statistical approaches will need to be addressed explicitly in the development of studies to extrapolate fluxes either through stratification or modeling approaches.

Finally, extrapolations using measured flux data or modeled estimates can only be as good as the ground-based or remote sensed regional and global data bases. At this point in time, extrapolations are probably more limited by a lack of flux data or estimates collected using a reasonable strategy (as discussed above) than by data bases with which to use them. However, as model development proceeds, data bases of important controlling factors (e.g., soil moisture) operating with sufficient detail may well become limiting.

WHAT ARE THE USES FOR, AND LIMITATIONS OF, CURRENT GLOBAL DATA BASES IN TRACE GAS FLUX EXTRAPOLATIONS? WHAT NEW DATA ARE REQUIRED?

Extrapolation of local measurements of trace gas fluxes to estimate regional and global emissions requires data bases that describe the geographic distribution and total areal extent of factors associated with flux. At present, there are global data bases of climatic variables, land surface characteristics, vegetation, and economic/agricultural activities. Tables 2 and 3 list a selection of currently available data bases. In the following paragraphs, we will briefly discuss their limitations and then assess the requirements for new data bases and handling of all data bases.

Because specific data bases are normally compiled with a specific use in mind (social, economic, political, or scientific), they often are not ideal for use by other groups for other purposes. For example, the FAO/UNESCO soils data bases are based on information from 1940–1960 and were collected primarily for use in land management. Some of the properties which would be of most interest from a trace flux gas point of view are not included (e.g., soil water holding capacity) or are compiled too coarsely. While a great deal of information is held within the classification scheme, much of it can only be extracted by expert pedologists who are knowledgeable about specific regions. This information may well be lost to users of digital data bases. Specific attribute files of the various soil classes should be made available with the data bases.

One of the most severe limitations for trace gas work is the lack of land-use data bases. Land-use information is not well recorded in map form (Matthews 1983). Information on land conversions is lacking, especially in tropical areas, and should be a major focus in the development of new data bases.

Data recovered from polar orbiting satellites currently provide information about atmospheric and surface characteristics (e.g., surface radiation temperature, cloud cover, ozone concentrations, etc.). The AVHRR (Table 3) has collected data on vegetation reflectance (NDVI, often assumed to be an index of photosynthetic activity) since 1979. However, these long-term data sets are not useful for trend analysis because of lack of instrument calibration, atmospheric correction, and viewing geometry correction. Instrument calibration should be a major focus of upcoming satellite missions. Promising but not yet operable remote sensing activities include monitoring of land-use change, inundation patterns, soil moisture, and fire intensity.

In our discussion of global data bases, it became clear that accessibility to the user is a major problem. Data bases should ideally be located in international holdings; Tables 2 and 3 show encouraging preliminary attempts in this direction. Moreover, it would be useful to have data bases referenced,

TABLE 2. Status of preliminary GRID global data sets as of 1 February 1988 (all data sets held at GRID-Processor, Geneva).

Parameter	Coverage	Source	Georef.	Projection	Avail.	D.Q.[a]
Boundaries (Land-water political)	Global	World Database II (WBDBII) —US State Department	Yes	Long-Lat	Yes	A-3
Elevation (10 min. grid)	Global	National Geophysical Data National Oceanic and Atmospheric Administration, USA (US–NOAA)	Yes	Long-Lat	Yes	A-3
Soils	Global	FAO/UNESCO 1.5 M soils map	Yes	Long-Lat	Yes	A-3
Vegetation	Global	Goddard Institute of Space Studies (GISS)	Yes	Long-Lat	Yes	
Cultivation Intensity	Global	Goddard Institute of Space Studies (GISS)	Yes	Long-Lat	Yes	
Albedo–4 Seasons	Global	Goddard Institute of Space Studies (GISS)	Yes	Long-Lat	Yes	
Vegetation Index (Weekly 4.82–4.87)	Global	US–NOAA	No	Polar Stereog.	Yes[b]	B-1
Precipitation Anomalies (Monthly 1985–on)	Global	Climate Anal. Center US–NOAA/WMO (digitized by GRID)	Yes	Long-Lat	Yes	A-2
Temperature Anomalies	Global	Climate Anal. Center US–NOAA/WMO (digitized by GRID)	Yes	Long-Lat	Yes	A-2

Surface Temperature	Global	NASA-Jet Propulsion Lab./ Goddard Space Flight Center (JPL/GSFC)	Yes	Long-Lat	No[c]	B-1
Ozone Distribution	Global	NASA, Total Ozone Measuring System (TOMS)	No	[—]	No[c]	B-1
Vegetation	Africa	DMA 1.2 M Topo Maps (digitized by ESRI for UNEP/ FAO)	Yes	Long-Lat	Yes	A-3
″	Africa	White's UNESCO/AETFAT map (digitized by GRID)	Yes	Long-Lat	Yes	A-2
″	Africa	FAO/Toulouse (UNESCO-class)		In preparation		
Vegetation Index (Seasonal 1982)	Africa	NASA-GSFC (from AVHRR 4 km Data)	No	Mercator	Yes	B-1
Watersheds	Africa	FAO data 1.5 M scale (digitized by ESRI for UNEP/ FAO)	Yes	Miller Obl. Stereog.	Yes	A-3
Rainfall (Mean Annual)	Africa	FAO data 1.5 M scale (digitized by ESRI for UNEP/ FAO)	Yes	Long-Lat	Yes	A-3
No. Wet Days (Mean Annual)	Africa	FAO data 1.5 M scale (digitized by ESRI for UNEP/ FAO)	Yes	Long-Lat	Yes	A-3
Windspeed (Mean Annual)	Africa	FAO data 1.5 M scale (digitized by ESRI for UNEP/ FAO)	Yes	Long-Lat	Yes	A-3

continued

TABLE 2. (Continued)

Parameter	Coverage	Source	Georef.	Projection	Avail.	D.Q.[a]
Protected Areas	Africa	IUCN/CMC (center points and areal extent)	Yes	Long-Lat	Yes	A-3
Biogeographical Province	Africa	IUCN paper no. 18 by UDVARDY	Yes	Long-Lat	Yes	
Species	Africa	IUCN/CMC (center points 20 endangered plants and animals)	Yes	Long-Lat	Yes	A-3

[a] D.Q. = Data set qualifiers; see Interim Data Release Policy.
[b] Data are held as received from NOAA and will not be georeferenced or transferred into long-lat. as standard archive practice unless there is a specific request.
[c] waiting modified data set from NASA.
[d] Awaiting recalibration by NASA.

TABLE 3. Global georeferenced data sets (available at the University of New Hampshire).

Base Maps	
World DataBank II Coastlines and International Boundaries, set 1.	Vector-based global template for registering other data sets: 100 m resolutions
World DataBank II Coastlines and International Boundaries, set 2.	Grid-cell conversion gridded at various resolutions: 4–50 km, AVHRR compatible
Physical Features Maps	
Mode Elevation	10 minute grid cells
Minimum Elevation	10 minute grid cells
Maximum Elevation	10 minute grid cells
Primary and Secondary Terrain Characteristics	Identifies significant features, e.g., marsh, rugged mountains
Global Slope and Aspect	Derived from global elevation and from FAO soils (see below)
Soil Maps	
FAO World Soils	Grid cell, resolution 0.5 degrees
USDA 7th Approximation World Soils	Grid cell, resolution 0.5 degrees
USSR World Soils Map	Vector digitized from 1:15 M scale maps from the USSR Inst. of Geogr. (in progress)
World Organic Soil Carbon and Nitrogen (source: Zinke et al. 1984)	Point data
Soil Texture and Phase	Grid cell, resolution 0.5 degrees
Vegetation Maps	
Holdridge Life Zones (source: Emanuel et al. 1985)	Grid cell, resolution 0.5 degrees, climatically based
World Ecosystem Complexes (source: Olson et al. 1983)	Grid cell, resolution 0.5 degrees, approx. carbon contents associated
Global Vegetation and Albedo Classification (source: Matthews 1983)	Grid cell, 1.0 degree converted to 0.5 degrees, Unesco
Primary and Secondary Land Cover (source: Henderson-Sellers et al. 1985)	Grid cell, 1.0 degree converted to 0.5 degrees
Hydrology and Climatic Data Sets	
World Rivers and Lakes	Vector format, contains several stream and river sizes
World Monthly Surface Station Climatology (source: Spangler and Jenne 1987)	NCAR met station point data historical ppt, temp.

continued

TABLE 3. (Continued)

Northern Hemisphere Historical Climate (source: Bradley et al. 1985)	Tabular data, 1851–1980
Global Temperature, Precipitation, soil moisture, and PET (Willmott et al. 1985)	Grid cell, 1 degree resolution
Global Cloud Cover	Grid cell, 0.5 degree resolution
Global Irradiance	Grid format, 0.5 degree, corrected for cloud cover
Soil Moisture	Grid format, 0.5 degree, South America, global in progress
Land Use and Demographic Data Sets	
World Bank Development Indicators	Integrated with International boundary file, used for region selection and definition
Fronts of Deforestation	Vector based, 50 km resolution time series maps of deforestation, Brazil, global in progress
Global Contemporary Land Use	Vector encoded from USSR map at 1:15 M scale, in progress

so that they are easily available to the international research community. As part of this procedure, definitional differences for categories (now a problem that makes intercomparison difficult) should be addressed. Finally, many of the ground-based data bases, especially land-use data sets, require better georeferencing and increased spatial resolution.

For remote sensing-based data bases, intercalibrations and intercomparisons of various sensors should be performed to facilitate interannual and long-term trend analysis of global change. Archiving of data is of utmost importance; the costs of this cannot be ignored in future planning and should be borne by the international community. Finally, development of improved algorithms that tie spectral data to biological and physical functioning, and development of new sensors that are designed explicitly for use in trace gas studies, should be carried out by this community in collaboration with remote sensing researchers.

It is clear that the status of global data bases is far from perfect. While the limitations cannot be ignored, it is likely that current data and upcoming improvements cannot be adequately utilized without a great deal more flux data and/or simulation models that take advantage of the data bases.

HOW DO WE EXTRAPOLATE AND TEST THE RESULTS?

A number of techniques can be used to measure trace gas fluxes, the advantages and disadvantages of which are discussed elsewhere in this volume (Vitousek et al.; Fowler and Duyzer, both this volume). Ecological (chamber) methods yield the most detailed information on the underlying processes but apply only to very small areas and are not without their own biases. In general, a method is needed to transfer the results of chamber studies to larger scales (i.e., from meters to hundreds of meters) and eventually to grid cells of global-scale numerical models (i.e., from hundreds of meters to hundreds of kilometers). The extrapolation must take into account the full range of complexity imposed by microsite and ecosystem variability at the smallest scales, to terrain and meteorology at the largest.

Figure 1 illustrates the time and space scales that can be addressed by the various measurement methods. In general, there are no two techniques that can be easily compared side-by-side. A major problem confronting global measurements is how to construct estimates of large-scale emission rates when the available measurement methods are so limited in their applicability. If chamber and/or tower measurements were carried out across vegetation or soil types, or by using some other stratification approach, combination with remote sensing of these types would allow fluxes to be estimated over large regions. Alternatively, if models of trace gas fluxes, as they vary with environmental and edaphic variables, were developed, then they too could be used to estimate fluxes on regional scales.

In concept, the space and time scale limitations evident in Fig. 1 could be overcome by judicious replication of measurements. Obviously, chambers can be used repetitively to extend their time coverage, and arrays can be used to provide spatial integration. Tower measurement methods address a larger spatial scale than do chambers (hundreds of meters rather than meters) and can also be applied in an array. However, the fundamental purpose of such flux measurements is to identify and formulate the processes that control the flux, not to quantify the fluxes for direct input into global budgets and models. Group discussions led to agreement that a critical step is to combine these process descriptions in models that will permit extrapolation based on limited field measurements of fluxes.

The appropriate area-averaged concepts include (*a*) descriptions of the surface sources; (*b*) description of where sources occur within the canopy, parameterized in micrometeorological terms that are gas-species specific; and (*c*) combination of these concepts in terms of some integrating model such as a multiple-resistance routine. All of the above have relevance only as area averages, over minimum areas typically about 100 m $\times$ 100 m. Once specified in this way, detailed scaling-up models can then be constructed. Some models of this kind have been developed, but further development

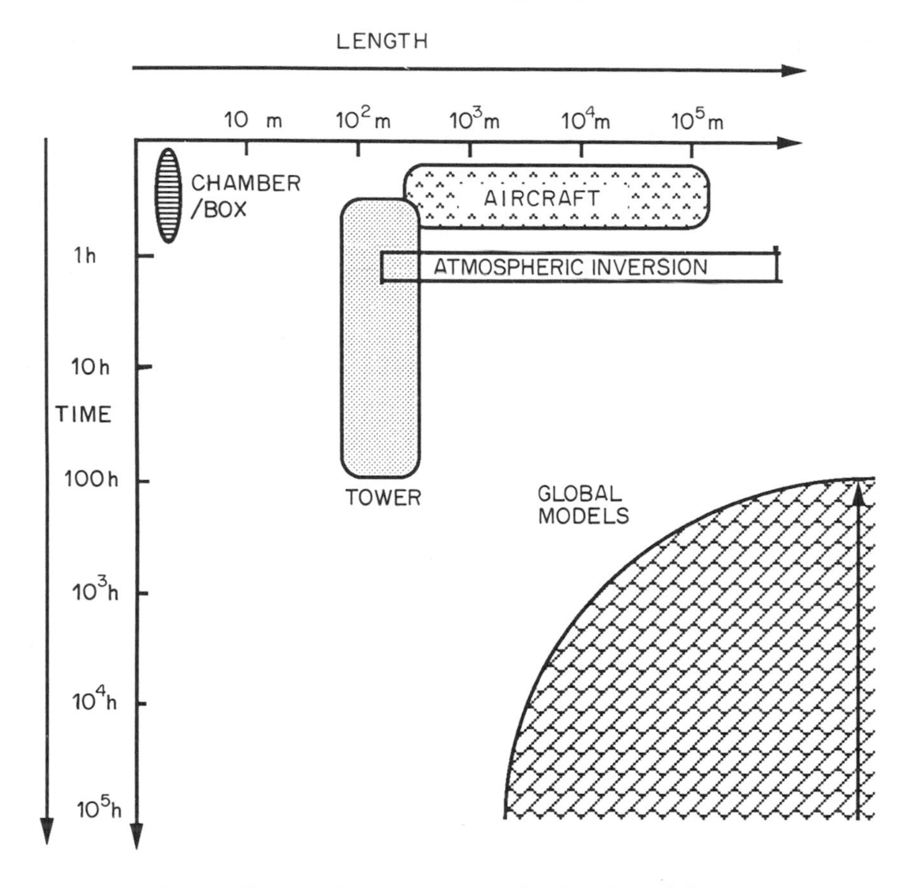

Fig. 1—Time and space aspects of estimation of fluxes.

and testing is necessary. There is, in fact, a hierarchy of models of various complexity that represents the current "state-of-the-art" for flux modeling. It is these integrated models that are especially well suited to computation of emission rates from natural landscapes.

Aircraft eddy correlation measurements have a special role, since they provide a direct measurement of the exchange over large areas and hence can be used to test predictions for global model grid-cell areas.

The group agreed that the answers that are required cannot be provided by measurement programs alone, nor only by models. It is critical, therefore, that measurement programs be conducted in close contact with model development programs.

The trace gas exchange models that are developed in programs of this kind are descriptions of the overall exchange process, intended to help compute area averages across terrain that is not always simple. These models will, by necessity, vary with the gas of interest. For example, the presence of a vegetation canopy strongly influences the effective emission of NO_x; micrometeorological and canopy exchange components in addition to soil microbiological and physical components will be necessary to model NO_x fluxes. On the other hand, canopy exchange is unlikely to have a dominant influence on long-term emission rates of soil-derived species, such as N_2O and CH_4, and can likely be ignored in their models.

Measurement Verification

The uncertainty associated with global emission estimates is intimately related to the uncertainty of individual field measurements of fluxes. The surface variability of emission rates might be large, leading to the generation of highly variable results from field programs. It should not necessarily be assumed that emission fluxes are normally distributed, nor that the arithmetic average of a limited number of samples is necessarily the appropriate quantity to multiply by surface area in order to estimate a global average. The matter is further complicated by the consequences of imperfect measurement. In general, the arithmetic average of limited flux data obtained in a specific circumstance may or may not be an acceptable basis for constructing estimates for larger areas.

There is a basic need to minimize errors of measurement, as well as to verify the large-area flux estimates produced by various extrapolation schemes (including specialized numerical models). Remote sensing is seen as a principal source of data to drive the models that are developed to compute emission rates for areas where no detailed flux data are available. As the various techniques are improved, it is anticipated that integrated, large-scale studies conducted over target areas of special interest are likely to become increasingly important.

There are three different kinds of field study that need to be considered. First, there are intensive studies of specific processes, intended to derive the formulations to be used in process models. Second, there are technique comparison studies, by which confidence is generated in the ability to measure fluxes under field conditions. Third, there are field tests of model predictions. These three kinds of study need to be considered in the context of extrapolation from localized chamber studies to areas addressed directly by micrometeorology, and from these areas to the larger grid-cell areas of global models.

For technique comparison studies, one can test the ability to measure fluxes by comparing results obtained over various scales; this frequently involves combining numerous samples obtained over small scales to arrive at an estimate appropriate for the larger scale addressed directly by another method. For example, the result derived from an array of chamber measurements over bare soil can be tested against tower-based eddy flux measurements. The evaluation of airborne flux measurements might well be carried out by selecting a relatively homogeneous area with a minimum size of about 4 km × 4 km and comparing against flux measurements made using ground-based systems.

Once measurement techniques have been tested, it becomes meaningful to use data collected at coarser scales as independent data sets against which to test model estimates of flux. For example, aircraft-based measurements of fluxes on a scale of 100 km might be used to test models that predict the distribution and quantity of fluxes as a function of vegetation, soil characteristics, climatic variables, or other variables obtained through remote sensing. The models can then be used with some assurance in areas where aircraft sampling is not possible. Other methods for evaluating area-averaged emission rates are available, such as budget calculations over some area containing an emission source of special interest, but these methods are less direct than aircraft eddy correlation methods and have yet to be fully tested. So far, they have been used mainly for studies of small test plots.

Independent Corroboration on the Global Level

Independent validation of estimates of regional and global fluxes must be attempted to identify the relative proportions of contribution from various sources. Sources of uncertainty not only include those associated with the flux measurements themselves (chambers, towers, aircraft) but also, and probably more importantly, those due to the difficulty in extrapolation to larger scales.

Balancing emission against global sinks, e.g., photochemical reactions in the atmosphere, is one obvious way of checking the estimates of the total source flux (cf. section on GLOBAL FLUX ESTIMATES). A somewhat more sophisticated approach is "inverse modeling," where information on systematic variations in the atmospheric concentration of a trace gas (height, latitude, season) is used in combination with an atmospheric transport model to derive information on surface sources and their distribution in space and time.

Additional information regarding the relative importance of different sources can be obtained from measurements of stable isotopes. Successful application of these techniques requires that there are predictable, consistent variations in the molecular or isotope signatures between various sources. In addition, all the partitioning and isotope effects have to be understood.

Isotope techniques offer tracer information at both the process level (e.g., microbiological production, consumption) and at the integrative, global level. This is based on the distinctive isotope signature of the various sources of gases, especially CH_4. In the case of CH_4 the combination of stable carbon and hydrogen isotopes with ^{14}C information (Wahlen et al. 1987, 1989) characterizes the various sources such as biogenic, thermogenic, and others (volcanic, geothermal). Current limitations to establishing the relative importance of these sources via isotope techniques are: (*a*) unknown isotope discrimination (e.g., D/H discrimination in the reaction of OH with CH_4), (*b*) a limited isotope data base on specific environments, (*c*) possible overlaps in isotope signals, and (*d*) the currently very small isotope data base for atmospheric gases.

Perhaps one of the most useful applications of isotopes is the potential to monitor change in the isotope ratios with time, and hence identify the source responsible for the absolute and relative changes in atmospheric concentration. However, limitations to this approach include an absence of long-term data bases for isotope abundances.

Radiogenic trace gases also offer the ability to distinguish between different sources and to evaluate their relative contributions. Such trace gases include ^{222}Rn, ^{220}Rn, and ^{212}Pb (Bonsang et al. 1985, 1988) or ^{85}Kr. In addition, the molecular composition of the minor trace gas components in atmospheric gas mixtures, such as nonmethane hydrocarbons (C_2–C_6 aliphatics or aromatics), can be used in concert with isotopes to assess the relative proportions of various sources, e.g., marine or terrestrial (Bonsang et al. 1988), bacterial or thermogenic (Whiticar et al. 1986). They are also valuable for information on transport processes.

CONCLUSIONS

Embodied in this report are many suggestions for improvement of current estimates of gaseous fluxes and for the extrapolation of flux measurements to regional and global scales. We would like to draw particular attention to the following priority research areas:

1. *More flux data is needed for all the trace gases that we considered, especially from areas such as wetlands and tropical forests which are extremely undersampled in relation to their importance in the global production of trace gases.*

In this connection, we discussed the stratification of sampling procedures and the use of variability in ecosystems to improve extrapolation approaches. The extent to which a selected environmental gradient incorporates the important controlling factors on gas production and utilizes ground-based or remote-sensed regional and global data bases needs to be clarified. We urge that these approaches be considered when point measurements are being made so that the valuable flux data collected at one or several sites may be extrapolated to a wider range. In this regard:

2. *We emphasize the need for the development of ecosystem models which recognize the relationship between fluxes and controlling factors and link them to data bases.*

   The verification and extension of this approach should be emphasized.

3. *Global data bases should (a) be improved and (b) be accessible to the user. This will require one or several central international holding agencies and an improvement of existing data bases.*

Existing data bases have to be rigorously examined with regard to their ability to provide information on processes that control gaseous production as well as to accurately depict current biome extent, soil, and vegetation properties. This may require the development of integrative models which utilize several data bases and incoming climatic data to predict fluxes. Several important data bases appear to be missing. Chief among these is data on land use and changes in land use with time. More emphasis could be placed on improving the resolution of satellite imagery. Instrument calibration should be a major focus of upcoming satellite missions. In this regard:

4. *We emphasize the need for intercalibration and intercomparisons of various sensors used in remote sensing, as well as the need to archive data.*

The latter aspect is important as we continue to develop algorithms that tie spectral data to biological and physical characteristics.

We also discussed the widely differing techniques which are used to measure trace gas fluxes and the difficulties encountered in comparing measurements from chambers, towers, or aircraft and transferring the results of point measurements to global-scale numerical models. Priority, therefore, should be given to:

5. *The development of methods which allow the result of chamber studies to be extrapolated to larger scales and which take into account the full range of complexity imposed by microsite and ecosystem variability at the smallest scales to terrain and meteorological variations at the largest.*

In this regard, we suggest that there is a great need for intercomparison across similar vegetation or soil types. We agreed that a critical step is to combine these process descriptions and models that will permit extrapolation based on limited field measurement of fluxes. It was obvious that the answers required cannot be provided by measurement programs alone nor only by models. It is critical, therefore, that

6. *Measurement programs be conducted in close contact with model development programs.*

With regard to the need to minimize errors of measurement, we recognized the potential of remote sensing as a principal source of data to drive models that compute emission rates for areas where detailed fluxes are not available. We anticipate integrated large-scale studies conducted over target areas of special interest. Therefore, we need intensive studies of specific processes, technique comparison studies, and field tests of model predictions. It is obvious that all types of measurements (chamber, tower-based eddy measurements, and airborne flux measurements) are necessary and can be used to test results of measurements and models at different scales.

7. *Finally, there is a need for independent validation of estimates of regional and global fluxes.*

This can be carried out by inverse modeling or by the use of isotope data. We suggest that both these approaches are promising but that they need to be further developed. We therefore suggest that the validity of these approaches will be improved by more intensive work in this area and the establishment of long-term data bases for isotope abundances.

REFERENCES

Bonsang, B., M. Kanakidou, G. Lambert, and P. Moufray. 1988. The marine source of C_2–C_6 aliphatic hydrocarbons. *J. Atmos. Chem.* 6:3–20.

Bonsang, B., and G. Lambert. 1985. Nonmethane hydrocarbons in an oceanic atmosphere. *J. Atmos. Chem.* 2:257–271.

Bradley, R.S., P.M. Kelly, P.D. Jones, C.M. Goodess, and H.F. Diez. 1985. A climatic data bank for Northern Hemisphere land areas 1851–1980. Report TR017. Washington, D.C.: US Dept. of Energy.

Cicerone, R.J., and R.S. Oremland. 1988. Biochemical aspects of atmospheric methane. *Glob. Biogeochem. Cyc.* 2:299–327.

Emanuel, W.R., H.H. Shugart, and M.P. Stevenson. 1985. Climatic change and the broad-scale distribution of terrestrial ecosystem complexes. *Clim. Change* 7:29–43.

Henderson-Sellers, A., M.F. Wilson, G. Thomas, and R.E. Dickinson. 1985. Current global land-surface data sets for use in climate in climate-related studies. NCAR Technical Note (TN-272+STR). Boulder, CO: National Center for Atmospheric Research.

Matthews, E. 1983. Global vegetation and land use: new high resolution data base for climatic studies. *J. Clim. Appl. Meteor.* 22:474–487.

Olson, J.S., J.A. Watts, and L.J. Allison. 1983. Carbon in live vegetation of major world ecosystems. Environmental Sciences Division Publ. 1997. Oak Ridge National Laboratory. Oak Ridge, TN: NTIS, US Dept. of Commerce.

Spangler, W., and R.L. Jenne. 1987. World monthly surface station climatology. Data set maintained by National Center for Atmospheric Research, Boulder, CO.

Wahlen, M., et al. 1987. ^{13}C, D and ^{14}C in methane. *Trans. Am. Geophys. U.* 68/44:1220.

Wahlen, M., N. Tanaka, R. Henry, B. Deck, J. Seglen, J.S. Vogel, J. Southon, A. Shemesh, R. Fairbanks, and W. Broecker. 1989. Carbon-14 in methane sources and in atmospheric methane: the contribution from fossil carbon. *Science* 245:286–290.

Whiticar, M.J., E. Faber, and M. Schoell. 1986. Biogenic methane formation in marine and freshwater environments. CO_2 reduction vs. acetate fermentation—isotope evidence. *Geochim. Cosmo. Acta* 50:693–709.

Willmott, C.J., C.M. Rowe, and Y. Mintz. 1985. Climatology of the terrestrial seasonal water cycle. *J. Climatol.* 5:589–606.

Zimmerman, P.R., J.P. Greenberg, S.O. Wandiga, and P.J. Crutzen. 1982. Termites: a potentially large source of atmospheric methane, carbon dioxide and molecular hydrogen. *Science* 218:563–565.

Zinke, P.J., A.G. Stangenberger, W.M. Post, W.R. Emanuel, and J.S. Olsen. 1984. World wide organic soil carbon and nitrogen data. Environmental Science Division Publ. 2212. Oak Ridge National Laboratory. Oak Ridge, TN: NTIS, US Dept. of Commerce.

Exchange of Trace Gases between Terrestrial Ecosystems and the Atmosphere
eds. M.O. Andreae and D.S. Schimel, pp. 175–187
John Wiley & Sons Ltd

Chamber and Isotope Techniques

A.R. Mosier

USDA-ARS, P.O. Box E
Fort Collins, CO 80522, U.S.A.

Abstract. Gas collection chambers on the soil surface are commonly used to quantify the flux of trace gases from the soil to the atmosphere. Even though chamber methods are imperfect, they must be used in most field situations because no other techniques are available. As chambers will be used in the future, we need to increase awareness of problems associated with their use. This paper discusses some of the theoretical and practical problems which cause variability in gas flux measurements using chambers and the effect of chambers on gas flux measurements using natural and enriched stable isotopes.

INTRODUCTION

Soil cover chamber techniques are routinely used to estimate the flux of a trace gas (CO_2, CO, CH_4, N_2O, NO, etc.) from the soil to the atmosphere. Investigators would prefer to quantify the exchange of gases between the soil and atmosphere by the most direct, unobtrusive techniques possible. Chamber methods do not meet this requisite because they disturb the soil–plant–atmosphere continuum. The problem is, in order to get any estimate of trace gas emissions from the soil to the atmosphere in most field situations, we must use some sort of soil surface gas concentration device. We are locked into this dilemma because analytical methods for many trace gases are either not sufficiently sensitive or response times are not rapid enough to measure directly the small upward or downward flux of gases between the soil and the atmosphere. Chamber measurements are useful, cheap, and efficient when applied to the right problem. For example, only chamber techniques can be used to measure CH_4 consumption from the atmosphere by soil or to isolate soil emission from other components of the system (M. Keller, pers. comm.). Chamber techniques can also be applied when systems are manipulated, i.e., by fertilization at scales smaller than those required for atmospheric techniques. Alternative methods have been used concurrently with chamber methods to measure emissions of a gas from the soil to the atmosphere in only a few cases (Hutchinson and Mosier 1979; Kaplan et al. 1988; Parrish et al. 1987).

Analytical technology is slowly progressing so that we may soon be able to use micrometeorological techniques to measure the flux of N_2O and CH_4 routinely from the soil. Tunable diode laser technology is advancing to provide a sensitive, fast response analytical technique for eddy correlation measurement of CH_4 flux. The possibility also exists for production of a tunable diode laser for N_2O, but routine availability is still a few years away.

In this chapter some of the theoretical and practical problems associated with chambers are discussed. I will also briefly delve into the use of stable isotope measurements to identify trace gas production mechanisms and to quantify emission rates. The emphasis of my discussion will be on N_2O measurements with some reference to other trace gases. Chamber techniques for measuring NO–NO_x and CH_4 fluxes are discussed by Johansson, and Schütz and Seiler, respectively (this volume).

CHAMBER METHODS

Basic design, theory, and limitations of chambers used to collect gases emitted from the soil have been discussed in detail by a number of authors (Denmead 1979; Jury et al. 1982; Hutchinson and Mosier 1981; Matthias et al. 1978; Sebacher and Harriss 1982). Some version of two basic chamber designs is used to estimate fluxes of CO_2, CH_4, NO, N_2O, and N_2 and other trace gases from soil and water surfaces. Both of the common chamber types enclose a distinct volume of air above a known area of soil and prevent or control emanating gas from mixing with the external atmosphere. The concentration of N_2O, for example, beneath the cover will increase or decrease whenever there is a positive or negative flux out of the soil. The two chamber types are those with forced flow-through air circulation, designated as "open soil covers," and those with closed-loop air circulation or no forced air circulation, designated as "closed soil covers."

Closed Chambers

Gas flux from the soil using closed chambers can be calculated by periodically collecting gas samples from the chamber and measuring the change in concentration of gas with time during the period of linear concentration change. Reasons for selecting the closed cover method are typically (*a*) very small fluxes can be measured, (*b*) no extra equipment requiring electrical supply is needed, (*c*) there is diminished disturbance of the site due to the short time a cover has to be in place for each gas flux estimate, (*d*) the chambers are simple to construct, (*e*) they are easy to install and to remove thus giving the opportunity to measure different locations at different times with the same equipment, and (*f*) they are relatively inexpensive to prepare.

Problems generally attributed to closed soil covers include:

1. Concentrations of gas in the enclosure atmosphere can build up to levels where they inhibit the normal emission rate. This problem can be limited by using short collection periods and correction equations (Jury et al. 1982; Hutchinson and Mosier 1981).
2. Closed covers either eliminate or alter the atmospheric pressure fluctuations which normally are found at the soil surface due to the natural turbulence of air movement. These fluctuations can pose a "pumping action" on the surface layer of soil which increases soil air movement, thus a totally closed cover may underestimate the flux that would have occurred without the cover in place. An appropriately designed vent does, however, allow pressure equilibration in and outside the chamber (Hutchinson and Mosier 1981).
3. Boundary layer resistance at the soil–atmosphere interface may be higher inside the chamber than outside the chamber. As a result, the flux rate of gas inside the chamber will be less than outside the chamber (Denmead, pers. comm.).
4. Pressure changes in the soil can be caused by inserting the chamber into the soil. This problem may be overcome by installing collars in the soil that are normally open to the atmosphere and to seal the cover to the collar when the chamber is used (Seiler and Conrad 1981; Duxbury et al. 1982). Alternatively, after initially inserting the chambers into the soil, the chambers may be removed for a brief time to allow dissipation of any gas released during the disturbance and then replaced (Livingston et al. 1988).
5. Temperature changes in the soil and atmosphere under the chamber can occur. Temperature differences within and outside the chamber can be reduced by insulating the chamber and covering it with reflective material.

Open Chambers

Open soil covers used by Ryden et al. (1978) and Denmead (1979) are coupled to the atmosphere via an air inlet through which outside air is continuously drawn into the cover and forced to flow over the enclosed soil surface. The gas flux from the soil surface can be calculated from concentration difference, flow rate, and area covered by the open soil cover. The main advantage of open soil covers is that they maintain environmental conditions close to those of the uncovered field. This implies that open systems are more applicable for a continuous long-term monitoring of gas flux if moisture conditions inside and outside the chamber remain the same. Open chambers are, however, sensitive to pressure deficits inside the chamber caused by the induced air flow which may cause artificially high fluxes. If even small pressure deficits occur, induced mass flow of the gas into the cover will lead to overestimation of the gas flux. This can be readily overcome by ensuring that the size of the inlet gas orifices are large compared to size of outlet (Denmead 1979). An additional consideration in open cover systems is the time required for gas concentration in the soil and chamber air to adjust to new equilibrium values. Since measurements assume this equilibrium flux between soil atmosphere and chamber atmosphere gas concentration, estimates will be erroneous during the time of equilibration (Denmead 1979).

PLANT EFFECTS

One aspect of chamber techniques that has been only superficially investigated is the effects of plants on transport of gases, produced in the soil, from the soil to the atmosphere. Many studies have omitted plants because it is difficult to include them in the gas collection system. The existence of growing plants in the soil unquestionably affects soil nutrient, water atmosphere, and temperature status (Conrad et al. 1983). The actual effect of plants on soil processes such as denitrification is still being debated (Haider et al. 1985). The physical effect of the existence of plants on the movement of soil gases through root channels or along root surfaces to the atmosphere has received little attention. Plants that grow under flooded conditions undoubtedly are involved in the evolution of gases from soil into the atmosphere. Cicerone and Shetter (1981) and Seiler et al. (1984) have shown that approximately 95% of the CH_4 emitted into the atmosphere by rice paddies was due to transport through rice plants. Nitrous oxide emissions from a rice field were greater when rice plants were included in collection chambers compared to the amount evolved into chambers positioned between the rice plants (Mosier et al. 1989).

MEASUREMENT VALIDATION

Another area of analytical inadequacy is in direct validation of chamber flux measurements. Method comparisons have been made in some instances and improved analytical methods for some gases will provide new opportunities for such comparisons. For example, Kaplan et al. (1988) have shown that NO emissions from a tropical forest soil measured using soil enclosures were in good agreement with flux estimates derived from vertical profiles of NO and O_3. Parrish et al. (1987) also found that NO emissions from a grassland measured by chambers and atmospheric gradient techniques agreed "very well." Hutchinson and Mosier (1979) reported that N_2O emissions from a cornfield measured by a micrometeorological method were close to the average of closed soil cover measurements made at the same time from the same field. For example, on July 20, 1978, N_2O emissions measured by the chamber technique averaged 258 ± 163 ng N m^{-2} s^{-1} (samples were collected within and between plant rows to make up the average value). The within cornrow flux was 373 ± 20 and between rows was 143 ± 30 ng N m^{-2} s^{-1}. The N_2O measured by micrometeorological methods, flux computed as the product of an eddy diffusivity and the vertical concentration gradient in the lower atmospheric boundary layer, was 350 ng N m^2 s^{-1}.

Does the apparent good agreement between gas fluxes measured by chamber and nonpoint source techniques demonstrate that chamber flux measurements represent gas efflux from a system or is the agreement merely coincidental? I ask this question because the variability among the individual chamber measurements is large and the assumptions that must be made in the atmospheric gradient techniques provide for significant uncertainty.

COMPARISON OF GAS PRODUCTION RATE AND FLUX RATE

Jury et al. (1982) suggest that surface measurements of N_2O flux probably are not quantitatively related to production rate in the soil. For a quantitative relationship between N_2O production rate and surface flux to exist, a steady state must be reached in the entire soil atmosphere. For very wet conditions, this time period is likely to be long, compared to the time over which denitrification is expected to occur, because gas diffusion in water is about 10,000 times slower than in air. For higher values of the diffusion coefficient, in drier conditions, steady state is generally reached rapidly enough for the flux out of the soil surface to represent the amount produced. In drier soils, open covers could successfully relate the amount released to the amount produced. However, using a closed cover for periods longer than one or two hours, without flushing the chamber, would likely result in an error and poor interpretation of both flux and production rate.

Estimates of field trace gas production are difficult to make in wet soils. One method that has been used to estimate field production rates is to take intact soil cores (Ryden et al. 1987; Robertson and Tiedje 1987). Soil core methods allow one to take a "relatively undisturbed" piece of a field soil into the laboratory where N_2O gas production rates can be estimated under controlled conditions. Little evidence exists, however, to show that these values relate to field gas production rates and then to chamber emission rates measured in the field.

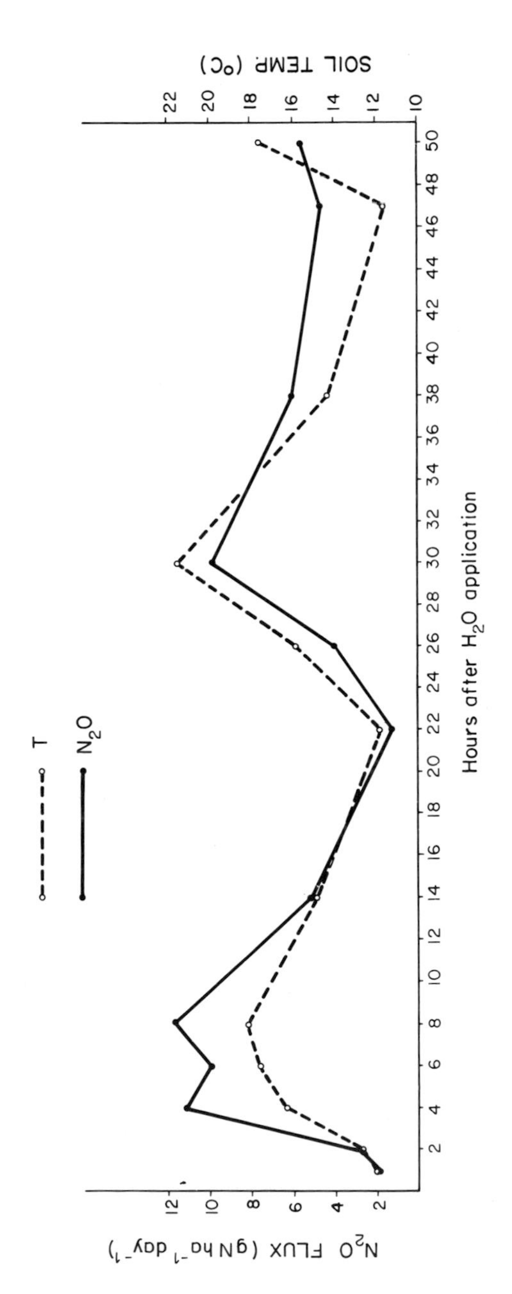

Fig. 1—Diurnal variation of N_2O flux and soil temperature (5 cm) in a maize field, where the change in soil temperature and N_2O flux are correlated.

ANNUAL ESTIMATES

Estimating annual or long-term production rates and quantities of gas flux from soil to the atmosphere is yet another problem. Any chamber method creates such a change in soil environmental conditions that long-term placement of a chamber is not possible. We should also question the value of periodic spot measurements, or weekly measurements at a site for that matter, in quantifying the amount of a given gas evolved from a site during a long period of time.

Time Sampling Variability

Sampling schemes and data collection frequency using chambers and other methods are inadequate to quantify annual emissions in a number of published studies. To determine the number and frequency of sampling necessary to estimate annual trace gas flux, we need to consider the types of variability involved in measuring gas fluxes and how these variations affect gas flux rates. There are typically two types of temporal variability associated with chamber measurements: diel variability, i.e., occurring over a 24-hour period that is generally associated with soil temperature change, and day-to-day variability. The following three sets of data (Figs. 1–3) illustrate time scale variability problems.

Diel Variability

Figure 1 represents a relatively regular gas efflux period from a Colorado maize field site. These diel variations in N_2O flux are very closely correlated ($r = 0.77$; N_2O flux $= -7.55 + 0.91$ T) to soil T. In this type of example one could readily predict at what soil T or time of day a mean daily N_2O flux could be measured with one measurement. Blackmer et al. (1982) have shown a similar data set for an Iowa soil.

Figure 2 shows an example of commonly encountered diel variability that is not predictable as a function of time or soil T [$r = 0.15$, (N_2O flux $= -48.4 + 3.32$ T)]. Using a very similar data set, Blackmer et al. (1982) concluded that "there is no short time during a 24-hour period that is always satisfactory for assessing the amount of N_2O evolved during that period." For net CH_4 flux from tundra soils, Whalen and Reeburgh (1988) found "no relationship between CH_4 flux, soil temperature, and time." In order to minimize errors in flux estimates due to temporal variability we have been using a sampling strategy based upon what we have learned about factors that cause large changes in N_2O or CO_2 gas production in cultivated and grassland soils. Increases in N_2O and CO_2 production are predicated upon the addition of water to the system. When water is added, elevated gas efflux occurs for a time (hours to days), then returns to very low, more

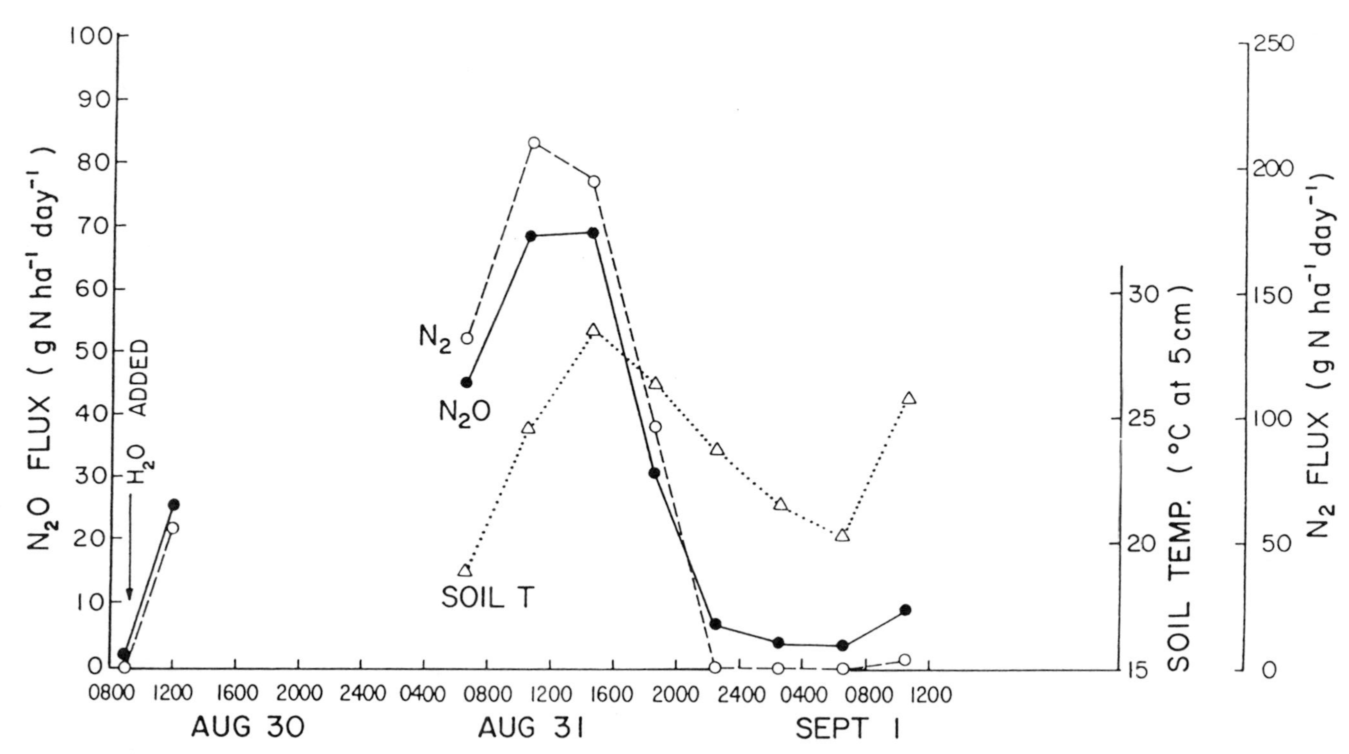

Fig. 2—Diurnal variation of N_2O and N_2 flux and soil temperature (5 cm) in a maize field, where the changes in temperature gas flux are not correlated.

or less baseline levels that vary relatively regularly, similar to Fig. 1. As a result, I suspect that relatively infrequent spot sampling at a more or less mean daily T gives a reasonable estimate when gas flux is low and variability is small. However, when water is added to the soil, N_2O production and efflux events similar to Fig. 2 are common. This type of event requires frequent measurements to characterize the event, as noted in the next paragraph.

Day-to-Day Variability

Figure 3 demonstrates sampling problems caused by day-to-day variability. These data show that it is not possible to characterize the effect of irrigation on N_2O or N_2 flux by weekly or biweekly measurements. Each of the data points in Fig. 3 is the mean of three 1-hour measurements made each day. With this sort of time variability in mind, it seems unlikely that annual gas emissions can be estimated by a few seasonal measurements.

SPATIAL VARIABILITY

Spatial variability is probably the major problem in making estimates of a given gas flux from a field, watershed, ecosystem, etc., based upon chamber

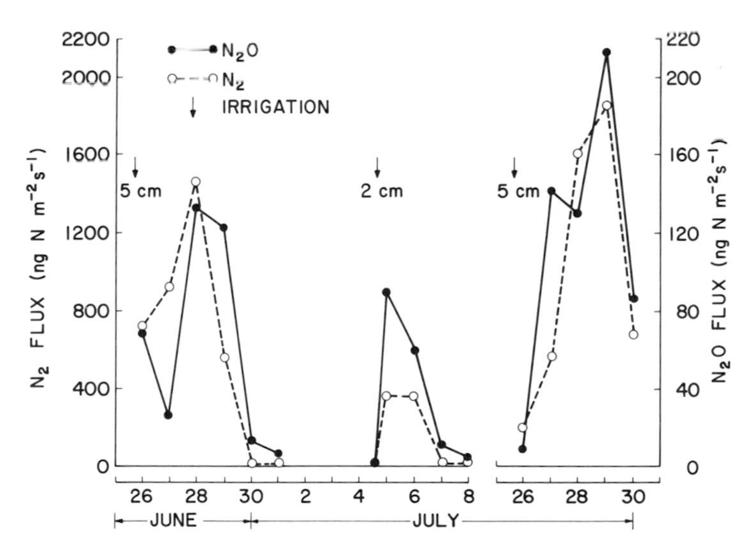

Fig. 3—Temporal variation in N_2O and N_2 flux from the soil following irrigation events. Reproduced from *Soil Science Society of America Journal*, vol. 50, No. 2, March-April 1986, pp.344–348, by permission of the Soil Science Society of America, Inc.

measurements. Whalen and Reeburgh (1988) found coefficients of variation for net CH_4 flux within uniform tundra sites to range between 50 and 100%. Studies of variability of N_2O emissions from the soil indicate that within an agricultural field having a single soil type, measured variability is extremely large over a relatively small space. Matthias et al. (1980) found that within a 100 m^2 area of an Iowa field, N_2O emissions had coefficients of variation ranging from 31 to 168%. In California field plots (3 $\times$ 36 m), Folorunso and Rolston (1984) calculated coefficients of variation from 161 to 508% for N_2O + N_2 emissions. They also found, as have most studies, that N_2O fluxes were ln-normal rather than normally distributed. Folorunso and Rolston (1984) calculated the number of sample replicates needed for various degrees of precision at the 95% confidence interval. These calculations serve very well to show how difficult it is to take chamber measurements over a given soil system and conclude correctly that these measurements truly represent the gas emissions from that system. They show that if we want a $\pm$ 10% estimate of the true mean, 156–4117 observations would have been required on the 3 $\times$ 36 m plots. If we could have been satisfied with an estimate $\pm$ 100% of the true mean then 2, 4, 42, 4, 9, or 4 replicate samples would have been required from each of the 6 plots.

STABLE ISOTOPIC TECHNIQUES FOR IDENTIFYING PRODUCTION MECHANISMS OF TRACE GASES IN THE SOIL AND QUANTIFYING EMISSION RATES

Natural Abundance Isotopic Studies

Several studies have recently reported that natural abundance isotopic fractionation can be used to determine the mechanism of production of N_2O and CH_4. I will briefly relate some examples of these studies and then pose questions as to the effect of using soil chamber enclosures to measure gas efflux from soil.

First, Wahlen and Yoshinari (1985) show that $^{18}O/^{16}O$ isotopic ratios of N_2O can be sensitively measured with high-resolution infrared spectroscopy using tunable diode lasers. They reported $\delta^{18}O$ values varying quite widely from 45.4 for ambient air, 35.9 for soil atmosphere, 59.3 for coal-fueled power plant stack gas to 93.0 for lake water sampled 13 m below the surface. Yoshinari and Wahlen (1985) show that $\delta^{18}O$ differs from N_2O derived from nitrification and denitrification, 22.4 and 27.9, respectively. It is assumed that $\delta^{18}O$ values in N_2O produced from nitrification have a relatively light oxygen isotope composition relative to the O from denitrification, since the two O atoms incorporated into NO_2^{-1} in the oxidation of ammonium are equally derived from O_2 and H_2O and since N_2O appears to be formed mostly from the reduction of NO_2^{-1} when O_2 concentration is low.

Assuming that these isotopic differences are uniform among different systems, given the isotopic composition of the N_2O, one might directly determine the source of the gas. For example, using ^{15}N techniques, Yoshida and Matsuo (1983) concluded that bacterial reduction of N_2O was the major consuming process of N_2O on the Earth's surface.

In a further example, Alperin and Reeburgh (1984) show how stable carbon isotopes may be used as natural tracers of geochemical processes. They used the general concept that a chemical bond containing a lighter atom will have a lower activation energy and therefore will react at a faster rate than the same bond involving a heavier atom. A chemical reaction will result in products enriched in ^{12}C (lighter) and reactants depleted in ^{12}C (heavier). They show that methane oxidation occurs in sediments under anaerobic conditions. Considering the sizeable differences in isotopic composition of the CO_2 produced during methane oxidation in the sediment, it seems possible to define production mechanisms in a specific system by measuring the isotopic composition of CO_2 and CH_4 evolving from the soil.

Enriched Stable Isotope Studies

Nitrogen fertilizers enriched with ^{15}N are typically used in agricultural research as tracers to monitor the utilization of the fertilizers by crop plants and to investigate specific soil N reactions. Use of N fertilizers enriched with ^{15}N also permits measurement of N_2O and N_2 emissions from soil and permits ascribing quantitative N isotopic values to the source of the gases evolved. This method involves applying a highly ^{15}N-enriched source to the soil (> 20 atom % ^{15}N) and then, using a chamber to cover the ^{15}N-fertilized plot, isolating the atmosphere above the soil for a specified time period. This permits one to determine the rate of change of ^{15}N atoms (N-gases) in the chamber atmosphere over time. The method's calculations utilize the fact that the soil N-gases (N_2 and N_2O, principally) that evolve into the chamber headspace containing normal air do not randomly mix isotopically with the N-gases in the chamber. By analyzing the chamber contents on an isotope ratio mass spectrometer for the increases in m/e $^{30}N_2$ and $^{29}N_2$, the emissions of N_2 or N_2O may be quantified (Hauck et al. 1958; Mosier et al. 1986; Mulvaney and Kurtz 1982).

Effect of Chambers on Isotopic Gas Flux Measurement Methods

Because of the relatively small isotopic concentration changes that these methods deal with, it is necessary to use closed chambers. Earlier it was discussed that the buildup of a gas inside the chamber causes a reequilibrium of the gas in the soil profile and therefore changes the rate of diffusion from the soil to the atmosphere. Does the same problem exist for the individual isotopes of a gas themselves? For example, when measuring N_2O efflux from a ^{15}N-amended soil, does the increase in concentration of $^{46}N_2O$ inside the chamber alter the diffusion of this isotope relative to the $^{44}N_2O$ evolving from the soil at the same time? This question has not, to my knowledge, been investigated.

A different problem may exist for measuring N_2 emissions from the soil into a chamber headspace that is already about 78% N_2. The very small quantity of N_2 that evolves into the chamber should not affect the equilibrium concentration of N_2 in the soil and therefore should not affect flux. However, if the nitrate in the soil is highly labeled with ^{15}N, then the N_2 produced in the soil is predominantly $^{29}N_2$ and $^{30}N_2$. Initially in the chamber, only about 0.001% and 0.73% of the N_2 is mass 30 and 29, respectively. Rather small additions of mass 30 into the chamber may alter the $^{30}N_2$ concentration in the headspace gas and affect actual gas emissions from the soil. One recent study (Mosier et al. 1989) suggests that this may not be a significant problem. N-gas flux rates measured by a closed chamber maintained over the soil for 16 hours did not measurably change between 4- and 16-hour measurements.

REFERENCES

Alperin, M.J., and W.S. Reeburgh. 1984. Geochemical observations supporting anaerobic methane oxidation. In: Microbial Growth on C-1 Compounds, ed. R.L. Crawford and R.S. Hanson, pp. 282–289. Washington, D.C.: Am. Soc. Microbiol.

Blackmer, A.M., S.G. Robbins, and J.M. Bremner. 1982. Diurnal variability in rate of emission of nitrous oxide from soils. *Soil Sci. Soc. Am. J.* 46:937–942.

Cicerone, R.J., and J.D. Shetter. 1981. Sources of atmospheric methane: measurements in rice paddies and a discussion. *J. Geophys. Res.* 86:7203–7209.

Conrad, R., W. Seiler, and G. Bunse. 1983. Factors influencing the loss of fertilizer nitrogen into the atmosphere as N_2O. *J. Geophys. Res.* 88:6709–6718.

Denmead, O.T. 1979. Chamber systems for measuring nitrous oxide emissions from soils in the field. *Soil Sci. Soc. Am. J.* 43:89–95.

Duxbury, J.M., D.R. Bouldin, R.E. Terry, and R.L. Tate. 1982. Emissions of nitrous oxide from soils. *Nature* 298:462–464.

Folorunso, O.A., and D.E. Rolston. 1984. Spatial variability of field measured denitrification gas fluxes. *Soil Sci. Soc. Am. J.* 48:1214–1219.

Haider, K., A.R. Mosier, and O. Heinemeyer. 1985. Phytotron experiments to evaluate the effect of growing plants on denitrification. *Soil Sci. Soc. Am. J.* 49:636–641.

Hauck, R.D., S.W. Melsted, and P.E. Yankwich. 1958. Use of N-isotope distribution in nitrogen gas in the study of denitrification. *Soil Sci.* 86:287–291.

Hutchinson, G.L., and A.R. Mosier. 1979. Nitrous oxide emissions from an irrigated cornfield. *Science* 205:1225–1226.

Hutchinson, G.L., and A.R. Mosier. 1981. Improved soil cover method for field measurement of nitrous oxide fluxes. *Soil Sci. Soc. Am. J.* 45:311–316.

Jury, W.A., J. Letey, and T. Collins. 1982. Analysis of chamber methods used for measuring nitrous oxide production in the field. *Soil Sci. Soc. Am. J.* 46:250–256.

Kaplan, W.A., S.C. Wofsy, M. Keller, and J.M. DeCosta. 1988. Emission of NO and deposition of O_3 in a tropical forest system. *J. Geophys. Res.* 93:1389–1395.
Livingston, G.P., P.M. Vitousek, and P.A. Matson. 1988. Nitrous oxide flux and nitrogen transformations across a landscape gradient in Amazonia. *J. Geophys. Res.* 93:1593–1599.
Matthias, A.D., A.M. Blackmer, and J.M. Bremner. 1980. A simple chamber technique for field measurement of emission of nitrous oxide from soil. *J. Env. Qual.* 9:251–256.
Matthias, A.D., D.N. Yarger, and R.S. Weinbeck. 1978. A numerical evaluation of chamber methods for determining gas fluxes. *Geophys. Res. Lett.* 5:765–768.
Mosier, A.R., A.B. Chalam, S.P. Chakravorti, and S.K. Mohanty. 1989. Gaseous N loss from intermittently flooded rice measured directly by a ^{15}N technique and indirectly by ^{15}N balance. *Soil Sci. Soc. Am. J.*, in press.
Mosier, A.R., W.D. Guenzi, and E.E. Schweizer. 1986. Field denitrification estimation by nitrogen-15 and acetylene inhibition techniques. *Soil Sci. Soc. Am. J.* 50:831–833.
Mulvaney, R.L., and L.T. Kurtz. 1982. A new method for determination of ^{15}N-labeled nitrous oxide. *Soil Sci. Soc. Am. J.* 46:1178–1184.
Parrish, D.D., E.J. Williams, D.W. Fahey, S.C. Liu, and F.C. Fehsenfeld. 1987. Measurement of nitrogen oxide fluxes from soils: intercomparison of enclosure and gradient measurement techniques. *J. Geophys. Res.* 92:2165–2171.
Robertson, G.P., and J.M. Tiedje. 1987. Nitrous oxide sources in aerobic soils: nitrification, denitrification, and other biological processes. *Soil Biol. Biochem.* 19:187–193.
Ryden, J.C., L.J. Lund, and D.D. Focht. 1978. Direct in-field measurements of nitrous oxide flux from soil. *Soil Sci. Soc. Am. J.* 42:731–738.
Ryden, J.C., J.H. Skinner, and D.J. Nixon. 1987. Soil core incubation system for the field measurement of denitrification using acetylene-inhibition. *Soil Biol. Biochem.* 19:753–757.
Sebacher, D.I., and R.C. Harriss. 1982. A system for measuring methane fluxes from inland and coastal wetland environments. *J. Env. Qual.* 11:34–37.
Seiler, W., and R. Conrad. 1981. Field measurement of natural and fertilizer induced N_2O release rates from soils. *ACPA J.* 31:767–770.
Seiler, W.A., A. Holzapfel-Pschorn, R. Conrad, and D. Scharite. 1984. Methane emission from rice paddies. *J. Atmos. Chem.* 1:241–248.
Wahlen, M., and T. Yoshinari. 1985. Oxygen isotope ratios in N_2O from different environments. *Nature* 313:780–782.
Whalen, S.C., and W.S. Reeburgh. 1988. A methane flux time series for tundra environments. *Glob. Biogeochem. Cyc.* 2:399–409.
Yoshida, N., and S. Matsuo. 1983. Nitrogen isotope ratio of atmospheric N_2O as a key to the global cycle of N_2O. *Geochem. J.* 17:231–239.
Yoshinari, T., and M. Wahlen. 1985. Oxygen isotope ratios in N_2O from nitrification at a wastewater treatment facility. *Nature* 317:349–350.

Exchange of Trace Gases between Terrestrial Ecosystems and the Atmosphere
eds. M.O. Andreae and D.S. Schimel, pp. 189–207
John Wiley & Sons Ltd

Micrometeorological Techniques for the Measurement of Trace Gas Exchange

D. Fowler[1] and J.H. Duyzer[2]

[1] *Institute of Terrestrial Ecology*
Bush Estate, Penicuik EH26 OQB, U.K.

[2] *TNO, Division of Technology for Society*
2600 AE Delft, The Netherlands

Abstract. The exchange of trace gases between natural surfaces and the atmosphere may be measured in the field using a range of different micrometeorological methods. The theory and practical restrictions of each of the methods are described and examples of field measurements for fluxes of NO, NO_2, O_3, SO_2, NH_3, HNO_3, HCl, and PAN are provided. Restrictions and instrumental difficulties currently limiting a wider application of these methods are discussed. Some of these restrictions may be overcome by selecting appropriate surface and atmospheric conditions to minimize errors, while for other gases, more sensitive or faster response gas analyzers are required before fluxes may be detected.

INTRODUCTION

The exchange of sensible heat, water vapor, and momentum between terrestrial ecosystems and the atmosphere has been the subject of extensive field study during the last four decades. Throughout this period there has been a requirement for corresponding flux measurements in the field, and the development of theory, techniques, and equipment for these measurements has represented the core of micrometeorological research.

At the fringes of these central interests, studies of trace gas and particle exchange above aerodynamically rough surfaces have been conducted. The techniques used are extensions of those developed for heat, water vapor, and momentum transfer and have become quite widely used during the last decade. There is now literature appropriate for a number of trace gases over a range of surfaces. These studies include SO_2, O_3, NO, NO_2, NH_3, I_2, PAN, HNO_3, HCl, and a range of particles and droplets.

The use of micrometeorological techniques has several major advantages over enclosure methods. First, the fluxes are averaged over large areas and therefore are not subject to the sampling problems associated with enclosure methods. Second, the measurements do not disturb the vegetation or soil within the "measurement area" since the equipment and disturbance is located downwind of the area where exchange is occurring. Third, the sampling times necessary for micrometeorological measurements (typically 15 to 60 minutes) permit studies of the changes in rates of trace gas exchange with changing atmospheric and surface conditions.

These methods therefore offer an excellent opportunity to improve our understanding of trace gas exchange. They are, however, subject to a series of theoretical and practical restrictions which limit the range of sites and gases suitable for measurements. The objective of this chapter is to outline, briefly, the main methods available and their applicability for a range of trace gases. In this way we will identify opportunities for further measurements and ways of overcoming some of the restrictions.

PRINCIPLES

For gases absorbed at the ground, transport through the free atmosphere to within a mm or so of the absorbing surface is by turbulent diffusion, in which the displacement of individual eddies is the basic transport process. Very close to the absorbing surface, typically over distances <1 mm, turbulence is damped by viscous forces, and transport over these very short distances and through the stomatal apertures relies on molecular diffusion. For the distance scales over which measurements are practical, turbulent transport is therefore the dominant mechanism. In the simplest of the micrometeorological methods, the flux may be measured by sensing the concentrations and velocities of components of the turbulence. The method termed "eddy correlation" provides vertical fluxes from measurements of the covariance of vertical wind velocity fluctuations with gas concentration fluctuations and is developed formally in the following section.

This and other "equilibrium" methods require extensive uniform areas of ground, a flux to the surface which is uniform throughout the upwind area influencing the sample point (fetch), and constant atmospheric conditions during each measurement period. In flat, homogeneous terrain the flux measured at the chosen sampling point above the surface provides the average vertical flux over the upwind fetch, provided that the sample point is in the height range in which the vertical flux is constant with height. The depth of the constant flux layer increases with fetch at a rate that varies with vegetation height and atmospheric stability but is usually about 0.5% of the fetch. This layer is therefore about 1 m deep at the downwind edge of a 200 m uniform field (Monteith 1973).

Within the constant flux layer, gradient measurements of wind velocity, air temperature, and concentrations of trace gases provide the other main method of measuring vertical fluxes over extensive surfaces, and are generally termed flux-gradient techniques.

More recently, micrometeorological methods have been applied in nonequilibrium conditions to measure trace gas fluxes from small plots of land using a mass balance method (Denmead et al. 1977). In this case, the areas over which fluxes are measured may be much smaller than is required for use of the equilibrium methods.

SCALE

The scales over which micrometeorological flux measurements may be made extend from a few square meters of ground or vegetation, as in the case of mass balance or field wind tunnel methods (Colbourn et al. 1987), to 10^4 ha with eddy flux measurements from aircraft. The emphasis, however, is on the average fluxes at a field scale (10–100 ha) so that micrometeorological methods therefore complement the enclosure techniques which work on much smaller scales and only provide overlap for the largest practical enclosures.

EDDY CORRELATION

The vertical flux density, F_g, of a trace gas may be written as

$$F_s = \overline{w\rho_s} \tag{1}$$

where w is the vertical velocity and ρ_s is the density of the trace gas. This may be considered as the sum of two components, the product of mean vertical windspeed $\overline{w}$ and trace gas density $\bar{\rho}_s$ and fluctuations about the means of the same quantities w' and ρ_s':

$$F = \overline{w}\,\overline{\rho}_s + \overline{w'\rho_s'} \tag{2}$$

where ρ_s is the gas density and w' and ρ_s' are the instantaneous vertical wind velocity and the departure from the mean concentration of the trace gas, respectively. Typical instrumentation for eddy correlation trace gas measurements is shown in Fig. 1. It has been assumed by some that there is no mean vertical flow of air to complicate the above, very simple relation.

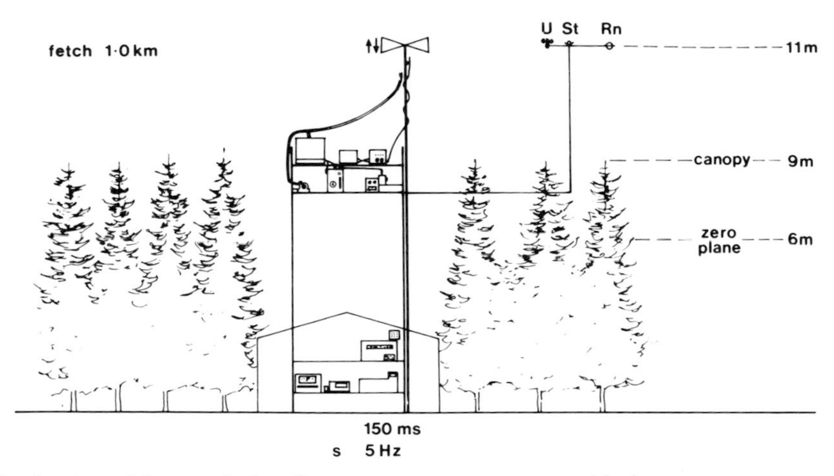

Fig. 1—An eddy correlation flux measurement system with fast response trace gas analyzer, vertical wind velocity, and temperature sensor on main tower. Additional measurements include horizontal wind speed (U), global short wave radiation (St), and net radiation (Rn).

However, in practice, sensible and latent heat fluxes cause vertical gradients in air density which result in an apparent vertical flow of air. This effect, described by Webb et al. (1980), has important implications for fluxes of some trace gases. The mean w signal resulting from these effects is too small to be detected in field measurements. However, the effect on trace gas fluxes has been estimated by Denmead (1983) from the work of Webb et al. (1980). Equation 2 now becomes

$$F = \overline{w'\rho_s'} + (\rho_s/\rho_a)\,[\mu/1 + \mu\sigma(1 + \mu\sigma)]E + (\rho_s/\rho)H/c_pT \tag{3}$$

where H is the sensible heat flux, E the latent heat flux, μ equals the ratio of molecular weight of dry air to that of water vapor, σ is the ratio of density of water vapor to the mean density of air, c_p is the specific heat of air at constant pressure, and T is the air temperature. The corrections are large for any of the trace gas species whose vertical flux is small in relation to the ambient concentration. Vertical fluxes are frequently normalized for ambient concentrations at a reference height $S_{(z)}$. The resulting quantity $F_s/S_{(z)}$, which is identical to the deposition velocity (v_d) (Chamberlain 1975), has been widely applied in trace gas studies, and the error in determining v_d in conditions of moderate sensible and latent heat fluxes becomes significant for deposition velocities smaller than 5 mm s^{-1} (Fig. 2). These problems, however, may be overcome by the measurement system. For example, if the sample gases are dried and brought to the same temperature

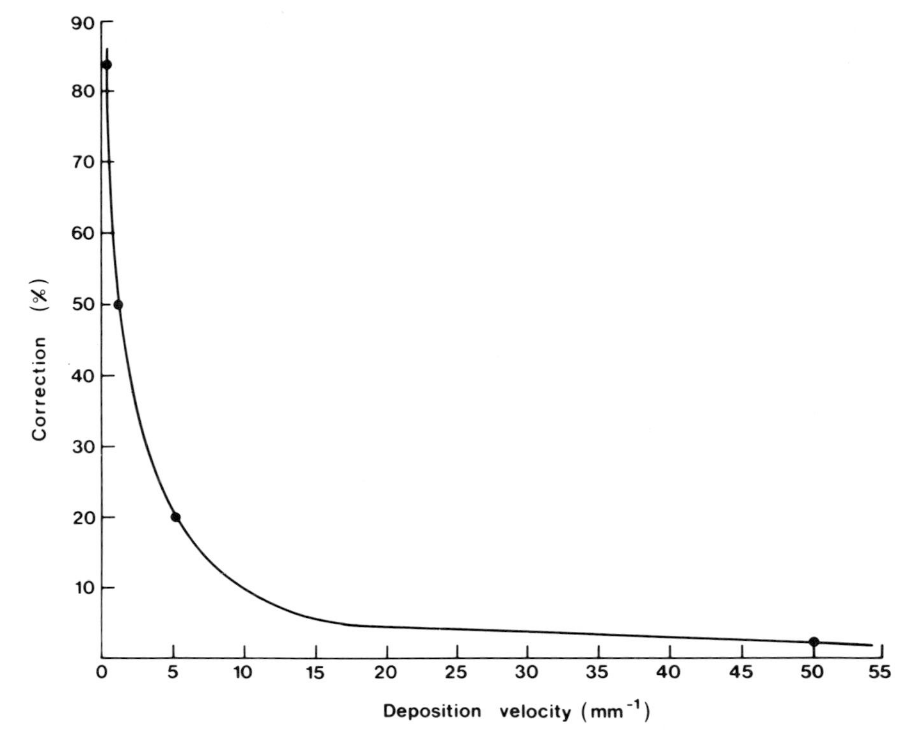

Fig. 2—Corrections to measured trace gas fluxes from profile measurements *in situ* (from Denmead 1983).

or if the mixing ratios are determined for each sampling point, density problems are eliminated.

FLUX-GRADIENT METHODS

The Aerodynamic Method

The vertical transport of an entity towards the surface may be written as

$$F_s = -\rho K_s \partial_s / \partial_z \tag{4}$$

where K_s is the transfer (diffusion) coefficient for the trace gas s and ∂_s/∂_z is the vertical gradient in air concentration in the constant flux layer. If the concentration decreases towards the surface, ∂_s/∂_z is negative by convention and the flux is towards the surface, and vice versa.

The flux density for momentum, F_m (more commonly denoted τ), may be written as

$$F_m = -\bar{\rho} K_{m(z)}(\partial_u/\partial_z) \tag{5}$$

and for heat

$$F_h = c_p \rho K_{H(z)}(\partial\theta/\partial_z) \tag{6}$$

where θ is the potential temperature.

In neutral atmospheric conditions the eddy diffusivities for heat, water vapor, trace gases, and momentum are equal ($K_{m(z)} = K_{H(z)}$, etc.). In these conditions the eddy diffusivity may be determined from the wind profile equation, where

$$U_{(z)} = \frac{U_*}{k} \ln \frac{z - d}{z_o} \tag{7}$$

and

$$K_{m(z)} = k\, U_* \,(z - d) \tag{8}$$

where U is the windspeed at height z, U_* is the friction velocity, k is von Karman's constant, d is the zero plane displacement, and z_o the roughness length. The measurement of a concentration gradient (∂_s/∂_z) then enables an estimate to be made of the trace gas flux from Eq. 4, provided the distribution of sources and sinks is such that the value of d is the same for momentum and for the trace gas in question.

The zero plane displacement (d) is in principle an unknown quantity. Over low vegetation (<0.5 m) d is normally small relative to the measurement height. However, over forests the uncertainty in d may be a major limitation for the analysis and interpretation of field measurements.

The magnitude of the gradient in most circumstances is small and requires an accuracy in measurement of concentration of 1–4% in most cases to be able to detect fluxes.

Stability Corrections

The presence of vertical temperature gradients in excess of the adiabatic lapse rate (i.e., a decrease > 0.01°C m^{-1}), known as "unstable" conditions, increases the rates of vertical exchange. In these conditions, eddy diffusivities for momentum and heat (or mass) transfer differ. When air temperatures increase with height, the atmosphere is termed stable and vertical mixing is damped. The stability may be quantified using a Richardson number (R_i):

$$R_i = [(g/\theta)(\partial_\theta/\partial_z)/(\partial_u/\partial_z)^2] \tag{9}$$

where g is the acceleration due to gravity. Positive R_i refer to stable conditions (commonly found during night and winter), while negative R_i (unstable conditions) generally occur whenever there is a significant sensible heat flux from the surface into the atmosphere and are largest on summer days.

The influence of stability on flux measurements is quantified using the stability functions ϕ_m, ϕ_H, and ϕ_s for momentum, heat, and trace gases, respectively. The trace gas fluxes are then calculated according to

$$F_s = -\frac{\rho K_s \partial_s/\partial_z}{\phi_s} \tag{10}$$

where ϕ_s is assumed identical to ϕ_H in all stabilities and also equal to ϕ_m in stable and neutral conditions, where

$$\phi_m = \phi_H = \phi_s = (1 - 5R_i)^{-1} \tag{11}$$

while in unstable conditions

$$\phi_m = (1 - 16R_i)^{-0.25} \tag{12}$$

$$\phi_H = \phi_s = (1 - 16R_i)^{-0.5}. \tag{13}$$

These semiempirical relationships were derived from the work of Dyer and Hicks (1970).

Fluxes may therefore be estimated from vertical gradients in gas concentration, wind velocity, and air temperature in a wide range of conditions. An example of the instrumentation necessary for such measurements is provided in Fig. 3.

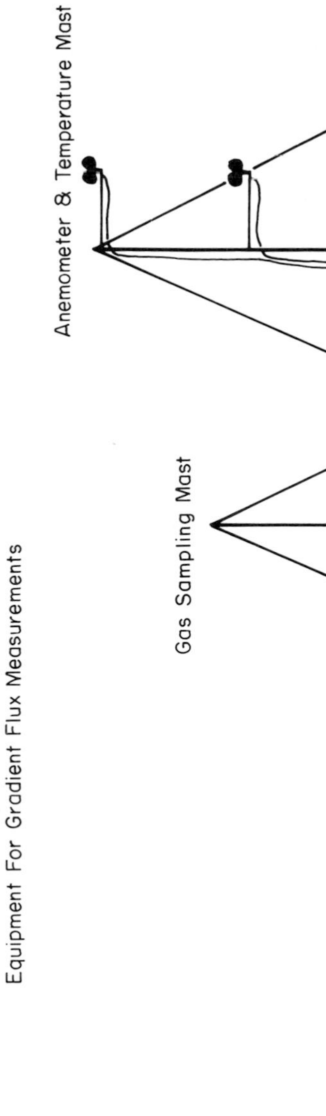

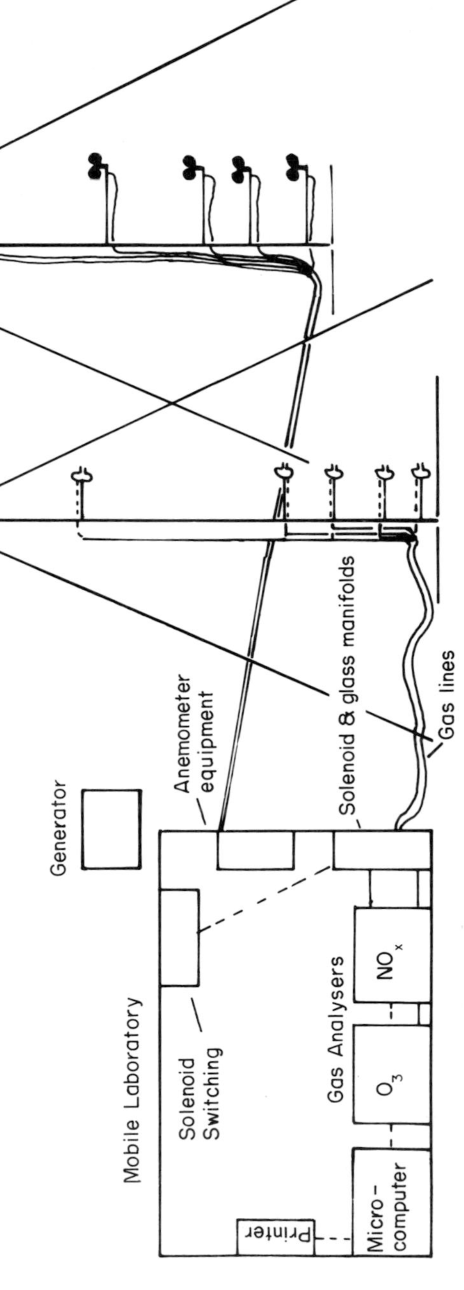

Fig. 3—Typical instrumentation for flux gradient measurements in field conditions.

Bowen Ratio Methods (Energy Balance)

Another gradient technique which does not require wind velocity profile measurements is that based on the energy balance at the ground. The incoming net radiation (R_n) is partitioned between sensible heat flux (H), latent heat flux (λE [the ratio $H/\lambda E$ is the Bowen ratio]), and the soil heat flux (G). A further term corresponding to heat storage in the vegetation and air below the height of measurement can become significant for a forest canopy, so that Bowen ratio methods are best suited to low vegetation (grass, agricultural crops, etc.).

$$R_n = H + \lambda E + G \tag{14}$$

Rearranging Eq. 14 and substituting for H and λE using equations equivalent to 4 and 6 yields

$$R_n - G = -K_s\,(\rho\, c_p \partial\theta/\partial_z + \lambda \rho_a \partial_w/\partial_z) \tag{15}$$

where the gradient ∂_w/∂_z is the gradient in absolute humidity (x). Defining an effective temperature, T_e, for sensible and latent heat transport, where

$$T_e = \theta + (\lambda/\rho_p)\, x/(1 + \gamma) \tag{16}$$

and γ represents the ratio of the mean densities of water vapor and air, the eddy diffusivity for mass and heat transfer may be obtained as

$$K_s = -\,(R_n - G)/(\rho\, c_p\, \partial T_e/\partial_z) \tag{17}$$

and the trace gas flux F_s is given by

$$F_s = \frac{R_n - G}{c_p\,(1 + \gamma)}\,\frac{\partial s}{\partial T_e}\,. \tag{18}$$

Energy balance methods therefore require measurements of vertical gradients of gas concentration, temperature, and humidity, and these provide estimates of the fluxes of sensible heat, water vapor, and the trace gas without the complication and uncertainty introduced by stability corrections. The greatest drawback, and a most important one for trace gas fluxes, is that substantial net radiation fluxes are required for the term (R_n − G), as the error in the flux is directly proportional to the accuracy of the (R_n − G) measurement. In cloudy, night, or winter conditions the available net radiation is frequently too small (<50 W m^{-2}) to permit satisfactory flux estimates. There are many variations in the formulation of Bowen ratio methods, some of which are described by Woodward and Sheehy (1983) and Thom (1975).

In the most commonly applied modified Bowen ratio technique, the sensible heat flux is measured directly by eddy correlation. This permits measurements of trace gas fluxes from vertical gradients in the trace gas concentration and air temperature and, again, is not subject to stability correction. The zero plane displacement (d) does not have to be evaluated for this method, but the measurements are limited by the assumption that sources and sinks for the trace gases and heat fluxes are distributed similarly.

Mass Balance Methods

The eddy correlation and flux-gradient methods are generally used to provide average fluxes over large areas of vegetation, generally with a fetch of betweeen 100 and 1000 m. These methods require uniform surface and atmospheric conditions at the site. In contrast, mass balance methods measure the horizontal flux across a vertical plane from an emission area and can be considered as a plume measurement (Denmead 1983). The flux F_s into the atmosphere is given by

$$F_s = 1/x \int_o^z \overline{u\rho_s}\, d_z \tag{19}$$

where x is the fetch, ρ_s is the density of gas in excess of background, and the term $u\rho_s$ is the time-averaged horizontal flux density at any level in the plume.

In practice, the flux in Eq. 19 is the result of two processes: the horizontal flux in the plume, which is given by the product of mean wind velocity and gas density ($\bar{u}\ \bar{\rho}_s$), and the mean product of the turbulent diffusion $u'\rho_s'$. However, the turbulent flux has been shown to be less than 10% of the convective flux. Equation 19 may therefore be simplified to

$$F_s = 1/x \int_o^z \bar{u}\ \bar{\rho}_s\, dz. \tag{20}$$

The measurements of u and ρ_s must cover the vertical zone where air concentrations are modified by the emission patch, and as a rule of thumb the upper height of measurement should be at about 0.1x. Atmospheric stability has a strong influence on the depth of the plume: it increases the upper boundary of the affected layer in unstable conditions by the additional vertical mixing and decreases it in stable conditions as shown by Denmead (1983).

Complications in the use of this method caused by irregular plot dimensions and varying wind direction are overcome using a circular plot, with the profile mast at the center (Wilson et al. 1983). A further advantage of this method is that it does not demand such precision in gas concentration measurement as the gradient technique.

APPLICATIONS OF MICROMETEOROLOGICAL METHODS TO DATE

The three measurement methods (eddy correlation, flux gradient, and mass balance) have all been used to measure trace gas fluxes, but practical restrictions have restricted some gases to just one technique. Examples of the uses of the different techniques for a range of trace gases are provided in Table 1.

Practical and Theoretical Limitations to Micrometeorological Trace Gas Fluxes

With the exception of the mass balance method, micrometeorological methods are based on the assumption that the flux to or from the surface is either identical to the vertical flux measured at a reference level some distance above the surface or sufficiently near this value that corrections can be applied with confidence. These two fluxes may differ as a consequence of three processes: (*a*) chemical reaction within the air column between the measurement level and the ground (Q), (*b*) changes in concentration with time and therefore changes in the storage of the trace gas within the air column (∂_s/∂_t), and (*c*) horizontal gradients in air concentration leading to advection ($U\partial_s/\partial_x$). These effects may be simplified to

$$\frac{\partial F_s}{\partial_z} = Q - U\frac{\partial_s}{\partial_x} - \frac{\partial_s}{\partial_t}\,. \tag{21}$$

TABLE 1. Trace gas fluxes measured using micrometeorological methods.

Gas	Surface	Method	Reference
NO_2	Grassland	Eddy correlation	Wesely et al. 1982
HNO_3	Grassland	Gradient	Huebert and Robert 1985
SO_2	Cereal	Gradient	Fowler and Unsworth 1979
SO_2	Forest	Eddy correlation	Galbally et al. 1979
NH_3	Grassland	Mass balance	Denmead et al. 1977
HCl	Grassland	Gradient	Rapsomanikis and Harrison 1987
PAN	Grassland	Gradient	Dollard et al. 1986
Cloudwater	Moorland	Gradient	Gallagher et al. 1988

Chemical Production (Q)

This term is of importance only for relatively fast reactions, such as those taking place in the photochemical reaction cycle of O_3, NO, and NO_2, or the growth of hygroscopic aerosol as a result of water vapor condensation. The error from this mechanism may be approximated as

$$F_s \simeq \pm Z k_c s \tag{22}$$

where k_c is the first-order rate constant for the production of the compounds and s is its concentration in air.

The production of NO_2 from NO and O_3 is very rapid ($k_s \simeq 25$ ppm^{-1} min^{-1}). In an environment where O_3 is depositing rapidly, a marked vertical gradient in O_3 is produced. The air chemistry may generate a gradient in NO with concentrations increasing towards the ground, and this gradient in NO may be mistakenly interpreted as the result of an emission flux. The problem of modified gradients in NO, NO_2, and O_3 due to chemistry/deposition interaction has been considered by Duyzer et al. (1983). In cases where fluxes of NO and NO_2 are small (5 to 50 ng m^{-2}s^{-1}) in relation to the O_3 flux (0.5 to 1μg m^{-2}s^{-1}), independent measurements of the NO flux using an enclosure at the site of the micrometeorological measurements is necessary.

Changes in Concentrations with Time

This term is sometimes linked with advection errors but can be considered independently. Considering an imaginary unit cube of air above the ground (Fig. 4), the upper surface represents the measurement reference level and the lower surface represents the site of uptake (or release). Any change in concentrations with time will result in a change in storage of the trace gas in the air below the measurement level. The change may also be associated with adsorbed surface gas uptake in equilibrium with the air concentration, representing additional storage. At its simplest, the divergence between these two heights (z_1, z_2) may be written as

$$\Delta F_s = \int_{z_1}^{z_2} \frac{\partial_s}{\partial_t} d_z \tag{23}$$

which may be simplified to

$$\Delta F_s = \partial_s/\partial_t \, (z_2 - z_1) \tag{24}$$

and which shows that for a flux of 20 ng m^{-2}s^{-1}, a change in concentration of 20 μg m^{-3}h^{-1} leads to an error in the measured flux 1 m above ground

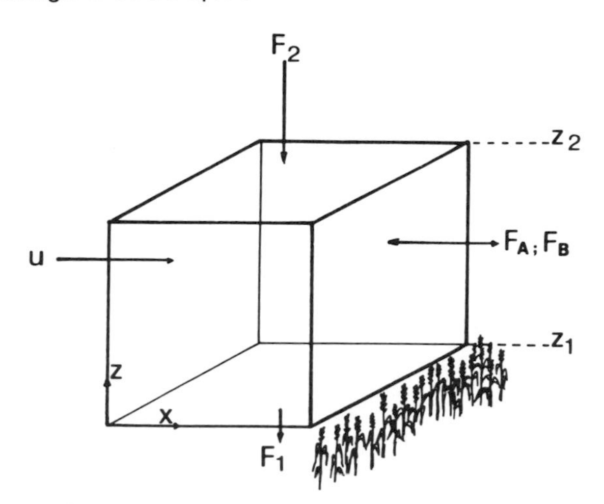

Fig. 4—Notation for discussion of advection (F_A) and storage (F_B) errors in measurements of vertical fluxes (F_1).

TABLE 2. Advection and storage errors (see Fig. 4 for notation).

Advection

Height range (z_1 to z_2) m	Horizontal conc. gradient $\mu g\ m^{-3}\ km^{-1}$	Actual flux (F_1) $\mu g\ m^{-2}\ s^{-1}$	Advection error (F_A) $\mu g\ m^{-2}\ s^{-1}$	$F_A/F_1 \times 100$ %
1	1	0.1	0.002	2
1	10	0.05	0.02	40
5	5	0.05	0.05	100

Storage

Height range (z_1 to z_2) m	Rate of change in concentration $\mu g\ m^{-3}\ h^{-1}$	Actual flux F_1 $\mu g\ m^{-2}\ s^{-1}$	Storage F_s $\mu g\ m^{-2}\ s^{-1}$	$F_S/F_1 \times 100$ %
1	6	0.1	0.0017	2
1	60	0.05	0.017	40
5	60	0.1	0.085	85

of 30%. Table 2 provides the magnitude of storage flux errors for a range of conditions.

An example of errors in deposition velocity (v_d) resulting from concentration changes with time is provided by the changes in mixing layer height in the absence of changes in concentration in the higher layer.

$$\Delta v_d = \frac{z}{H} \frac{\partial H}{\partial_t} \tag{25}$$

where H is the mixing layer height. The term $1/H\ \partial H/\partial_t$ may reach 100% h^{-1}. Van Aalst (1989) used a data set consisting of six years of hourly observations of O_3, NO_2, and O_3 at 4, 100, and 200 m heights above the ground to calculate oxidant ($NO_2 + O_3$) fluxes at two levels. The result (Fig. 5) shows evidence of marked flux divergence in the morning attributed by Van Aalst to the changes in mixing layer height.

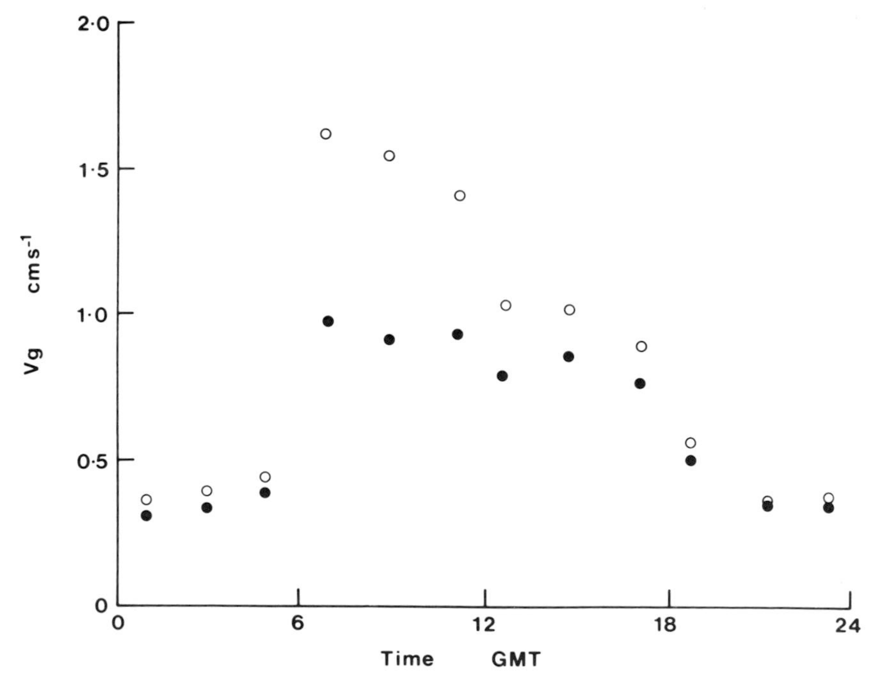

Fig. 5—The average diurnal variation in deposition velocity at Cabauw on grassland with time from six years' data (1980–1986), derived from measurements at 4 m and 100 mn (o) and from 100 m and 200 m (●).

Horizontal Gradients in Trace Gas Concentration: Advection Errors

Considering again the imaginary unit cube of atmosphere above the surface (Fig. 4), if a horizontal gradient in a trace gas concentration exists, then the quantity entering the upwind face of the cube will differ from that leaving the downwind face. This may be written as

$$\Delta F_s = \int_{z_1}^{z_2} U \, \partial_s/\partial_x \, d_z. \tag{26}$$

Defining the mean windspeed over the height interval z_2 to z_1 as $\bar{u}$ and the mean horizontal gradient in concentration ∂_s/∂_x, Eq. 26 may be simplified to

$$\Delta F_s = u \, \partial_s/\partial_x \, (z_2 - z_1). \tag{27}$$

The error introduced through advection may be very large for any trace gas which is also a common pollutant (e.g., NO, NO_2, SO_2). For a vertical flux of 100 ng m^{-2} s^{-1} a horizontal gradient of 10 μg m^{-3} km^{-1} leads to a 20% error in the flux to the ground.

Consider, for example, the case of NO_x flux measurements in the proximity of roads. The traffic density is 5000 vehicles day^{-1} and the site of measurements is between 500 and 1000 m from the road. A Gaussian plume dispersion model has been used to estimate concentrations (C) at the receptor point according to

$$C = \frac{2q \exp(-H^2/2\sigma z^2)}{2 \pi z \, u \sin \theta} \tag{28}$$

where H is emission height, z is the vertical dispersion coefficient, and θ is the angle the wind (u) makes with the road. In this case the flux divergence at the measurement site was estimated to be of the same order as the vertical flux 60 ng $m^{-2}s^{-1}$. Thus, although attractive from a logistical aspect, the site proved unsuitable for flux measurements of NO and NO_2. Again, Table 2 provides further examples of the scale of advection errors for different conditions.

Measurements over Tall Vegetation

The foregoing arguments allow fluxes to be estimated from measurements within the constant flux layer above the ground either from gradient or eddy correlation methods. In practice, gradient methods have been used mainly over short vegetation because for the same flux, gradients are then much larger than over forests. An additional problem over very tall vegetation (>10 m) is that flux-gradient relationships break down close to the canopy top. These problems led to a reluctance within the micrometeorological community to apply any gradient methods close to the canopy surface and recent work by Raupach and Finnigan (1989) provides a physical basis for the problems encountered. They show that flux-gradient measurements over forests are satisfactory provided that the measurement height range is between a lower limit of 1.5 h (where h is the canopy top) and the upper extent of the constant flux layer provided by the fetch. For a forest 10 m tall with 2 km of fetch, the measurements would need to be between 15 m and 28 m. Over such a height range the concentration differences for trace gas fluxes are generally very small and difficult to detect. Wherever possible, research groups are applying eddy correlation methods over such surfaces. The main practical limitation to eddy correlation studies is the response time of the sensor (which should be better than <1 s and ideally better than 0.1 s). As mean eddy size increases with height above the ground and with aerodynamic roughness of the surface, sensors detect a larger fraction of the flux over very rough surfaces and at higher levels.

Fast response gas analyzers have been developed for SO_2, O_3, NO_2, and CO_2, but their response times are generally in the range 0.1 to 2.0 s. Such response times lead to the sensor detecting only a fraction of the eddy sizes present and the resulting "high frequency loss" which varies with site and atmospheric conditions is generally in the range 10 to 50% in the measurements reported (Garratt 1975). These losses can be corrected for when the sensor characteristics are known.

RESEARCH OPPORTUNITIES

The description of measurement methods and restrictions, while very familiar to those engaged in the work, presents a daunting list of problems. These problems and the requirement for large amounts of capital equipment have restricted the application of micrometeorology in the trace gas field. There are instrumental solutions to some of the problems and ways of making field measurements which avoid some of the problems introduced by air density variation with height and which are nonstationary.

Instrumentation

1. Several groups are currently developing sensitive optical techniques for a range of trace gases. Such instruments would be suitable for eddy flux methods, having response times in the range 0.01–0.1 s.

2. The development of eddy accumulation methods are needed in which the upward and downward moving eddies are separated and accumulated into enclosures in which small differences may be detected.
3. The sampling of gases for gradient measurement, using a time-shared gas analyzer in which all levels are measured at the same temperature, obviates the need to correct flux gradient data for the density effects produced by temperature gradients.

Site Conditions

1. The effects of large heat fluxes on trace gas exchange highlights the importance of daytime sunny conditions. In the majority of micrometeorological flux measurements of CO_2 and H_2O fluxes above crop canopies, the bulk of the transfer occurs during the day and fluxes are often approximately proportional to R_n. The fluxes of trace gases are in many cases largely independent of net radiation, though emission of soil-derived gases does vary with soil temperature. Valuable data may therefore be obtained in a wide range of conditions for which the corrections are very small. There are good reasons to believe that fluxes in cloudy, windy, and dark conditions may be more representative of Northern Europe!
2. Flux-gradient methods have recently been applied in gently undulating terrain and have provided precisely logarithmic wind velocity profiles (Gallagher et al. 1988). These data show that many sites which have previously been regarded as nonideal are in fact perfectly adequate for measurements. Similar arguments apply to fetch/height ratios, where field studies in conditions with fetch height ratios of 100:1 have provided satisfactory fluxes.

In measurements at nonideal sites one of the main requirements of the investigators is that the shapes of wind velocity profile be determined by measurement rather than be assumed. Further, energy balance measurements of sensible and latent heat fluxes must be measured whenever possible in parallel with trace gas fluxes, so that in the event of peculiar trace gas fluxes the data may be examined to show whether components of the energy balance appear satisfactory.

The careful selection of sites should also enable the worst cases of advection and storage errors to be avoided. The development of sensors with very low detection thresholds is an important step in this direction.

Lastly, in designing field studies of trace gas fluxes, the integration of micrometeorological measurements with independent measurements of exchange rates using enclosure methods offers considerable promise. This combination provides an important contribution to quality asssurance in the measurements and an opportunity to study the fine detail while obtaining fluxes representative on a much larger scale.

Acknowledgements. The authors acknowledge discussion with ITE and TNO colleagues, the very helpful comments and additions from Bruce Hicks (and the reminder that these problems have been known for many years). This work was supported financially by the U.K. Department of the Environment and the Commission of the European Communities.

REFERENCES

Chamberlain, A.C. 1975. The movement of particles in plant communities. In: Vegetation and the Atmosphere, ed. J.L. Monteith, vol. 1, pp. 155–201. New York: Academic.

Colbourn, P., J.C. Ryden, and G.J. Dollard. 1987. Emission of NO_x from urine treated pasture. *Envir. Poll.* 46:253–261.

Denmead, O.T. 1983. Micrometeorological methods for measuring gaseous losses of nitrogen in the field. In: Gaseous Loss of Nitrogen from Plant-soil Systems, ed. J.R. Freney and J.R. Simpson, pp.133–157. Amsterdam: W. Junk.

Denmead, O.T., J.R. Simpson, and J.R. Freney. 1977. A direct field measurement of ammonia emission after injection of anhydrous ammonia. *Soil Sci. Soc. Am. J.* 41:1001–1004.

Dollard, G.J., J.T. Davies, and J.P.C. Lindstrøm. 1986. Measurements of the dry deposition rates of some trace gas species. In: Physico-chemical Behaviour of Atmospheric Pollutants, ed. G. Angletti and G. Restelli, pp. 470–476. Dordrecht: Reidel.

Duyzer, J.H., G.M. Meyer, and R.M. van Aalst. 1983. Measurement of dry deposition velocities of NO, NO_2, and O_3 and the influence of chemical reactions. *Atmos. Envir.* 17:2117–2120.

Dyer, A.J., and B.B. Hicks. 1970. Flux-gradient relationships in the constant flux layer. *Q. J. Roy. Meteor. Soc.* 96:715–721.

Fowler, D., and M.H. Unsworth. 1979. Turbulent transfer of sulphur dioxide to a wheat crop. *Q. J. Roy. Meteor. Soc.* 105:767–784.

Galbally, I.E., J.A. Garland, and M.J.E. Wilson. 1979. Sulphur uptake from the atmosphere by forest and farmland. *Nature* 280:49–50.

Garratt, J.R. 1975. Limitations of the eddy correlation technique for the determination of turbulent fluxes near the surface. *Bound.-Lay. Meteor.* 8:255–259.

Gallagher, M.W., T.W. Choularton, A.P. Morse, and D. Fowler. 1988. Measurements of the size dependence of cloud water deposition at a hill site. *Q. J. Roy. Meteor. Soc.* 114:1291–1304.

Huebert, B., and C.H. Roberts. 1985. The dry deposition of nitric acid to grass. *J. Geophys. Res.* 90:2085–2090.

Monteith, J.L. 1973. Principles of Environmental Physics. London: E. Arnold.

Rapsomanikis, S., and R.M. Harrison. 1987. Dry deposition of HNO_3 and HCl onto some vegetative surfaces. In: Acid Rain, Scientific and Technical Advances, ed. R. Perry, J.N.B. Bell, and R.M. Harrison, pp. 201–204. London: Selper.

Raupach, M.R., and J.J. Finnigan. 1989. Stand overstay processes. *Phil. Trans. R. Soc. Lon. B*, in press.

Thom, A.S. 1975. Momentum mass and heat exchange. In: Vegetation and the Atmosphere, ed. J. Monteith, pp. 57–109. New York: Academic.

van Aalst, R.M. 1989. Ozone and oxidants in the planetary boundary layer. In: Atmospheric Ozone Research and its Policy Implications, ed. T. Schneider et al. Amsterdam: Elsevier, in press.

Webb, E.K., G.I. Pearman, and R. Leuning. 1980. Correction of flux measurements for chemistry effects due to heat and water vapour transfer. *Q. J. Roy. Meteor. Soc.* 106:85–100.

Weseley, M.L., J.A. Eastman, D.H. Stedman, and E.D. Yelvac. 1982. An eddy correlation measurement of NO_2 flux to vegetation and comparison to ozone flux. *Atmos. Envir.* 16:815–820.

Wilson, J.D., V.R. Catchpole, O.T. Denmead, and G.W. Thurtell. 1983. Verification of a simple micrometeorological method for estimating ammonia losses after fertilizer application. *Agric. Meteor.* 29:283–290.

Woodward, F.I., and J.E. Sheehy. 1983. Principles and Measurements in Environmental Biology, pp. 129–133. London: Butterworths.

Exchange of Trace Gases between Terrestrial Ecosystems and the Atmosphere
eds. M.O. Andreae and D.S. Schimel, pp. 209–228
John Wiley & Sons Ltd

Methane Flux Measurements: Methods and Results

H. Schütz and W. Seiler

Fraunhofer-Institut für Atmosphärische Umweltforschung
8100 Garmisch-Partenkirchen, F.R. Germany

Abstract. CH_4 fluxes from (or to) soils, sediments, and the digestive tract of animals and humans have been studied during the last decade. The most frequently employed technique in soil/sediment studies is the closed (static) chamber (box) technique. Open chambers are primarily used in animal studies. High areal and temporal variations of CH_4 flux rates (over 5 orders of magnitude) have been found in wetland environments; therefore, comparative and continuous measurements are required to obtain reliable average flux rates. From the data presently available, biogenic CH_4 from natural wetlands, rice paddies, and ruminants represents the predominant global source for atmospheric CH_4.

INTRODUCTION

Methane is a radiatively active trace gas (IR-absorption), and plays an important role in numerous chemical reactions in the atmosphere. It is well documented that the atmospheric mixing ratio of CH_4 is increasing at a rate of approximately 1% per year (Blake and Rowland 1988). This increase is most probably due to an increasing global source strength for atmospheric methane (Seiler 1984). As a consequence, studies on the atmospheric CH_4 cycle have focused on the determination of the magnitude of CH_4 fluxes from a great variety of potentially important CH_4 sources. As shown in Table 1, recent estimates indicate that CH_4 is predominantly produced biogenically in anoxic soils of natural wetlands and rice paddies and in the digestive tract of ruminants. Biogenic CH_4 is also released from organic waste degradation sites, e.g., landfills. The most important abiological source of atmospheric methane is its production by biomass burning. In addition, thermogenic methane is released into the atmosphere from natural reservoirs such as coal mines and natural gas. The main sink of atmospheric CH_4 is its oxidation by OH radicals in the troposphere and stratosphere. Methane

TABLE 1. Estimates of the global contribution of sources to atmospheric CH_4.

CH_4 source	Global annual emission (Tg)	Reference
Rice paddies	70 – 170	Holzapfel-Pschorn and Seiler (1986)
Natural wetlands	110	Matthews and Fung (1987)
Digestive tract	70 – 100	Crutzen et al. (1987)
Organic waste	30 – 70	Bingemer and Crutzen (1987)
Other biogenic	6 – 45	Cicerone and Oremland (1988)
Total biogenic	286 – 495	
Biomass burning	55 – 100	Bolle et al. (1986)
Natural gas	30 – 40	Bolle et al. (1986)
Coal mining	35	Bolle et al. (1986)
Other nonbiogenic	1 – 2	Bolle et al. (1986)
Total nonbiogenic	121 – 177	
Total of sources	407 – 672	

is also taken up by microorganisms in natural soils. At the current atmospheric CH_4 mixing ratios, this sink appears to be of minor importance for the global CH_4 budget (Bolle et al. 1986).

Methane produced in wetland soils and sediments may enter the water column or the atmosphere via different pathways, including molecular diffusion or pressurized ventilation (resulting from thermo-osmosis and Knudsen diffusion), through the aerenchym of aquatic plants, ebullition of gas bubbles, and/or molecular diffusion of dissolved CH_4 across the water/atmosphere boundary layer. Because of these different pathways, different techniques have to be employed for determinations of CH_4 fluxes. These techniques are described below. Analysis of CH_4 is performed either by gas chromatography (FID or TCD) or by infrared-absorption (IR) measurement.

TECHNIQUES APPLIED FOR CH_4 FLUX MEASUREMENTS

Closed Chamber

The chamber technique is a common method for the determination of the flux of atmospheric trace gases (emission or deposition) from or to soil and water (e.g., Conrad and Seiler 1985). This method employs an open bottom chamber which is placed over the measuring site; changes in the trace gas concentration are monitored. One of the disadvantages of the chamber technique is the exclusion of the natural air turbulence, which may have a significant influence on the exchange of air between soil and atmosphere. To get an impression of this influence, experiments using wind generators have been performed in closed chamber studies. Temperature and humidity alterations due to the enclosure must be minimized, which is best achieved using open and dynamic chambers or controlled environment closed chambers. In the latter case, it is possible to reduce heating of the chamber air by shading of the box. However, the effect of shading on soil temperatures must be considered. It is very important that the measuring period be kept as short as possible and that the measuring site be reexposed to natural conditions at regular intervals. When measuring over vegetation, the chamber must be built of material causing only minor changes in the spectral composition or radiation intensity in the physiologically active range. Materials often used for building the chambers are aluminum, glass, and plexiglass. In some cases, plastic bags are used for the enclosure of plants. Plexiglass (3 mm thick), the most commonly used material for chamber constructions, absorbs radiation effectively at wavelengths lower than 400 nm but does not significantly influence the active radiation spectrum for photosynthesis of higher plants. However, physiological reactions induced by bluelight may be altered.

The most common method for measuring CH_4 fluxes from soil or water surfaces into the atmosphere is the closed or static chamber technique. This method is applicable in view of the long lifetime of CH_4 (about 10 years), the low solubility of CH_4 in water (appr. 24 mg/l at 20°C and ambient pressure), and the low interaction of CH_4 with common chamber materials. In order to evaluate the CH_4 flux, the CH_4 mixing ratios in the air enclosed by the chamber are recorded as a function of time. The effect of the altered CH_4 concentration gradients in the upper soil layer mediated by the increasing CH_4 concentrations in the enclosed air can be minimized by using boxes with a large chamber volume as compared to the enclosed soil area (Matthias et al. 1978). A vertical CH_4 gradient within the box can be avoided by using a fan. The CH_4 soil flux (in mg/m^{2}h) in a static chamber is calculated by the equation

$$E = h\,p\,\frac{dm}{dt},$$

where h is the height of the box, p is the density of CH_4 at the pressure and temperature recorded inside the box, and dm/dt is the linear increase in the CH_4 mixing ratio inside the box during the measuring period.

In order to quantify CH_4 fluxes caused by gas bubble ebullition, gas collecting devices (traps) are placed over the sediment–water or soil–water interface. Ebullition gas is collected over time intervals. The flux rate is calculated by determination of the gas volume's CH_4 concentration.

Open and dynamic chambers

The term "open chamber" is used in this context for boxes flushed with ambient air at a constant flow rate during the measuring period. In contrast to static chambers, the continuous air flow in an open chamber hinders temperature and humidity increase within the chamber. If high flow rates are applied and the design of the chamber is appropriate, uniform mixing of the chamber air results and short residence times for gas molecules in the chamber (<1 min) are achieved. Relative humidity and mixing ratios are kept near ambient conditions, and the relatively short measuring periods are advantageous for, e.g., undisturbed soil conditions. The term "dynamic chamber" is applicable when high flow rates are applied so that the soil flux (F) is determined from the mass balance of the chamber by the equation

$$F = Q/A\ p\ dc,$$

where Q is the air flow rate through the chamber, A is the soil area covered by the chamber, p is the density of methane at the temperature and pressure recorded inside the chamber, and dc is the difference in concentration of methane between chamber outlet and chamber inlet.

Application of CH_4/CO_2 Ratios

For the global source evaluation of CH_4 produced from biomass burning, air samples in fire plumes at soil surface and at different altitudes were taken and the CH_4/CO_2 emission ratio factor was determined (Crutzen et al. 1979). In the same way, the global contribution of termites to CH_4 emission was determined from data on their carbon ingestion applying species-specific ratios of measured CH_4/CO_2 fluxes (Seiler et al. 1984). This technique assumes the CO_2 flux is well quantified, which may be questionable.

Diffusive Flux

The diffusive flux of CH_4 across the water/atmosphere boundary layer can be determined by using a static chamber and covering the emission site with a fine-mesh screen which impedes CH_4 bubble emission. The diffusional flux can also be determined indirectly using a model for gas exchange across the water/atmosphere boundary layer. Applying, for example, the stagnant film model described by Liss and Slater (1974), the concentration of CH_4 dissolved in the surface water below the laminar film (C) and the atmospheric mixing ratio of CH_4 (m) can be measured (and multiplied with its solubility

α). The flux (F) can be calculated if the transfer velocity (K_L) of CH_4 across the water–air interface is known:

$$F = K_L (C - \alpha m)$$

Sediment/Water Flux by Vertical Gradient Measurement

In order to determine the flux of CH_4 from sediments to the overlying water column, the vertical gradient of dissolved CH_4 at the sediment–water interface (dC/dz) is measured. The gradient supported flux (J) of CH_4 can be calculated using Fick's relationship, provided the porosity (Φ) is known and the diffusivity (D_s) is assumed to be molecular (i.e., 2×10^{-5} cm^2 s^{-1}):

$$J = \Phi D_s dC/dz.$$

However, the actual diffusion coefficient is subject to spatial and temporal variations such as open bubble tubes or benthic animals. Furthermore, the shape of the gradient may be influenced by local CH_4 oxidation reactions in the sediment.

Micrometeorological Methods

Chamber methods for flux determination are certainly not free from interference. It is obvious that enclosure of the measuring site can cause changes in the natural conditions and, thus, in the calculated flux rates of CH_4. Alternative methods for studying trace gas fluxes under "nondestructive" conditions can be carried out using micrometeorological flux monitoring, especially the gradient technique and the eddy correlation technique (e.g., Baldocchi et al. 1988).

Gradient technique. The vertical flux (j_c) of a trace gas per unit area is determined by measuring its vertical concentration gradient $\delta c/\delta z$ caused by turbulent transport in the atmosphere. The flux is calculated by the equation:

$$j_c = K_c \, \delta c/\delta z,$$

where the diffusion coefficient K_c has to be determined by micrometeorological measurements (wind gradient, temperature gradient) and parameters derived from these measurements. The trace gas concentration has to be measured simultaneously at at least two different heights from the ground. Analytical instruments with high resolution, high precision, and fast response times (<1 s) have to be employed. Although the gradient technique has been used to measure turbulent fluxes of CH_4 by using grab-bag air samplers or bubblers as chemical sensors (Baldocchi et al. 1988), a direct and fast detecting instrument is a limiting factor to its applicability to flux measurements of CH_4. In addition, the method depends on areas of homogeneous surfaces with fetch/measuring height approximately 100/1, so that spatial differences in CH_4 fluxes on a small scale, as reported for some wetland ecosystems, are not detectable. A He–Ne laser technique has recently been developed for CH_4 flux measurements using eddy correlation methods or gradient methods (R.C. Harriss, pers. comm.).

Eddy correlation technique. The eddy correlation technique determines the vertical transport of trace gases and presumes measurement of the vertical wind speed (w) and the trace gas concentration (c) over a sufficiently large time interval at one height above the measuring site. The flux is calculated by the equation

$$j = c\,w\,.$$

In contrast to the gradient technique, the eddy correlation technique requires only one measuring instrument and the flux is directly determined from the two parameters measured. Therefore, better results may be obtained by the eddy correlation method than by the gradient method, provided the instrumentation requirements are met. However, the same limitations as for the gradient technique apply to its use in CH_4 flux monitoring due to the analytical instrumentation and the areal characteristics of the measuring sites. The present problems which result from the lack of instrumentation for rapid and high resolution CH_4 analysis may be solved in the future through use of the diode laser technique for CH_4 analysis. G. Sachse, NASA Langley Research Center, obtained excellent results with the tunable diode laser technique for CH_4 flux by eddy correlation from an aircraft platform in summer 1988 over Alaskan tundra (R.C. Harriss, pers. comm.).

APPLICATION OF DESCRIBED TECHNIQUES

Natural Wetland Studies

The occurrence of CH_4 in natural wetlands was first observed by Alessandro Volta in the 18th century (describing the discovery of inflammable air in a letter to his friend, Pater Campi, from 14 November 1776). Natural wetlands are globally distributed over the world's surface (appr. 5.3×10^{12} m^2 [Matthews and Fung 1987]) and, therefore, are likely to be an important source for atmospheric methane. During recent years measurements of CH_4

fluxes from these ecosystems employing the chamber techniques have been intensified.

Closed chamber. A continuous sampling and analyzing system for CH_4 was developed by Sebacher and Harriss (1980) and used for measurements of CH_4 fluxes from soil or water interfaces to the atmosphere from various sites, including natural wetlands. This system is illustrated in Fig. 1. It consists of a gas collecting chamber, a gas filter correlation analyzer (IR) sampling cell, a signal processor, and a recorder. The system is portable and floatable, and can be operated at remote locations using batteries. The chamber air is uniformly mixed and continuously recycled in a closed cycle from the chamber through a silica gel filter (to remove water vapor which interferes with the IR-detection of CH_4) to the IR-analyzer and back to the chamber, so that any change in CH_4 concentration is continuously monitored. Thus, for example, CH_4 emission by gas bubbles is easily documented by the typical stepwise increase of the CH_4 mixing ratios over time. The system has a fast response time (<1 s) and high analytical sensitivity (0.01 ppmv CH_4). The minimum detectable flux is as low as 0.005 mg CH_4/m^2h, which is at the lower end of the range of values commonly found in wetland ecosystems (Table 2). Flux measurements can be performed with this system within minutes. In a modified version, Sebacher and Harriss (1982) equipped the closed chamber with a blower in order to study the effects of air velocity from 0 to 4.4 m/s on CH_4 fluxes from water–atmosphere interfaces.

The contribution of individual plants to the emission of methane from wetland and aquatic environments has been analyzed by enclosing the plant in a plastic bag and monitoring the temporal increase of the CH_4 mixing

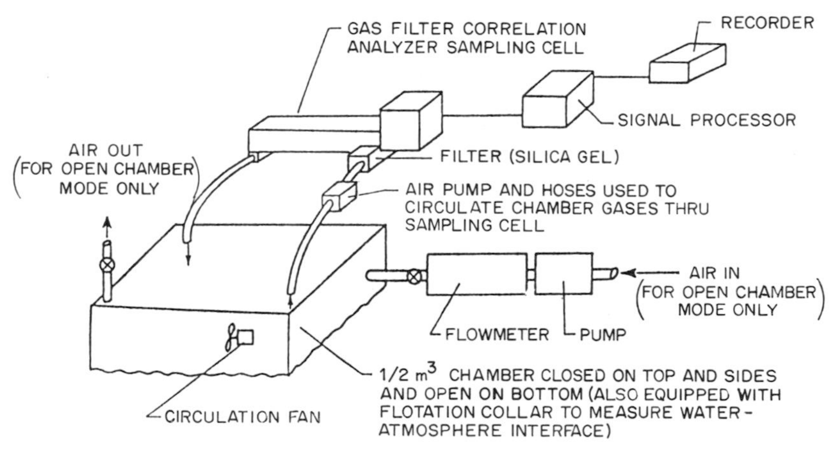

Fig. 1—Schematic outline of a system used to measure methane fluxes (figure taken from Sebacher and Harriss [1980]).

ratios inside the bag volume (Dacey 1981; Sebacher et al. 1985). This technique has also been employed to determine the mechanism by which gas passes the aerenchymal parts of plants by following the distribution of tracers (e.g., ethane) injected into the bag (Dacey and Klug 1982).

Open chamber. The open chamber mode was employed by Sebacher and Harriss (1980) in wetland studies. At the flowrate of 6.5 l/min ambient air applied and a chamber volume of 0.5 m^3, the residence time of the chamber air volume was approximately 77 min and, thus, much longer than time values of <1 minute needed for dynamic chambers. A comparative study of both chamber techniques, closed and open, resulted in the same CH_4 flux from a swamp environment.

Rice Paddy Studies

Because of the large areal extent of approximately 1.5×10^6 km^2, rice paddies are likely to be one of the most important individual sources of atmospheric methane. It is also likely that the methane fluxes from rice paddies are not only influenced by regional differences in climate, soil conditions, and plant variety, but are also subject to variation due to differences in the agricultural management.

Closed chamber. The first, discontinuously performed field measurements in California (Cicerone and Shetter 1981), Spain (Seiler et al. 1984), and Italy (Holzapfel-Pschorn and Seiler 1986) using mainly static chambers confirmed the assumption that high temporal variations in CH_4 fluxes from rice paddies exist. Thus, in order to monitor the temporal variations of CH_4 emission rates as well as the influence of fertilizer application, a continuously sampling and analyzing system was developed and used to monitor CH_4 emission rates from rice paddies in Italy (Schütz et al. 1989) and China. This system allows the semicontinuous determination of CH_4 emission rates at 16 individual field sites using gas collector boxes (static chambers) made of plexiglass (colorless, smooth, 3 mm thick). The boxes are covered with a removable lid which is opened by a pneumatic pressure cylinder. In the field, the gas collector boxes are mounted on stainless steel frames fixed in the ground prior to flooding and which remain during the entire vegetation period. The height of a frame is adjusted in such a way that approximately 5 cm of the box is below water. In this way the air volume inside the box is separated from the ambient atmosphere, but the water beneath the box can exchange with the surrounding water body, avoiding temperature increases in the flooded soil which would lead to overestimated CH_4 emission rates.

This system, illustrated in Fig. 2, consists of two completely separate units each with a gas chromatograph (Mini2, Shimadzu) connected to a dual-

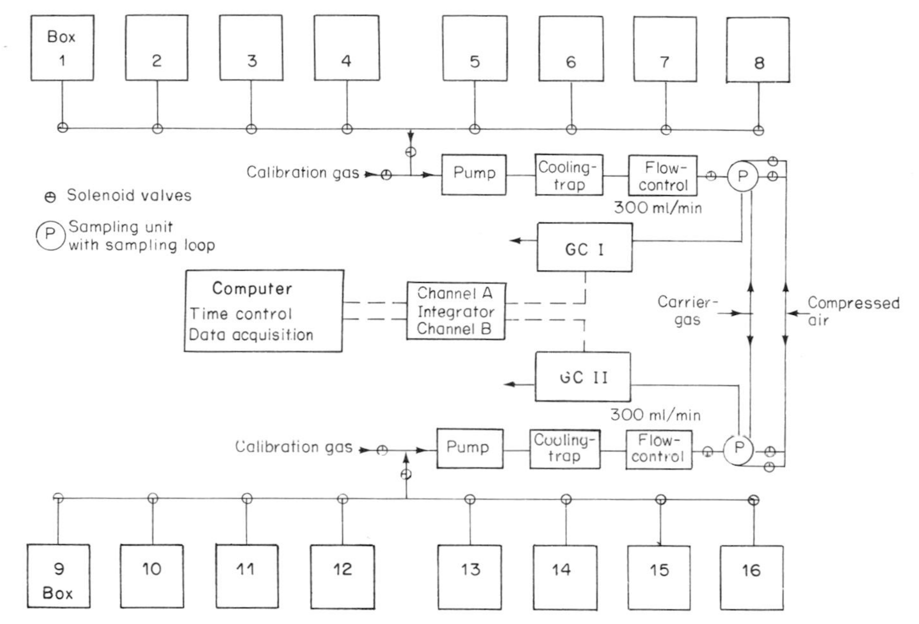

Fig. 2—Schematic of the automatic system employed in continuous CH_4 flux measurements in rice paddies.

channel integrator. Magnetic valves controlling the air flow in the system are operated by a programmable micro-computer (Hewlett-Packard HP 9835) which also stores the raw data output of the GC-systems. Air samples from inside the individual closed boxes are taken periodically and analyzed for CH_4 using gas chromatography (FID). This system allows automatic determination of the CH_4 emission rates at each field plot eight times per day.

Open chamber. Using an open chamber (flow-through saran bag at 1l/min ambient air), Cicerone and Shetter (1981) measured CH_4 fluxes in Californian rice paddies which were in the same range as values measured by using two different closed chambers.

CH_4 Flux at Soil–Atmosphere Interface

The static chamber method has also been used to measure deposition fluxes of CH_4 from the atmosphere to the soil (e.g., Seiler et al. 1984). For these measurements a glass box (8 to 18 cm high, 800 cm^2 covered soil area) was equipped with a septum on top for extraction of air samples and with an open glass coil to guarantee pressure equilibration of air inside and outside the box. Measurements were started at ambient atmospheric CH_4 mixing ratios, which declined with time and finally reached an equilibrium. The finding of such an equilibrium indicates that deposition and emission are both the result of dynamic exchange processes of methane between biosphere and atmosphere. Flushing the chamber with CH_4-free air would have resulted in CH_4 emission contrary to the actual observation under ambient conditions. Using the static box technique, Hao et al. (1988) did observe emission of CH_4 from dry tropical savanna soils.

CH_4 Flux from Freshwater and Saltwater Sediments

The techniques employed for the measurement of CH_4 fluxes from sediments of freshwater lakes and saltwater marshes include the chamber technique, the use of ebullition gas traps, and the measurement of dissolved CH_4 for diffusive flux calculation. A static chamber was also used by Martens and Klump (1980) to measure the CH_4 flux from oceanic sediments into the oceanic water. For this purpose a benthic chamber was placed at the oceanic sediment–water interface and the temporal increase of dissolved methane in the water body enclosed by the chamber was measured.

Digestive Tract Studies

Closed chamber. The closed chamber technique has often been used to study CH_4 emission by exhalation, eruction, and flatus from the digestive tract of animals and humans. A closed rebreathing system was developed (Rodkey et al. 1972; Fleming 1980) for CH_4 flux measurements of smaller animals (rats, guinea pigs), in which the animal is placed in a closed chamber and the chamber air is circulated by a pump. Samples of chamber air were taken through a sampling port and were analyzed for CH_4. Moisture and CO_2 were absorbed on dry ice and calcium hydroxide, respectively. The resulting negative pressure inside the chamber allowed oxygen to enter the chamber on demand to equalize the pressure (Fleming 1980). In another study (Rodkey et al. 1972), oxygen was monitored continuously in the air flow and added to the system on demand from a tank.

Field studies of CH_4 fluxes from termite nests into the atmosphere (Seiler et al. 1984) have been performed using the static chamber technique with a Tedlar bag placed over an aluminium frame. The chamber was separated from the atmosphere with a layer of sand on the plastic hood over the soil. Air samples from inside the chamber were taken by hand with syringes. Quantification of the total flux of CH_4 from termites was achieved by applying the rate of CH_4 to the rate of CO_2 produced (determined by simultaneous flux measurement) from the consumption of biomass.

CH_4 flux measurements in humans have been carried out in clinical studies. For the determination of pulmonary methane excretion, a closed (chamber) system was used (Bond et al. 1971). For this purpose, the test person's head was enclosed in a polyvinyl hood. The gas inside the hood was circulated by a pump through a CO_2 absorber and an ice bath (to remove water), and through an O_2 monitor back to the hood. O_2 was continuously monitored and added to the system on demand. Breath methane concentrations were also collected by exhalation through a plastic mouth piece into a collector bag. Methane excretion by infants was measured in a similar closed system by enclosing the entire infant in a polyurethane bag and regulating the O_2 influx. A constant gas perfusion technique was employed for the site and rate of CH_4 production in the bowel (Levitt and Ingelfinger 1968). In these experiments a triple lumen polyvinyl tube was placed in the test persons terminating in the terminal ileum, the proximal ileum, and the mid-jejenum. N_2 or air was constantly infused and gas samples were collected via the intestinal tube or via a rectal tube (Calloway and Burroughs 1969). Methane concentrations were also determined in fecal specimens after homogenization and collection of the evolved gas.

Continuous measurements. In order to quantify the emission by exhalation and eruction from sheep, Murray et al. (1976) employed a continuously monitoring gas collecting and analyzing apparatus. A fiberglass mask was attached to the animal's head and ambient air was drawn across the muzzle back to the room by a pump at 60 l/min. From the mask, gas samples were drawn to the CH_4 analyzer (IR). Flatus gas was collected in plastic bags by using an anus mask.

To determine the total CH_4 flux from animals, respiration chambers (appr. 9 m^3 volume) have been developed (Gädcken and Fliegel 1974) that are continuously flushed with ambient air (2 to 20 m^3/h for large animals such as cattle, sheep, and pigs; 0.3 to 3 m^3/h for smaller animals). Air samples are taken automatically and continuously by a special gas collecting device and analyzed for CH_4 by using the IR technique. The chamber climate can be controlled with regard to temperature and humidity.

RESULTS

Table 2 gives a summary of CH_4 emission/deposition rates from the most important measuring sites. Flux measurements of CH_4 have been carried out predominantly on natural wetland soils (swamps and ponds in the U.S.A., Amazon floodplain, boreal peatlands of the U.K., Sweden, and Alaska), on irrigated rice paddies (California, Spain, Italy, and P.R. China), on dry soils (subtropical savanna, South Africa; tropical savanna, Venezuela; drought swamp, U.S.A.; tropical forest, South and Middle America; deciduous forest, U.S.A. and F.R. Germany), on freshwater lakes (U.S.A., Canada, and Amazon region), and on saltwater marshes (U.S.A.) by measuring the emission by exhalation, eruction, and flatus from ruminants, other herbivorous fauna (such as termites), and humans, and by measuring the CH_4 concentrations in fire plumes. Measurements of CH_4 emission rates on landfills and similar sites (e.g., pits) as well as determination of the emission rates of CH_4 from natural reservoirs are still scarce.

Natural Wetlands

As shown in Table 2, CH_4 flux rates from natural wetlands varied over a wide range depending on the environmental characteristics and the location. Generally, low rates of CH_4 emission (<10 mg/m^{2}h) have been found in the dryer, nutrient-poor sites of northern and midlatitudinal peatlands. However, wet peatland ecosystems exhibited sites of high CH_4 emission reaching values of up to approximately 80 mg CH_4/m^{2}h. Average rates range from 0.06 to 27.7 mg/m^{2}h. CH_4 flux rates from northern peatlands are positively correlated with both increasing water depth and soil temperatures. Higher CH_4 emission rates than in boreal peatlands are found in swamps and ponds of subtropical and tropical regions reaching individual values of up to 216 mg CH_4/m^{2}h and average values of up to 51 mg/m^{2}h. According to the estimate given by Matthews and Fung (1987; see also Table 1) natural wetlands contribute significantly to the global emission of CH_4 with northern peatlands being the most important contributor (appr. 60%) to this source.

Rice Paddies

Discontinuous measurements of CH_4 flux rates from irrigated rice paddies have been performed in the U.S.A., Spain, and Italy. However, extensive field measurements have become possible only after development of the automatic system described here. The system was employed in measurements of CH_4 fluxes from rice paddies in Italy and in the People's Republic of China. The design of the system enabled a comparative study of flux rates in fertilized and unfertilized Italian paddy fields. The data indicated that fertilization with ammonium sulfate or urea depresses the CH_4 emission by up to 50%. Fertilization with organic matter (rice straw), however, enhances the CH_4 emission up to a factor of two (Schütz et al. 1989). High diurnal as well as seasonal variations in the flux rates of CH_4 have been observed in the course of discontinuous measurements in the Italian rice paddies (Holzapfel-Pschorn and Seiler 1986; Fig. 3). Similar results have been obtained most recently by measurements in China (Hangzhou, Province Zhejiang). These measurements are the first extensive set of data on the CH_4 emission obtained in rice paddies of Asia, where more than 90% of the global rice growing area is located. A first evaluation of the data shows that even higher CH_4 flux rates than in Italy are monitored and that in contrast to the results obtained in Italy, the CH_4 fluxes are not correlated

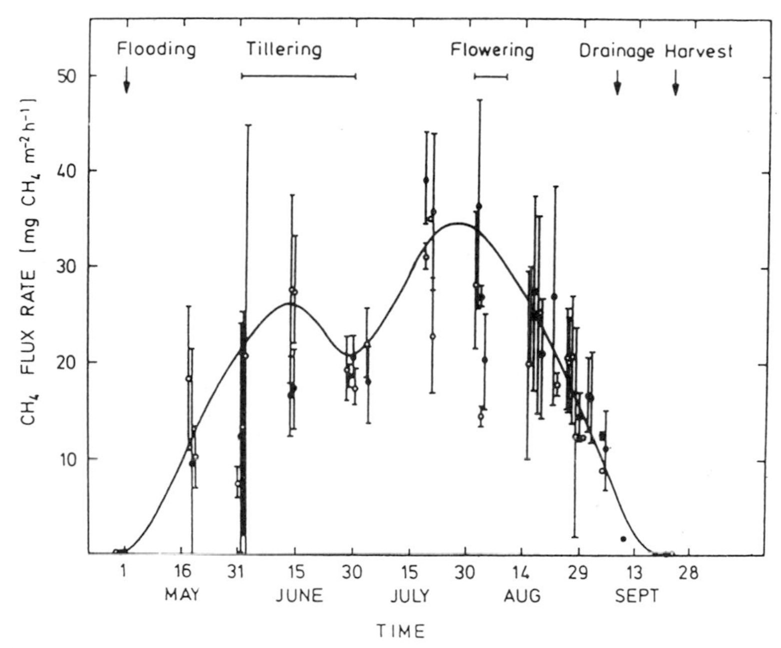

Fig. 3—Diurnal and seasonal variations of CH_4 emission rates in a rice paddy (unfertilized [o] and fertilized with $CaCN_2$ [•]) in Vercelli, Italy (figure taken from Holzapfel-Pschorn and Seiler [1986]).

with soil temperature. It is, therefore, likely that the global contribution of rice paddies to atmospheric methane is even higher than the estimate given by Holzapfel-Pschorn and Seiler (1986; Table 1). Rice paddies, therefore, are likely to be the most important individual source for atmospheric methane.

Soil

Dryland soils of temperate, subtropical, and tropical regions are sinks for atmospheric methane yielding a strength of approximately 20 Tg/yr (Bolle et al. 1986). CH_4 deposition on soils follows first-order kinetics, and deposition rates at atmospheric mixing ratios of CH_4 are rather low (mostly $< 100\ \mu g/m^2h$) compared to rates of CH_4 emission from wet soils. Factors influencing the CH_4 uptake by soils have not been studied intensively. Most recently, however, it was found that dry tropical savanna soils are a source for atmospheric methane. Hao et al. (1987) measured CH_4 emission rates which ranged from approximately 3 to 100 $\mu g/m^2h$ (average appr. 41 $\mu g/m^2h$) on undisturbed soils, providing an insignificant source for atmospheric methane. More measurements are needed to elucidate the role of dryland soils in the atmospheric cycle of methane.

TABLE 2. CH_4 flux values measured in various ecosystems.

	Range of CH_4 flux rates	Average rates	Reference
A. *Wetland Soils*			
Natural wetland soils	−0.44–216.7 mg/m²h	0.1–51	Baker-Blocker et al. (1977), Bartlett et al. (1985), Barber et al. (1987), Sebacher et al. (1983), Harriss & Sebacher (1981), Harriss et al. (1982), Bartlett et al. (1988), Devol et al. (1988), Thathy et al. (1988), Harriss et al. (1988), Clymo & Reddaway (1971), Harriss et al. (1985), Svensson (1980), Svensson & Rosswall (1984), Sebacher et al. (1986), Whalen & Reeburgh (1988)
Irrigated rice paddies	0.02–220 mg/m²h	1.3–38	Cicerone & Shetter (1981), Cicerone et al. (1983), Seiler et al. (1984), Holzapfel-Pschorn & Seiler (1986), Schütz et al. (1989)
B. *Dry Soils*	−3634–140 μg/m²h	−52–41	Seiler et al. (1984), Hao et al. (1988), Harriss et al. (1982), Devol et al. (1988), Keller et al. (1983)
C. *Sediments*			
Freshwater	0.01–327 mg/m²h	0.2–28	Keller et al. (1986), Cicerone & Shetter (1981), Strayer & Tiedje (1978), Molongoski & Klug (1980), DeLaune

			et al. (1983), Barber et al. (1987), Dacey & Klug (1979), Miller & Oremland (1987), Kelly & Chynoweth (1980), Rudd & Hamilton (1978), Crill et al. (1988)
Saltwater	−0.1–90 mg/m^2h	0.1–11	Bartlett et al. (1985), Atkinson & Hall (1976), Martens & Klump (1980), Crill & Martens (1983), Sansone & Martens (1981), DeLaune et al. (1983), Cicerone & Shetter (1981), King & Wiebe (1978), Barber et al. (1987)
D. *Digestive Tract*			
Ruminants	1.6–10% CH_4 of energy intake	3–10	Krishna et al. (1978), Czerkawski (1969), Ritzman & Benedict (1938), Blaxter & Clapperton (1965), Moe & Tyrell (1979), Wainman et al. (1978), Murray et al. (1978), Blaxter & Czerkawski (1966), Kempton et al. (1976), Schneider et al. (1974)
Nonruminants	0.25–2% CH_4 of energy intake	0.25–2	Schneider & Menke (1982), Fleming (1980), von Engelhardt et al. (1970), Kempton et al. (1976)
Termites	0.001–19.3 mg CH_4/nest h	0.07–10.3	Seiler et al. (1984)
Humans	<0.02–0.67 ml CH_4/min	0.02–0.06	Calloway & Burroughs (1969), Levitt & Ingelfinger (1968), Bond et al. (1971), Tadesse et al. (1980)
E. *Biomass Burning*	1–2.2% (v/v; CH_4/CO_2)	1.6	Crutzen et al. (1979)

Freshwater Lakes and Saltwater Marshes

CH_4 flux rates measured in freshwater lakes in the U.S.A. and Canada range from 0.01 to 327.5 mg/m^{2}h with average values between 0.2 and 28.2 mg/m^{2}h. From these studies, the global contribution of freshwater lakes to CH_4 emission is only of minor importance (e.g., Bolle et al. 1986).

Although high CH_4 emission rates of up to 90 mg/m^{2}h have been measured on some sites in saltwater marshes, the average rates range betwcn 0.1 to 11 mg CH_4/m^{2}h. The low average rates of CH_4 emission from these ecosystems are probably due to the inhibition of CH_4 production in sediments with high salt (especially sulfate) concentrations. Therefore, the contribution of salt marshes to global methane seems to be negligible. CH_4 emission over the oceanic water–atmosphere interface is also a less important factor in the global CH_4 budget.

Digestive Tract

Measurements of CH_4 release from the digestive tract of ruminants have been reported mainly for domestic animals, like cattle and sheep. In order to estimate the global contribution of this source to CH_4 emission, the CH_4 release has been related to the gross energy intake of the animals. Values ranging from 3 to 10% (CH_4 of energy intake) have been obtained by this approach. According to the estimate given by Crutzen et al. (1986), CH_4 emission from the digestive tract is one of the three most important sources for atmospheric methane. It is governed by CH_4 production in ruminants, whereas CH_4 production in other herbivorous fauna and humans seems to be of minor importance. However, field measurements from other herbivorous fauna, especially termites, should be intensified.

Fire

Crutzen et al. (1979) determined the volumetric relation of CH_4 to CO_2 (vCH_4/vCO_2) in several fire plumes and found values in the range of 1 to 2.2% (average 1.6%), leading to an estimate of the global contribution of biomass burning to atmospheric methane (Table 1), which is comparable to the contribution of the three dominating biogenic sources. However, the proportion of carbon volatilized by forest fires and clearing burns in tropical forest areas is unknown.

CONCLUSIONS

Fluxes of CH_4 from various sources into the atmosphere, summarized in Table 2, indicate that biogenic production of CH_4 is the dominant source of CH_4 (Table 1). The values (given as ranges), however, indicate large uncertainties resulting from the large variations of flux rates in space and time. Consequently, more systematic measurements over longer time periods are necessary to arrive at a better estimate of the individual global source strengths for atmospheric methane. Modern techniques for CH_4 flux determinations, e.g., eddy correlation, should be introduced to determine averaged fluxes over large areas under natural environmental conditions. In addition, the chamber technique still provides a valuable tool for the study of the effects of several parameters on CH_4 fluxes on a smaller scale. Furthermore, measurements of stable isotopes ($^{13}CH_4$) should be intensified to obtain a data set which allows a comparative estimation of the global budget of methane based on its isotopic composition.

Acknowledgements. We thank Michael Nestlen and Heinz Rennenberg for valuable discussions, Pat Haug for linguistic corrections of the manuscript, and Robert C. Harriss for his kind permission to use Fig. 1.

REFERENCES

Atkinson, L.P., and J.R. Hall. 1976. Methane distribution and production in the Georgia salt marsh. *Est. Coast. Mar. Sci.* 4:677–686.

Baker-Blocker, H., T.M. Donahue, and K.H. Mancy. 1977. Methane flux from wetland areas. *Tellus* 29:245–250.

Baldocchi, D.D., B.B. Hicks, and T.P. Meyers. 1988. Measuring biosphere–atmosphere exchanges of biologically related gases with micrometeorological methods. *Ecology* 69:1331–1340.

Barber, T.R., R.A. Burke, Jr., and W.M. Sackett. 1987. Methane flux across the air–water interface from warm wetland environments. Paper presented at the 193rd ACS Natl. Meeting, Denver, CO.

Bartlett, K.B., P.M. Crill, D.I. Sebacher, R.C. Harriss, J.O. Wilson, and J.M. Melack. 1988. Methane flux from the Central Amazonian floodplain. *J. Geophys. Res.* 93:1571–1582.

Bartlett, K.B., R.C. Harriss, and D.I. Sebacher. 1985. Methane flux from coastal salt marshes. *J. Geophys. Res.* 90:5710–5720.

Bingemer, H.G., and P.J. Crutzen. 1987. The production of methane from solid wastes. *J. Geophys. Res.* 92:2181–2187.

Blake, D.R., and F.S. Rowland. 1988. Continuing worldwide increase in tropospheric methane, 1978 to 1987. *Science* 239:1129–1131.

Blaxter, K.L., and J.L. Clapperton. 1965. Prediction of the amount of methane produced by ruminants. *Br. J. Nutr.* 19:511–522.

Blaxter, K.L., and J. Czerkawski. 1966. Modifications of the methane production of the sheep by supplementation of its diet. *J. Sci. Food Agric.* 17:417–421.

Bolle, H.-J., W. Seiler, and B. Bolin. 1986. Other greenhouse gases and aerosols. In: The Greenhouse Effect, Climatic Change, and Ecosystems, ed. B. Bolin, B.R. Döös, J. Jäger, and R.A. Warrick, pp. 157–203. SCOPE 29. Chichester: Wiley.

Bond, J.H., Jr., R.R. Engel, and M.D. Levitt. 1971. Factors influencing pulmonary methane excretion in man. *J. Exp. Med.* 133:572–588.

Calloway, D.H., and S.E. Burroughs. 1969. Effect of dried beans and silicone on

intestinal hydrogen and methane production in man. *Gut* 10:180–184.
Cicerone, R.J., and R.S. Oremland. 1988. Biogeochemical aspects of atmospheric methane. *Glob. Biogeochem. Cyc.* 2:299–327.
Cicerone, R.J., and J.D. Shetter. 1981. Sources of atmospheric methane: measurements in rice paddies and a discussion. *J. Geophys. Res.* 86:7203–7209.
Cicerone, R.J., J.D. Shetter, and C.C. Delwiche. 1983. Seasonal variation of methane from a California rice paddy. *J. Geophys. Res.* 88:11022–11024.
Clapperton, J.L. 1979. Changes with time in the effect of methane inhibitors on the rumen fermentation of sheep. *EAAP Publ.* 14:99–102.
Clymo, R.S., and E.J.F. Reddaway. 1971. Productivity of *Sphagnum* (bog-moss) and peat accumulation. *Hydrobiol.* 12:181–192.
Conrad, R., and W. Seiler. 1985. Feldmessung von Emission und Deposition atmosphärischer Spurengase in Boden und Wasser. *GIT Suppl. Umweltsch.-Umweltanal.* 3/85:74–78.
Crill, P.M., K.B. Bartlett, J.O. Wilson, D.I. Sebacher, R.C. Harriss, J.M. Melack, S. MacIntyre, L. Lesack, and L. Smith-Morrill. 1988. Tropospheric methane from an Amazonian floodplain lake. *J. Geophys. Res.* 93:1564–1570.
Crill, P.M., and C.S. Martens. 1983. Spatial and temporal fluctuations of methane production in anoxic coastal marine sediments. *Limnol. Ocean.* 28:1117–1130.
Crutzen, P.J., I. Aselmann, and W. Seiler. 1986. Methane production by domestic animals, wild ruminants, other herbivorous fauna, and humans. *Tellus* 38B:271–284.
Crutzen, P.J., L.E. Heidt, J.P. Krasnec, W.H. Pollock, and W. Seiler. 1979. Biomass burning as a source of atmospheric gases CO, H_2, N_2O, NO, CH_3Cl and COS. *Nature* 282:253–256.
Czerkawski, J.W. 1969. Methane production in ruminants and its significance. In: World Review of Nutrition and Dietetics, vol. 11, pp. 240–282. Basel/New York: Karger.
Dacey, J.W.H. 1981. Pressurized ventilation in the yellow waterlily. *Ecology* 62:1137–1147.
Dacey, J.W.H., and M.J. Klug. 1979. Methane efflux from lake sediments through water lilies. *Science* 203:1253–1254.
Dacey, J.W.H., and M.J. Klug. 1982. Tracer studies of gas circulation in Nuphar: $^{18}O_2$ and $^{14}CO_2$ transport. *Physiol. Plant.* 56:361–366.
DeLaune, R.D., C.J. Smith, and W.H. Patrick. 1983. Methane release from Gulf Coast wetlands. *Tellus* 35B:8–15.
Devol, A.H., J.E. Richey, W.A. Clark, S.L. King, and L.A. Martinelli. 1988. Methane emissions to the atmosphere from the Amazon floodplain. *J. Geophys. Res.* 93:1583–1592.
Fleming, S.E. 1980. Measurement of hydrogen production in the rat as an indicator of flatulence activity. *J. Food Sci.* 45:1012–1018.
Gädeken, D., and H. Fliegel. 1974. Aufbau und Funktionsweise der neuen Respirationsanlage für Rinder, Schweine und Kleintiere. *EAAP Publ.* 14:241–244.
Hao, W.M., D. Scharffe, and P.J. Crutzen. 1988. Production of N_2O, CH_4, and CO_2 from soils in the tropical savanna during the dry season. *J. Atmos. Chem.* 7:93–105.
Harriss, R.C., E. Gorham, D.I. Sebacher, K.B. Bartlett, and P.A. Flebbe. 1985. Methane flux from northern peatlands. *Nature* 315:652–654.
Harriss, R.C., and D.I. Sebacher. 1981. Methane flux in forested freshwater swamps of the southeastern United States. *Geophys. Res. Lett.* 8:1002–1004.
Harriss, R.C., D.I. Sebacher, and F.P. Day, Jr. 1982. Methane flux in the Great

Dismal Swamp. *Nature* 297:673–674.

Harriss, R.C., et al. 1988. Sources of atmospheric methane in the South Florida environment. *Glob. Biogeochem. Cyc.* 2:231–243.

Holzapfel-Pschorn, A., and W. Seiler. 1986. Methane emission during a vegetation period from an Italian rice paddy. *J. Geophys. Res.* 91:11803–11814.

Keller, M., T.J. Goreau, S.C. Wofsy, W.A. Kaplan, and M.B. McElroy. 1983. Production of nitrous oxide and consumption of methane by forest soils. *Geophys. Res. Lett.* 10:1156–1159.

Keller, M., W.A. Kaplan, and S.C. Wofsy. 1986. Emission of N_2O, CH_4, and CO_2 from tropical soils. *J. Geophys. Res.* 91:11791–11802.

Kelly, C.A., and D.P. Chynoweth. 1980. Comparison of *in situ* and *in vitro* rates of methane release in freshwater sediments. *Appl. Env. Microbiol.* 40:287–293.

Kempton, T.J., R.M. Murray, and R.A. Leng. 1976. Methane production and digestibility measurements in the grey kangaroo and sheep. *Aust. J. Biol. Sci.* 29:209–214.

King, G.M., and W.J. Wiebe. 1978. Methane release from soils of a Georgia salt marsh. *Geochim. Cosmo. Acta* 42:343–348.

Krishna, G., M.N. Razdan, and S.N. Ray. 1978. Effect of nutritional and seasonal variations on heat and methane production in Bos indicus. *Ind. J. Anim. Sci.* 48:366–370.

Levitt, M.D., and F.J. Ingelfinger. 1968. Hydrogen and methane production in man. *Ann. N.Y. Acad. Sci.* 150:75–81.

Liss, P.S., and P.G. Slater. 1974. Flux of gases across the air–sea interface. *Nature* 247:181–184.

Martens, C.S., and J.V. ValKlump. 1980. Biogeochemical cycling in an organic-rich coastal marine basin. I. Methane sediment–water exchange process. *Geochim. Cosmo. Acta* 44:471–490.

Matthews, E., and I. Fung. 1987. Methane emission from natural wetlands: global distribution, area, and environmental characteristics of sources. *Glob. Biogeochem. Cyc.* 1:61–86.

Matthias, A.D., D.N. Yarger, and R.S. Weinbeck. 1978. A numerical evaluation of chamber methods for determining gas fluxes. *Geophys. Res. Lett.* 5:765–768.

Miller, L.G., and R.S. Oremland. 1987. Methane flux from stratified lakes. Paper presented at the 193rd ACS Natl. Meeting, Denver, CO.

Moe, P.W., and H.F. Tyrell. 1979. Methane production in dairy cows. *J. Dairy Sci.* 62:1583–1586.

Molongoski, J.J., and M.J. Klug. 1980. Anaerobic metabolism of particulate organic matter in the sediments of a hypereutrophic lake. *Freshwat. Biol.* 10:507–518.

Murray, R.M., A.M. Bryant, and R.A. Leng. 1976. Rates of production of methane in the rumen and large intestine of sheep. *Br. J. Nutr.* 36:1–14.

Murray, R.M., A.M. Bryant, and R.A. Leng. 1978. Methane production in the rumen and lower gut of sheep given lucerne chaff: effect of level of intake. *Br. J. Nutr.* 39:337–345.

Ritzman, E.G., and F.G. Benedict. 1938. Nutritional physiology of the adult ruminant. Carnegie Institution, Washington, D.C., pp. 14–30.

Rodkey, F.L., H.A. Collison, and J.D. O'Neal. 1972. Carbon monoxide and methane production in rats, guinea pigs, and germ-free rats. *J. Appl. Physiol.* 33:256–260.

Rudd, J.W.M., and R.D. Hamilton. 1978. Methane cycling in an eutrophic shield lake and its effect on whole lake metabolism. *Limnol. Ocean.* 23:337–348.

Sansone, F., and C.S. Martens. 1981. Methane production from acetate and associated methane fluxes from anoxic coastal sediments. *Science* 211:707–709.

Schneider, W., R. Hauffe, and W. von Engelhardt. 1974. Energie- und Stickstoffumsatz beim Lamm. *EAAP Publ.* 14:127–130.

Schneider, W., and K.H. Menke. 1982. Untersuchungen über den energetischen Futterwert von Melasseschnitzeln in Rationen für Schweine. *Z. Tierphysiol. Tierern. Futter.* 48:233–240.

Schütz, H., A. Holzapfel-Pschorn, R. Conrad, H. Rennenberg, and W. Seiler. 1989. A three-year continuous study on the influence of daytime, season, and fertilizer treatment on methane emission rates from an Italian rice paddy field. *J. Geophys. Res.*, in press.

Sebacher, D.I., and R.C. Harriss. 1980. A continuous sampling and analysis system for monitoring methane fluxes from soil and water surfaces to the atmosphere. Presented at the 73rd annual meeting of the Air Pollution Control Assn. in Montreal, June 22–27, 1980.

Sebacher, D.I., and R.C. Harriss. 1982. A system for measuring methane fluxes from inland and coastal wetland environments. *J. Env. Qual.* 11:34–37.

Sebacher, D.I., R.C. Harriss, and K.B. Bartlett. 1983. Methane flux across the air–water interface: air velocity effects. *Tellus* 35B:103–109.

Sebacher, D.I., R.C. Harriss, and K.B. Bartlett. 1985. Methane emissions to the atmosphere through aquatic plants. *J. Env. Qual.* 14:40–46.

Sebacher, D.I., R.C. Harriss, K.B. Bartlett, S.M. Sebacher, and S.S. Grice. 1986. Atmospheric methane sources: Alaskan tundra bogs, an alpine fen, and a subarctic boreal marsh. *Tellus* 38B:1–10.

Seiler, W. 1984. Contribution of biological processes to the global budget of CH_4 in the atmosphere. In: Current Perspectives in Microbial Ecology, ed. M.J. Klug and C.A. Reddy, pp. 468–477. Washington, D.C.: Am. Soc. Microbiol.

Seiler, W., R. Conrad, and D. Scharffe. 1984. Field studies of methane emission from termite nests into the atmosphere and measurements of methane uptake by tropical soils. *J. Atmos. Chem.* 1:171–186.

Seiler, W., A. Holzapfel-Pschorn, R. Conrad, and D. Scharffe. 1984. Methane emission from rice paddies. *J. Atmos. Chem.* 1:241–268.

Strayer, R.F., and J.M. Tiedje. 1978. *In situ* methane production in a small, hypereutrophic, hard-water lake: loss of methane from sediments by vertical diffusion and ebullition. *Limnol. Ocean.* 23:1201–1206.

Svensson, B.H. 1980. Carbon dioxide and methane fluxes from the ombrotrophic parts of a subarctic mire. In: Ecology of a Subarctic Mire, ed. M. Sonesson. *Ecol. Bull.* 30:235–250.

Svensson, B.H., and T. Rosswall. 1984. *In situ* methane production from acid peat in plant communities with different moisture regimes in a subarctic mire. *Oikos* 43:341–350.

Tadesse, K., D. Smith, and M.A. Eastwood. 1980. Breath hydrogen (H_2) and methane (CH_4) excretion patterns in normal man and in clinical practice. *Q. J. Exp. Physiol.* 65:85–97.

Thathy, J.P., R. Delmas, B. Cros, A. Marenco, J. Levant, and M. Labat. 1988. Methane emissions from flooded forest in Central Africa. *EOS* 69:1066.

von Engelhardt, W., S. Wolter, H. Lawrenz, and J.A. Hemsley. 1970. Production of methane in two non-ruminant herbivores. *Comp. Biochem. Physiol.* 60:309–311.

Wainman, F.W., J.S. Smith, and P.J.S. Dewey. 1978. The predicted and observed metabolizable energy values of mixtures of maize silage and barley fed to cattle. Rowett Research Institute, Aberdeen, Scotland, pp. 55–58.

Whalen, S.C., and W.S. Reeburgh. 1989. A methane flux time-series for tundra environments. *Glob. Biogeochem. Cyc.*, in press.

Exchange of Trace Gases between Terrestrial Ecosystems and the Atmosphere
eds. M.O. Andreae and D.S. Schimel, pp. 229–246
John Wiley & Sons Ltd

Fluxes of NO_x above Soil and Vegetation

C. Johansson

Department of Meteorology
University of Stockholm
106 91 Stockholm, Sweden

Abstract. This review is concerned with measurements and modeling the exchange of NO_x. The importance of meteorological and biological factors for both uptake and emission is well recognized but consistent quantitative information on the effects of NO_x fluxes is lacking. Measurements have shown a strong dependence of the NO emission on a number of environmental variables including temperature and soil moisture. Largest emissions from natural, uncultivated ecosystems are reported for soils in tropical climates. There is no clear evidence of NO_x emission from plants grown under natural conditions.

There are very few field studies on NO_x deposition at the levels encountered in clean continental air. Data from many micrometeorological measurements show large scatter, which has been attributed to inhomogeneity due to local sources of NO_x, interference from other oxidized nitrogen compounds on NO_2 detection, and alteration of fluxes and concentrations by rapid photochemical interactions.

Dry deposition, occurring mainly as NO_2, is regulated by physiological variables including both stomatal and mesophyllic resistance. Chamber studies indicate that the uptake of NO_2 increases linearly with concentration, but recent measurements on pine trees indicate essentially zero uptake at levels found in clean continental air.

Modeling of atmosphere–surface exchange has been mainly restricted to gases with no surface source, i.e., SO_2 and O_3, which is due to lack of information on the effect of environmental variables on the NO_x exchange. These models identified parameterization of surface resistance as major control for exchange but one subject to large uncertainties. There is a need for more emission and deposition data coupled with simultaneous information on physiological and meteorological variables to allow more rigorous large-scale assessment of NO_x fluxes above plant–soil surfaces.

INTRODUCTION

Surface atmosphere exchange of NO_x constitutes a major pathway for input and removal of atmospheric nitrogen oxides. The exchange includes simultaneous emission from and dry deposition to soils and plants. Most measurements of NO_x exchange made so far have addressed only one of these processes at a time. As NO and NO_2 are rapidly interchanged in the atmosphere, it is the net flux of NO_x to or from a surface that is of interest in atmospheric chemistry. Knowledge of which factors control NO_x fluxes is a prerequisite to allow proper generalization of measurements taken in a limited area within an ecosystem during a short time period.

The purpose of this chapter is to review the current literature on NO_x exchange, to examine the main processes involved in both the uptake and emission by terrestrial surfaces, and to identify controversial issues and open questions.

METHODOLOGIES USED TO STUDY NO_x EXCHANGE

NO_x exchange has been studied on a range of scales from a few tenths of square meters on individual trees up to a few thousands of square meters of ground area. This discussion will be focused on the methodological problems encountered during measurements of NO_x exchange. A more detailed discussion of chamber methods is given by Mosier (this volume). The different micrometeorological methodologies and their advantages and disadvantages are discussed in detail by Fowler and Duyzer (this volume).

Chamber Methods

The majority of field measurements of NO_x exchange have utilized chambers of different designs and sizes, with or without mechanical stirring of the air inside the chambers. Open chamber systems have a continuous air flow-through, either of "dry zero" air or ambient air. Open chambers typically cover the plant/soil for several days or hours whereas closed chambers only cover the soil/plant intermittently during measurements with enough air flow-through for the analyzers.

There is very little information on how the different systems of operation relate. The effect of a flow-through chamber on environmental conditions is dependent on air flow rate, which should be adjusted to mimic the natural environment as close as possible. The effect of closed chambers depends on the time that the chamber covers the surface. Comparisons of chamber methods run in closed vs. open mode of operation have shown systematic differences depending on the air flow rate in the open system and the time of coverage during measurements using a closed system (Denmead 1979;

Johansson and Granat 1984). The net exchange of NO above the surface has been shown to be dependent on the concentration, so that for concentrations above a certain level (compensation point), there will be uptake of NO, whereas below there will be emission (Galbally and Roy 1978). It is therefore important that the concentration inside the chamber is as close as possible to the ambient level.

There are a number of other potential problems associated with chamber meaurements of NO_x exchange:

1. Uptake or artificial production of gases by chamber wall material (Johansson 1987; Galbally et al. 1987). This effect needs to be corrected for if the fluxes of NO or NO_2 are very small.
2. The aerodynamic mixing close to the surface where NO production occurs is altered. Johansson and Sanhueza (1988) found an error of up to 30% introduced into the flux measurements on a dry sandy soil by different aerodynamic mixing in the chamber compared to outside.
3. Photochemical reactions involving NO, NO_2, and O_3 occur on time scales similar to that of the residence time in the chambers (a few minutes). These reactions need to be corrected for as they introduce uncertainties if the fluxes of the individual gases NO and NO_2 are to be measured using an open chamber system (Galbally et al. 1987). They are not important if a closed system is used or if the net NO_x (NO plus NO_2) exchange is the desired quantity.

The importance of these three effects will depend on the chamber design and may differ from site to site depending on soil characteristics and meteorological conditions.

Micrometeorological Methods

Micrometeorological measurements are often preferred as they represent a minimum of disturbance to the surface and as the flux obtained represents an average over a large surface area. A restriction of most micrometeorological methods is that they can only be applied in carefully selected areas with uniform, level terrain.

As will be discussed further below, a number of problems have been encountered in micrometeorological measurements that specifically relate to NO_x exchange. These include (*a*) inhomogeneity due to local sources of NO_x, i.e., emission from soils and plants or in some cases local traffic (Hicks and Matt 1988), (*b*) interference from other oxidized nitrogen species on the NO_2 detection (measurements so far have used either thermal or chemical converters that are known to reduce a number of other oxidized N species (Fehsenfeld et al. 1987), and (*c*) modification of the fluxes and concentrations due to photochemical reactions in the atmosphere (Lenschow and Delany 1987).

Recently, Parrish et al. (1987) and Kaplan et al. (1988) derived the NO emission from measured nighttime gradients of NO and O_3 in the surface layer of the atmosphere. During steady state and with negligible horizontal divergence, the emission of NO from the soil surface should equal the integrated loss of NO due to the reaction with O_3. The method is limited to nighttime, when there are no atmospheric sources or sinks for NO other than the reaction with O_3.

Comparison of Different Methods

Only three papers report on comparisons of NO fluxes obtained from chamber measurements with fluxes calculated from measured nighttime vertical concentration gradients of O_3 and NO (Parrish et al. 1987; Kaplan et al. 1988; Johansson and Sanhueza 1988). Both Parrish et al. (1987) and Kaplan et al. (1988) found no systematic differences between the two methods and the fluxes agreed within 20%. The intercomparison by Parrish et al. (1987) covers fluxes ranging from 0.2 to 20 ng N m^{-2} s^{-1} under a variety of different soil conditions.

Johansson and Sanhueza (1988) derived fluxes from the two methods within 30%. Calculations indicated that part of the discrepancy could be due to different mixing in the chamber compared to ambient air. Taking one particular occasion as an example, the aerodynamic resistance in the chamber was estimated to be a factor of ten smaller than the aerodynamic resistance in ambient air during the night when the measurements were made (Johansson and Sanhueza 1988).

NO_x EMISSION FROM TERRESTRIAL ECOSYSTEMS

Controlling factors and mechanisms for NO_x production in soils are addressed in more detail elsewhere in this volume (see Galbally; Firestone and Davidson, both this volume). A short discussion is included for the completeness of this review.

Mechanism for NO_x Production in Soils

Laboratory studies have shown three mechanisms that may lead to production of NO in soils: microbial nitrification, denitrification, and chemodenitrification (chemical decomposition of nitrite). Photolysis of nitrite has been suggested as a mechanism for NO production in ocean surface waters (Zafiriou and McFarland 1981) and in flooded rice fields (Galbally et al. 1987).

The mechanism responsible for the NO or NO_2 production in the actual field situation has not in any case been clearly assessed. There is, however, circumstantial evidence that nitrification is the main process in many soils (Johansson and Granat 1984; Slemr and Seiler 1984; Anderson and Levine 1987; Colbourn et al. 1987; Kaplan et al. 1988; Johansson et al. 1988).

Factors Controlling Emissions from Soils

Temperature. Measurements of NO emission from agricultural land show a rather strong dependence of the NO emission on temperature, which is what one would expect for a microbiological process (Johansson and Granat 1984; Slemr and Seiler 1984; Anderson and Levine 1987). Similarly, Williams et al. (1987) observed a strong temperature dependence of the NO emission from uncultivated grassland in Colorado. Measurements on a very dry tropical savanna soil showed no clear diurnal variation despite a 20°C change in temperature (Johansson et al. 1988). Only after soil wetting, fertilization, or burning of the surface biomass (all three processes causing increased nutrient levels) could a clear covariation with temperature be observed. The reason could be that microbial activity was inhibited by the extremely dry conditions, that the mobility of nutrients for the bacteria was restricted due to the dry conditions, or, finally, that there is another (presumably abiological) process responsible for the NO emission during dry conditions.

Soil moisture. As indicated earlier, the NO_x emission from soils is strongly dependent on the moisture content of the soil. The emission decreases drastically as soil moisture approaches saturation (Johansson and Granat 1984; Anderson and Levine 1987; Williams et al. 1987). When the soil surface is suddenly wetted after a period of drought, the emission has been observed to increase by a factor of 4 to 20 (e.g., Johansson et al. 1988). Still, there is no simple relation between soil moisture content and emission rate.

Plant cover. The presence of plants has been shown to reduce the emission of NO and NO_2 (Johansson and Granat 1984; Slemr and Seiler 1984; Williams et al. 1987). Both NO and NO_2 may be absorbed by soils and plants (see section on MODELS OF NO_x EXCHANGE). In addition, one might expect different rates of NO production from plant-covered soils compared to bare soils as several physical and chemical factors change: (*a*) nutrient levels will be different as plants assimilate nitrate and ammonium and may deposit carbon-rich compounds in the root zone; (*b*) temperature and moisture may differ between plant-covered and bare soil sites. These factors have a strong impact on microbial nitrification and denitrification. Certain grasses and tree species may even be inhibitory to the nitrification process.

Soil nutrient content. Galbally and Roy (1981) report enhanced fluxes from grazed pasture compared to ungrazed. Several studies have shown increased NO emission rates following application of nitrate, ammonium, or urea (Johansson and Granat 1984; Slemr and Seiler 1984; Anderson and Levine 1987; Colbourn et al. 1987). Loss rates of fertilizer nitrogen as NO range between 0.02 and 5.4%, with the highest rates for urea and ammonium and the lowest for nitrate. The loss occurs within the first few days after application.

There is, however, very little information on how "natural" variations in nutrient levels affect NO production. Williams et al. (1988) reported a strong correlation between NO fluxes and soil nitrate level. They found no obvious correlation between NO emission and ammonium levels. Equally high emission rates were observed from a savanna soil despite four times higher levels of nitrate and ammonium during the wet season compared to the dry season (Johansson and Sanhueza 1988). However, in both cases the soil was very dry (2 and 0.5% for the wet and dry season, respectively) and it was suggested that the NO production was limited by the dry conditions as discussed earlier (see section on *Temperature*).

NO Emission Rates from Terrestrial Ecosystems

NO emission from soils may contribute a large fraction to the global sources of NO_x in the atmosphere. Tables 1 and 2 summarize earlier measurements of NO emission from terrestrial ecosystems and represent an updated version of Table 4 in Johansson et al. (1988). Most measurements have been performed on cultivated soils where the emission rates varied by more than an order of magnitude from site to site (Table 1). In some measurements there was a net uptake of NO. Although part of this variation has been attributed to the environmental factors discussed above (mainly soil moisture and temperature), most of it was associated with the application of nitrogen fertilizers, which may increase the flux by two orders of magnitude.

As regards uncultivated, natural soils, the highest emission rates are reported for tropical ecosystems (Table 2). The measurements in the tropical rain forest of Brazil (Kaplan et al. 1988) and the savanna in Venezuela (Johansson et al. 1988; Johansson and Sanhueza 1988) show mean fluxes of a factor 3 to 160 times larger than fluxes observed in temperate ecosystems. The largest emission rates have been observed in the savanna during the rainy season, with emission rates immediately following rain of up to 250 ng N m^{-2} s^{-1} (Johansson and Sanhueza 1988). The estimated NO emission from the soil during the whole rainy season exceeded the input of nitrogen as wet deposition of nitrate by at least a factor of 2. It should be noted, however, that recent data at a different site in the savanna have shown much lower emission rates than those described here (E. Sanhueza, pers.

comm.). Soil analyses have shown an order of magnitude lower nitrate and ammonium concentrations at this site compared to that studied by Johansson et al. (1988) and Johansson and Sanhueza (1988).

NO_2 Emission

Most measurements show $NO_2^{(1)}$ production to be only a small fraction of the NO production (Galbally and Roy 1978; Johansson and Granat 1984; Johansson and Galbally 1984; Williams et al. 1987; Johansson et al. 1988). Only two studies report on substantial emissions of NO_2 from soils (Slemr and Seiler 1984; Colbourn et al. 1987). Slemr and Seiler (1984) found very large NO_2 emission rates from bare soil in Spain and postulated that the NO_2 emission was due to chemodenitrification (see also Galbally, this volume). The NO_2 emission was correlated with light intensity rather than temperature and was substantially less from plant covered soil, indicating a significant plant uptake of NO_2. Colbourn et al. (1987) suggested that the NO_2 they detected was due to oxidation of NO either in the soil air or in their chamber system.

NO_x Emission from Plants

Nitrogen compounds may be emitted from plants, either in reduced form as NH_3 or amines, or in oxidized form as NO_x or N_2O (Farquhar et al. 1983). Loss as reduced N seems to be the most quantitatively important pathway for plants. Klepper (1979) found substantial amounts of NO as well as some NO_2 emitted from herbicide treated plants. The herbide causes accumulation of nitrite in the cell tissue, which may be reduced to NO and NO_2. Nitrite levels during natural conditions are much lower, and it seems to be an open question whether oxidized N compounds are released from plants grown under natural conditions.

NO_x DEPOSITION TO TERRESTRIAL SURFACES

Mechanism for Uptake of NO_x by Plants

The exact mechanism for uptake of NO or NO_2 by plants is not known. Rogers et al. (1979) showed that $^{15}NO_2$ was incorporated into reduced nitrogen compounds of plants. The NO_2 absorbed formed nitrite, which was assimilated via nitrite reductase. Other studies have demonstrated a stimulating effect of NO_2 on the nitrite and nitrate reductase activity (e.g., Wingsle et al. 1987).

(1) The NO_2 detector may in many cases include organic nitrates and HNO_2.

TABLE 1. Measurements of NO emissions from cultivated soils.

Climate/System Location	Past use	Time period	Flux, ng N m^{-2} s^{-1}		Reference
			Range	Mean	
Temperate/Grassland Australia	grazed	Nov. to May	1.5–7.3	3.5	Galbally & Roy (1978)
Temperate/Cropland Sweden	unfertilized	Apr. to Sept.	0.3–17	0.6[a]	Johansson & Granat (1984)
Temperate/Cropland Sweden	fertilized	Apr. to Sept.	0.1–62	1.9[a]	ibid.
Subtropical/Cropland Spain	fertilized	Sept. to Oct.	−2–250	–	Slemr & Seiler (1984)
Temperate/Grassland Australia	grazed	April	0.6–124	–	Galbally et al. (1985)
Temperate/Cropland United States	–	June to July	−9–28	–	Delany et al. (1986)
Temperate/Flooded rice Australia	fertilized	Dec.	<0.2–0.95	–	Galbally et al. (1987)
Temperate/Cropland United States	unfertilized	May to June	0.003–67	1.7[a]	Anderson & Levine (1987)
Temperate/Cropland United States	fertilized	1 year	–	6.7[a]	ibid.
Temperate/Pasture United Kingdom	fertilized	July to Aug.	0–36	8	Colbourn et al. (1987)
Temperate/Sward United Kingdom	unfertilized	July to Aug.	−12–26	0.5[b]	ibid.
Temperate/Cropland United States	unfertilized	Aug.	0.2–3.8	1.2	Williams et al. (1988)
Temperate/Cropland United States	fertilized	Aug.	1.6–338	94	ibid.

[a] Weighted yearly average
[b] Calculated as NO_x–N (i.e., includes uptake of NO_2)

TABLE 2. Measurements of NO emission from natural ecosystems.

Climate/System Location	Time period	Flux, ng N m^{-2} s^{-1}		Reference
		Range	Mean	
Temperate/Grassland Australia	November	0.6–2.6	1.6	Galbally & Roy (1978)
Temperate/Coniferous forest, Sweden	June to Sept.	0.1–0.8	0.4[c]	Johansson (1984)
Temperate/Grassland Germany	August	−6–14	2.2	Slemr & Seiler (1984)
Temperate/Grassland United States	Aug. to Nov.	0.03–65	3	Williams et al. (1987)
Tropical/Evergreen rainforest, Brazil	July to Aug.[a]	9.2–16	11	Kaplan et al. (1988)
Tropical/Savanna Venezuela	Feb.[a]	3–15	8	Johansson et al. (1988)
Tropical/Cloud forest Venezuela	Feb.	0.1–2	0.5	ibid.
Temperate/Mixed forest, United States	July	0.2–4.1	1.2	Williams et al. (1988)
Tropical/Savanna Venezuela	Oct.[b]	2–250	10–65[d]	Johansson & Sanhueza (1988)

[a] Dry season
[b] Rainy season
[c] Estimated weighted mean for growing season
[d] Estimated mean for rainy season based on two different sites

Plants also assimilate NO, but several measurements have shown that the rate of uptake of NO is only a small fraction of that for NO_2 (e.g., Hill 1971; Galbally et al. 1985; Johansson 1987). As NO normally constitutes only on the order of 10% of the total amount of NO_x in the atmosphere, dry deposition of NO on plants can be neglected as far as removal of atmospheric NO_x is concerned.

Controls of NO_x Deposition to Plants

The uptake of NO_2, as well as of other trace gases, is usually characterized and modeled by the dry deposition velocity, v_d, defined as the flux divided by the concentration at some reference height above the surface. The deposition velocity is usually determined experimentally and depends both on meteorological conditions and surface and trace gas characteristics. To be able to apply deposition velocities under a variety of conditions, it is useful to interpret the deposition velocity as the inverse of an overall resistance to transport to the surface consisting of several consecutive resistances in series:

$$1/v_d = r_a + r_b + r_s$$

where r_a is the aerodynamic resistance to turbulent transport in the air above the surface, r_b is the boundary layer resistance to diffusion through the stagnant air layer close to the surface, and r_s is the resistance characteristic of the surface itself. For NO and NO_2, r_a and r_b usually constitute a small fraction of the total resistance; only during very calm or stable conditions might the influence on the deposition velocity become important. Therefore, this discussion will be limited to the characteristics of the surface resistance, which in turn consists of a resistance to uptake by plants through the stomata or onto the surface of the leaves and uptake on soil and litter.

The stomatal resistance has a major impact on the rate of uptake of NO_2 by plants, as demonstrated in studies on alfalfa (Hill 1971), soybean plants (Rogers et al. 1977; Wesley et al. 1982), and pine trees (Bengtsson et al. 1982; Johansson 1987). The degree of stomatal aperture shows a diurnal and seasonal cycle due to the strong influence of light intensity, leaf water potential, water vapor pressure deficit, and, to a minor degree, temperature.

Several studies have shown the existence of an internal or mesophyllic resistance that may contribute a substantial fraction to the total resistance. This resistance reflects either different path lengths for the diffusion of water vapor molecules compared to NO_2 and/or a metabolic/chemical limitation of the rate of uptake of NO_2. The mesophyllic resistance constituted up to 60% of the total surface resistance to uptake of NO_2 by pine and seemed to be independent of the stomatal resistance and NO_2 concentration (Johansson 1987). However, Bengtsson et al. (1982), also studying uptake by pine, found a mesophyllic resistance only following a period of NO_2 exposure at very high levels (around 100 ppbv). For soybeans, a value of at least 40% may be calculated based on the assumption that O_3 is solely removed via the stomata of the leaves (Wesley et al. 1982). The mesophyllic resistance derived in the latter study might be too high since the method cannot distinguish between deposition and emission of NO or NO_2 to/from plants or soil.

It is interesting to note that there seems to be no uptake on the cuticula of the leaves (needles). For other gases (HNO_3, SO_2) this uptake may be substantial (Fowler 1984; Huebert and Roberts 1985). However, data on the uptake by plants at higher concentrations (hundreds of ppbv NO_2) indicate an uptake on the surface of the leaves, in addition to the flux through stomata (see discussion by Farquhar et al. 1983). These results suggest we view with caution the extrapolation of measurements obtained at concentrations higher than those normally encountered in the atmosphere.

Furthermore, measurements on shoots of pine trees have shown that the uptake of NO_2 was linearly dependent on the concentration, but was essentially zero at concentrations around a few ppbv, indicating a compensation point of between 0.5 and 3 ppbv (Johansson 1987). A similar behavior has been observed earlier for exchange of NH_3 with plants (Farquhar et al. 1980).

NO_x Deposition Rates

There are very few studies on NO_x deposition at levels normally encountered in clean continental air (see recent review by Voldner et al. 1986). Most field studies have utilized micrometeorological techniques. Wesley et al. (1982), using an eddy correlation technique above a soybean field, reported deposition velocities ranging from around 0.5 mm s^{-1} at night up to 5.6 mm s^{-1} during the daytime. The range of surface resistances remaining after the influence of atmospheric mixing (aerodynamic resistance) is subtracted was 1.3 to 15 s cm^{-1}, the former value corresponding to daytime and the latter to night. A similar range (0.8 to 5.6 s cm^{-1}) was found above a cut grass surface by Delany and Davies (1983). The deposition velocities reported by Wesley et al. (1982) might be too large since the detection may have included some HNO_3, which would be deposited much more rapidly than NO_2 (Huebert 1983).

Hicks et al. (1983) found both positive and negative fluxes of NO_2 above a deciduous forest. On some occasions the deposition velocity approached the maximum allowed value permitted by turbulent transport. Subsequent measurements indicated that HNO_3 might have interfered with some of the NO_2 measurements. No simple relation between the flux and the concentration could be deduced from their data. Dollard et al. (1986) found no significant gradients for either NO or NO_2 above grass. Using an eddy correlation technique, Delany et al. (1986) report both positive and negative fluxes of NO_x to grass. Deposition dominated in the morning as the site experienced a diluted urban pollution plume and emission dominated in the afternoon as relatively clean air reached the site.

Results from many of the micrometeorological measurements of NO_x fluxes show much larger scatter than similar measurements of SO_2 and O_3 fluxes (Hicks and Matt 1988). This is probably related to the specific problems with micrometeorological measurements of NO_x exchange as discussed above (see section on MICROMETEOROLOGICAL METHODS) and to the fact that NO_x fluxes seem to be smaller than fluxes of SO_2 and O_3. The exchange of NO_x includes simultaneous emission and deposition. Up to a factor of 10, larger surface resistances for NO_2 compared to that for SO_2 and O_3 were recently reported for uptake by pine trees (Johansson 1987). The reason for the different uptake rates of NO_2 versus SO_2 and O_3 is not known.

NO_2 uptake during winter by pine trees and by snow has been shown to be small (see also review by Voldner et al. 1986). Deposition velocities of NO_2 to various soils are in the range 0.3 to 0.8 cm s^{-1}. NO is also absorbed by soils, mainly through microbial rather than physical processes (Johansson and Galbally 1984), but the rate of uptake is much lower than for NO_2. Both NO and NO_2 are sparingly soluble in water and the rate of uptake by surface waters should be very slow if there are no further reactions in the aqueous phase (Lee and Schwartz 1981). Accordingly, very small rates of deposition have been observed to fresh- and seawater (van Aalst 1982). However, Galbally et al. (1987), studying NO_x exchange above a flooded rice field, observed much more rapid uptake of NO_2 by floodwater than physical dissolution would suggest. From their data they derived a deposition velocity of around 3×10^{-4} m s^{-1} and suggested that NO_2 is efficiently scavenged by reactions occurring in the floodwater.

MODELS OF NO_x EXCHANGE

Many chamber studies of NO exchange have shown both emission and uptake by the soil/plant system. Galbally et al. (1985) examined two models for NO exchange above a grazed pasture. In the first, the net NO emission was treated as a freely evaporating surface with the difference in concentration between the atmosphere and an equilibrium concentration within the soil air pores as the driving force. However, this model failed to describe properly the features of data on NO exchange. In the other approach examined, the net flux was calculated as the difference between an emission flux and an uptake:

$$F = E - [NO]_s \, r_s^{-1}$$

where r_s is the surface resistance and $[NO]_s$ is the concentration of NO close to the surface. This model provided a good fit to data obtained by chamber measurements, indicating that the NO exchange consists of two separate processes: a deposition proportional to the NO level in the air and an emission that depends on the production and consumption in the soil, and on physical factors regulating the diffusion into the atmosphere. A similar conclusion was reached by Delany et al. (1986), who applied eddy correlation measurements above a wheat grass surface. The "compensation point" ($E \, r_s$) is generally higher than the ambient levels, i.e., there is usually a net emission of NO.

Many canopy dry deposition models are based on the resistance analog and treat the canopy as one single leaf with dry deposition occurring on one side of the leaf ("big-leaf" models), either in a single layer (Hicks et al. 1987) or in several layers within a canopy (Baldocchi 1988). These models do not directly apply to NO_x exchange since they only address gases without a surface source; however, the treatment of the dry deposition process will be the same.

As indicated earlier (see section on CONTROLS OF NO_x DEPOSITION TO PLANTS), one of the most critical parameters in the estimation of dry deposition is the surface resistance, which for many trace gases (SO_2, O_3, and to some extent also NO_2) is highly correlated with the stomatal resistance. Some recent models include empirical relationships to take into account the influence on the stomatal resistance of photosynthetically active radiation (PAR), temperature, humidity, and plant water status (Hicks et al. 1987; Baldocchi 1988; Meyers 1987). The stomatal resistance is derived according to the following expression:

$$r_s = r_{s,min}(1 + \beta/I)\, D_w\, F_T\, D_g$$

where $r_{s,min}$ is a minimum stomatal resistance and β is a light response factor, both parameters being dependent on plant species. F_T, D_w, and D_g are functions to correct for the influence on the stomatal resistance of temperature and water vapor deficit and a correction for the different diffusivity of the gas and water vapor, respectively. In the above expression, the stomatal resistance is calculated on a leaf area basis. An important problem concerns the scaling of this resistance from the leaf level to the whole canopy. For tall canopies like forests it is necessary to consider the variation of stomatal resistance within the canopy. One main factor that controls stomatal resistance is PAR, which varies with height in the canopy. Hicks et al. (1987) calculated the sunlit and shaded leaf fractions based on the assumption that the spatial leaf distribution is random and that the leaf inclination distribution is spherical.

Canopy deposition models based on either higher order closure theory or "K-theory" have shown the modeled deposition velocities (for SO_2) to be very sensitive to the parameterization used for the stomatal resistance, i.e., the values of $r_{s,min}$ and β (Meyers 1987; Baldocchi 1988). The leaf boundary layer resistance comprised only a small fraction of the total resistance and had therefore a minor influence on the dry deposition velocity calculated by the model (Meyers 1987). A factor of 4 decrease of $r_{s,min}$ caused up to a factor of 2 increase in v_d and a similar sensitivity was encountered for β (Baldocchi 1988). Both of these parameters may vary within the canopy as they are dependent on the age of the leaves. Furthermore, the minimum stomatal resistance may also depend on temperature and water stress.

As NO_2 has even higher surface resistance than SO_2, similar conclusions should hold for dry deposition of NO_2 as well. For NO_2, the surface resistance should include a mesophyllic resistance, which may constitute a large fraction of the total surface resistance (see above).

There is a need to characterize the NO_x emission rates in terms of meteorological, chemical, and biological variables in order to verify model calculations with measurements on soils under different conditions.

An understanding of the exchange of NO_x within a tall canopy needs to include the vertical distribution of NO and NO_2 levels and the distribution of sources and sinks for these compounds. The presence of a tall canopy will alter the photochemical equilibrium between O_3, NO, and NO_2; the photolysis of NO_2 will vary with height in the canopy. The reactions involved are:

$$NO + O_3 \rightarrow NO_2 + O_2$$

$$NO_2 + h\nu \rightarrow NO + O$$

$$O + O_2 + M \rightarrow O_3 + M$$

These reactions form a photostationary state, where the product of the concentration of NO and O_3 divided by the concentration of NO_2 is constant. As these reactions occur on the same (or even shorter) time scale as that of turbulent transport, they may have a substantial impact on the distribution of NO and NO_2 in the atmospheric surface layer.

Many observations show substantial deviations from the photostationary state set up by the reactions described above. Trainer et al. (1987) showed that this may be due to oxidation of NO by oxidized peroxyradicals, which are formed in high enough levels from naturally emitted hydrocarbons. The time scales for the reaction of NO with CH_3O_2, for example, is on the order of a few minutes (i.e., assuming a CH_3O_2 concentration of 20 pptv). The importance of these reactions for measurements of NO_x exchange has not yet been addressed. Another potentially important reaction during night is the formation of NO_3, which in turn rapidly reacts with hydrocarbons or NO_2, ultimately forming organic nitrates and HNO_3, respectively. Although the timescale for this reaction is much longer than for the reactions mentioned above (on the order of 10 to 15 hours), it may be important in, for example, coniferous forests where the surface removal of NO_2 during the night may be a very slow process (Johansson 1987).

CONCLUDING DISCUSSION

As the processes of deposition and emission are controlled by different physical and biological factors, it is important to know the contribution to the net NO_x flux from each individual gas in order to generalize results and to extrapolate to other areas and other seasons. Emission from soils seems to occur mainly as NO, whereas dry deposition of NO_x is dominated by NO_2. The net flux of NO_x may then be approximated as

$$F_{NO_x} = F_{NO} - V_d\,[NO_2]_s$$

where $[NO_2]_s$ is the NO_2 concentration at some reference height above the surface. For a canopy, the dry deposition velocity would consist mainly of a stomatal and internal resistance (i.e., $1/v_d = r_s + r_i$) and a compensation point for NO_2 may need to be considered. Emission rates from natural (uncultivated) soils typically range from 1 to 10 ng N m^{-2} s^{-1} (Table 2). There is very little information on the rate of uptake of NO_2. If we, as a first approximation, neglect the possible existence of a compensation point for NO_2 and use the daytime deposition velocities of 5.6 mm s^{-1}, as determined by Wesley et al. (1982), and NO_2 concentrations in clean continental air of 0.5 ppbv to 5 ppbv, then the dry deposition should range from 2 to 16 ng N m^{-2} s^{-1}, which is similar to the range of NO emissions. During many of the micrometeorological measurements, NO_2 levels were high enough for deposition to dominate over emission (Wesley et al. 1982; Delany et al. 1986). Most measurements show only slowly varying NO emission rates whereas atmospheric NO_2 levels may exhibit large variations resulting in net fluxes that may fluctuate from net emission to net deposition. Indications of such a pattern were obtained over a grass surface by Delany et al. (1986).

Opposing fluxes, which may be of the same order of magnitude, complicate measurements of the individual fluxes of NO and NO_2, since they are rapidly interconverted in photochemical reactions in the atmosphere. Both when micrometeorological techniques are applied and when flow-through chamber systems are utilized, these reactions need to be corrected for in order to obtain an estimate of the fluxes of the individual gases (Duyzer et al. 1983; Lenschow and Delany 1987; Galbally et al. 1987).

The estimation of NO_x fluxes relies on accurate measurements of NO and NO_2 concentrations. Usually NO_2 is thermally or chemically converted to NO, which is measured with a chemiluminescence analyzer. An important problem concerning the results of these measurements is that they suffer from severe interferences from other oxidized nitrogen compounds, such as HNO_3, HNO_2, PAN, or other organic nitrates (Fehsenfeld et al. 1987). Only a few measurements have used the more specific photolytic converter to convert NO_2 to NO (Williams et al. 1987; Galbally et al. 1987).

In order to make more reliable estimates of fluxes on regional and global scales, quantitative and qualitative information is needed to characterize the surface resistance of NO_2 and the emission of NO (and maybe other oxidized odd nitrogen species) in terms of the influence of biological, meteorological, and chemical factors. Such information would enable the development of models for predictive purposes, permitting more realistic extrapolations of data to obtain regional fluxes on annual or seasonal scales.

Acknowledgements. I would like to thank Ian Galbally, CSIRO Division of Atmospheric Research, and Arvin Mosier, USDA-ARS, Fort Collins, for reviewing the manuscript.

REFERENCES

Anderson, I.C., and J.S. Levine. 1987. Simultaneous field measurements of nitric oxide and nitrous oxide. *J. Geophys. Res.* 92:965–976.

Baldocchi, D. 1988. A multi-layer model for estimating sulphur deposition to a deciduous oak forest canopy. *Atmos. Envir.* 22:869–884.

Bengtsson, C., P. Grennfelt, C.-Å. Boström, E. Troéng, L. Skärby, Å. Sjödin, and K. Petersson. 1982. Deposition and uptake in Scots pine needles (Pinus sylvestris L.). Report IVL-B 647. Gothenburg, Sweden: Institute for Water and Air Pollution Research.

Colbourn, P., J.C. Ryden, and G.J. Dollard. 1987. Emission of NO_x from urine-treated pasture. *Envir. Poll.* 46:253–261.

Delany, A.C., and T.D. Davies. 1983. Dry deposition of NO_x to grass in rural East Anglia. *Atmos. Envir.* 17:1391–1394.

Delany, A.C., D.L. Fitzgerrald, D.H. Lenschow, R. Pearson, Jr., G.J. Wendel, and B. Woodruff. 1986. Direct measurements of nitrogen oxides and ozone fluxes over grasslands. *J. Atmos. Chem.* 4:429–444.

Denmead, O.T. 1979. Chamber systems for measuring nitrous oxide emission from soils in the field. *Soil Sci. Soc. Am. J.* 43:89–95.

Dollard, G.J., T.J. Davies, and P.J.C. Lindstrøm. 1986. Measurements of the dry deposition rates of some trace gas species. In: Physico-chemical Behaviour of Atmospheric Pollutants, ed. G. Angletti and G. Restelli, pp. 470–476. Dordrecht: Reidel.

Duyzer, J.H., G.M. Meyer, and R.M. van Aalst. 1983. Measurements of dry deposition velocities of NO, NO_2 and O_3 and the influence of chemical reactions. *Atmos. Envir.* 17:2117–2120.

Farquhar, G.D., P.M. Firth, R. Wetselaar, and B. Weir. 1980. On the gaseous exchange of ammonia between leaves and their environment: determination of the ammonia compensation point. *Plant Physiol.* 667:710–714.

Farquhar, G.D., R. Wetselaar, and B. Weir. 1983. Gaseous nitrogen losses from plants. In: Gaseous Loss of Nitrogen from Plant-soil Systems, ed. J.R. Freney and J.R. Simpson, pp. 158–180. The Hague: M. Nijhoff/W. Junk.

Fehsenfeld, F.C., R.R. Dickerson, G. Hubler, et al. 1987. A ground based intercomparison of NO, NO_x, NO_y measurement techniques. *J. Geophys. Res.* 92:14710–14722.

Fowler, D. 1984. Transfer to terrestrial surfaces. *Phil. Trans. R. Soc. Lon. B* 305:281–297.

Galbally, I.E, J.R. Freney, W.A. Muirhead, J.R. Simpson, A.C. Trevitt, and P.M. Chalk. 1987. Emission of nitrogen oxides (NO_x) from a flooded soil fertilized with urea: relation to other nitrogen loss processes. *J. Atmos. Chem.* 5:343–365.

Galbally, I.E., and C.R. Roy. 1978. Loss of fixed nitrogen by nitric oxide exhalation. *Nature* 275:734–735.

Galbally, I.E., and C.R. Roy. 1981. Ozone and nitrogen oxides in the Southern Hemisphere troposphere. In: Proceedings of the Quadrennial International Ozone Symposium, Boulder, Colorado, pp. 431–438.

Galbally, I.E., C.R. Roy, C.M. Elsworth, and H.A.H. Rabich. 1985. The measurement of nitrogen oxide (NO, NO_2) exchange over plant/soil surfaces. Technical paper no. 8, pp. 1–23. Aspendale, Australia: CSIRO Div. of Atmospheric Research.

Hicks, B.B., D.D. Baldocchi, T.P. Meyers, R.P. Hosker, Jr., and D.R. Matt. 1987. A preliminary multiple resistance routine for deriving deposition velocities from measured quantities. *Wat. Air Soil Poll.* 36:330–349.

Hicks, B.B., and D.R. Matt. 1988. Combining biology, chemistry and meteorology in modeling and measuring dry deposition. *J. Atmos. Chem.* 6:117–131.

Hicks, B.B., D.R. Matt, R.T. McMillen, J.D. Womach, and R.E. Shetter. 1983. Eddy fluxes of nitrogen oxides to a deciduous forest in complex terrain. In: The Meteorology of Acid Deposition, ed. P.J. Sampson, pp. 189–201. Pittsburgh: APCA.

Hill, A.C. 1971. Vegetation: a sink for atmospheric pollutants. *J. Air. Poll. Cont. Assn.* 21:341–346.

Huebert, B.J. 1983. Comment on: An eddy correlation measurement of NO_2 flux to vegetation and comparison with O_3 flux. *Atmos. Envir.* 17:1600.

Huebert, B.J., and C.H. Roberts. 1985. The dry deposition of nitric acid to grass. *J. Geophys. Res.* 90:2085–2090.

Johansson, C. 1987. Pine forest: a negligible sink for atmospheric NO_x in rural Sweden. *Tellus* 39B:426–438.

Johansson, C., and I.E. Galbally. 1984. Production of nitric oxide in loam under aerobic and anaerobic conditions. *Appl. Env. Microbiol.* 47:1284–1289.

Johansson, C., and L. Granat. 1984. Emission of nitric oxide from arable land. *Tellus* 36B:25–37.

Johansson, C., H. Rodhe, and E. Sanhueza. 1988. Emission of NO in a tropical savanna and a cloud forest during the dry season. *J. Geophys. Res.* 93:7180–7192.

Johansson, C., and E. Sanhueza. 1988. Emission of NO from savanna soils during rainy season. *J. Geophys. Res.* 93:14193–14198.

Kaplan, W.A., S.C. Wofsy, M. Keller, and J.M. da Costa. 1988. Emission of NO and deposition of O_3 in a tropical forest system. *J. Geophys. Res.* 93:1389–1395.

Klepper, L. 1979. Nitric oxide (NO) and nitrogen dioxide (NO_2) emission from herbicide-treated soybean plants. *Atmos. Envir.* 13:537–542.

Lee, Y.-N., and S.E. Schwartz. 1981. Evaluation of the rate of uptake of nitrogen dioxide by atmospheric and surface liquid water. *J Geophys. Res.* 86:11971–11983.

Lenschow, D.H., and A.C. Delany. 1987. An analytical formulation of NO and NO_2 flux profiles in the atmospheric surface layer. *J. Atmos. Chem.* 5:301–309.

Meyers, T.P. 1987. The sensitivity of modelled SO_2 fluxes and profiles within and above a vegetative canopy to leaf stomatal and boundary layer resistances. *Wat. Air Soil Poll.* 35:261–278.

Parrish, D.D., E.J. Williams, D.W. Fahey, S.C. Lui, and F.C. Fehsenfeld. 1987. Measurements of nitrogen oxides fluxes from soils: intercomparison of enclosure and gradient techniques. *J. Geophys. Res.* 92:2165–2171.

Rogers, H.H., H.E. Jeffries, E.P. Stahel, W.W. Heck, L.A. Ripperton, and A.M. Witherspoon. 1977. Measuring air pollutant uptake by plants: a direct kinetic technique. *J. Air Poll. Cont. Assn.* 27:1192–1197.

Rogers, H.H., H.E. Jeffries, and A.M. Witherspoon. 1979. Measuring air pollutant uptake by plants: nitrogen dioxide. *J. Env. Qual.* 8:551–557.

Slemr, F., and W. Seiler. 1984. Field measurements of NO and NO_2 emission from fertilized and unfertilized soils. *J. Atmos. Chem.* 2:1–24.

Trainer, M., E.Y. Hsie, S.A. McKeen, R. Tallamraju, D.D. Parrish, F.C. Fehsenfeld, and S.C. Liu. 1987. Impact of natural hydrocarbons on hydroxyl and peroxy radicals at a remote site. *J. Geophys. Res.* 92:11879–11894.

van Aalst, R.M. 1982. Dry deposition of NO_x. In: Air Pollution by Nitrogen Oxides, ed. T. Schneider and L. Grant, pp. 263–270. Amsterdam: Elsevier.

Voldner, E.C., L.A. Barrie, and A. Sirois. 1986. A literature review of dry deposition of oxides of sulphur and nitrogen with emphasis on long range transport modelling in North America. *Atmos. Envir.* 20:2101–2123.

Weseley, M.L., J.A. Eastman, D.H. Stedman, and E.D. Yalvac. 1982. An eddy

correlation measurement of NO_2 flux to vegetation and comparison with O_3 flux. *Atmos. Envir.* 16:815–820.
Williams, E.J., D.D. Parrish, M.P. Buhr, and F.C. Fehsenfeld. 1988. Measurements of soil NO_x emissions in Central Pennsylvania. *J. Geophys. Res.* 93:9539–9546.
Williams, E.J., D.D. Parrish, and F.C. Fehsenfeld. 1987. Determination of nitrogen oxide emissions from soils: results from a grassland site in Colorado, United States. *J. Geophys. Res.* 92:2173–2179.
Wingsle, G., T. Näsholm, T. Lundmark, and A. Eriksson. 1987. Induction of nitrate reductase in needles of Scots pine seedlings by NO_x and NO_3. *Physiol. Plant.* 70:399–403.
Zafiriou, O.C., and M. McFarland. 1981. Nitric oxide from nitrite photolysis in the central equatorial Pacific. *J. Geophys. Res.* 86:3173–3182.

Standing, left to right:
Christer Johansson, Leif Klemedtsson, Arvin Mosier, Franz Meixner, Helmut Schütz, Jürgen Kesselmeier

Seated, left to right:
Martin Wahlen, Tom Denmead, Peter Vitousek, Lucas Stal, David Fowler

Exchange of Trace Gases between Terrestrial Ecosystems and the Atmosphere
eds. M.O. Andreae and D.S. Schimel, pp. 249–261
John Wiley & Sons Ltd

Group Report
What Are the Relative Roles of Biological Production, Micrometeorology, and Photochemistry in Controlling the Flux of Trace Gases between Terrestrial Ecosystems and the Atmosphere?

P.M. Vitousek, Rapporteur
O.T. Denmead
D. Fowler
C. Johansson
J. Kesselmeier
L. Klemedtsson
F.X. Meixner
A.R. Mosier
H. Schütz
L.J. Stal
M. Wahlen

INTRODUCTION

Discussions within the group focused on the relative roles of biological production and consumption, micrometeorology, and photochemistry in the transfer of trace gases between vegetated surfaces and the atmosphere. Both the composition of the group and its charge precluded a serious discussion of photochemical sinks for trace gases in the stratosphere or troposphere. Within the range of interactions considered, we concluded that biological processes are overwhelmingly dominant in the control of N_2O, CH_4, and NO_x fluxes. For reactive gases (i.e., NO itself), micrometeorological and chemical processes can modify the extent to which biological processes control net exchanges between terrestrial surfaces and the atmosphere. Additionally, we concluded that the lack of micrometeorological methods for measuring integrated fluxes of N_2O and CH_4 limits our ability to understand their dynamics, and a dearth of information on the biological mechanisms involved in NO_x uptake limits our understanding of NO_x exchange.

In this report, we summarize the information and assumptions that underlie these conclusions. We describe constraints on methods for measuring trace gas fluxes, the importance of plants in modifying fluxes from soil, and the dynamics of NO_x within and above vegetation canopies as they are influenced by physiological mechanisms, photochemistry, and patterns of air movement. We conclude with a brief discussion of possible changes in the relative roles of these processes on a changing Earth and a discussion of what now limits our ability "to analyze the biological and physicochemical regulation of trace gas exchange for modeling on local to global scales, and to evaluate the importance of trace gas exchange to ecology, climate, and atmospheric chemistry."

ADVANTAGES AND DISADVANTAGES OF TECHNIQUES

In an ideal world, trace gas fluxes would be determined through an integrated set of measurements. Well-designed enclosures would be used to develop an understanding of the mechanisms controlling trace gas flux. Models would be used to summarize, predict, and integrate fluxes over space and time; such models would also feed back to the design of chamber-based experiments. The accuracy of the models in integrating fluxes over space and time would be determined from stationary (tower-based) flux measurements; coarser spatial scales would be analyzed by aircraft-based flux measurements. Both of these would out of necessity feed back to affect further development of models and of process measurements.

The current world offers a much richer spectrum of problems. For the longer-lived trace gases N_2O and CH_4, chambers have been the only choice for both process-level measurements and flux estimates. No sufficiently sensitive or fast-responding sensors have been available for either gas, although this is changing rapidly for CH_4. For NO_x, both atmospheric and chamber measurements have been possible—and as discussed later, the combination of both together has yielded much more useful information than either alone.

The reliance on chamber techniques for N_2O and CH_4 has made comparisons of results from different studies rather difficult. The micrometeorological community is relatively small and interactive. Its members know that they are working with difficult systems, and they have developed a more or less agreed upon standard for measurements. In contrast, chambers are deceptively simple. Moreover, there are many different kinds that are useful in different situations (Mosier, this volume), and practical considerations may be a controlling factor in important environments.

We decided that it would be useful to make a checklist of factors to consider in the selection of chambers for particular field applications. The list is drawn from a number of earlier studies (cf. Denmead 1979; Jury et al. 1982; Mosier, this volume). It includes:

1. Is the chamber material inert vis-à-vis the species of interest?
2. How uniform are gas concentrations within the chamber? Does the air within it need to be mixed?
3. How does the chamber affect the surface boundary layer resistance?
4. For closed chambers, do gas concentrations accumulate to the point where they could affect net gas flux?
5. Does the chamber modify the pressure gradient from atmosphere to soil? For closed chambers, does the absence of pressure fluctuations alter the flux? For open chambers, does circulation induce a pressure gradient?
6. How does the chamber alter the energy budget of the area enclosed? Does temperature change within chambers during the course of measurements?
7. Are soil moisture or nutrient concentrations within the chamber altered during (or between) measurements?
8. If plants are contained within the chamber, how do they function as compared with unconfined plants?
9. If plants are not included, has the possibility of flux through (or affected by) plants been considered?

Many of these problems can be minimized by using chambers only for a short time, although that itself can introduce other problems. The point of this list is not to suggest that conditions in chambers must reproduce exactly the conditions outside chambers; that is impossible. Rather, changes that alter fluxes of the gas of interest should be recognized and minimized.

Recently, there have been substantial advances in the development of sensors that allow micrometeorological measurements of CH_4 flux. A fast-response, high precision He–Ne laser detection system (C.E. Kolb, pers. comm.) has been used to perform eddy correlation measurements of CH_4 fluxes using towers as a platform. This laser system has low power requirements and is sufficiently robust for use in remote areas. For example, it was operated for thirty days in a remote tent camp in the Alaskan tundra during the NASA Arctic Boundary Layer Expedition (ABLE-3A) during July-August 1988. In addition, a recently developed tunable diode laser (TDL) technique for CH_4 detection (Sachse et al. 1989) has been combined with a turbulence measurement system on the NASA Electra research aircraft to perform airborne eddy correlation CH_4 flux measurements. The TDL uses a constant pressure white cell and a high flow venturi operated off the aircraft engine to achieve a measurement rate of approximately 5 Hz ($5s^{-1}$). This particular TDL system requires cryogenic cooling, considerable power for operation, and would not be suitable for operation at remote field sites.

During the NASA ABLE-3A expedition to the Alaskan tundra, these micrometeorological techniques were compared with chamber-based measurements. Field samples were stratified by plant community and results were extrapolated using remote sensing. The CH_4 source estimates produced by the two micrometeorological and one chamber-based methods were reasonably close (R.C. Harriss, pers. comm.). Perhaps similar techniques (whether eddy correlation or gradient-based) could be extended to N_2O in the future.

Future Research Directions

1. The development of CH_4 flux measurement systems offers the opportunity to combine process-level measurements, models, and integrated flux estimates for a relatively simple, stable gas. Further, integrated field campaigns like the ABLE-3A mission, and especially more intensive multidisciplinary collaborations like that carried out in FIFE (Sellers et al. 1988), can determine how well we can extrapolate from chamber-based methods to coarser-scale fluxes. More importantly, they will yield both a much improved understanding of the mechanisms controlling flux on a range of scales and a much better estimate for variation in net flux over space and time. If micrometeorological systems for N_2O prove viable, they will be equally useful.
2. The potential for use of mass balance techniques (within canopy, within inversion of planetary boundary layers, and with tower or aircraft platform defined "boxes") for estimating fluxes should be explored further.

EFFECTS OF PLANTS ON GAS TRANSPORT FROM SOILS

In this section, we are concerned with the extent to which plants modify the exchange of trace gases between soil and the atmosphere. The importance of trace gas uptake and release in the canopy will be discussed in the next section.

The importance of rice plants in the transfer of CH_4 from soil to the atmosphere is well documented (Schütz and Seiler, this volume; references therein). Up to 90% of the flux from a paddy passes through plants. Similar fluxes have been observed through native plants in natural wetlands, and the existence of species-specific differences within such sites has been documented. Moreover, there is some evidence for the transfer of N_2O through rice plants (Mosier et al. 1989). These observations raise a number of questions that have not been answered fully. To what extent does stomatal resistance influence gas flux through wetland plants? How important is passive diffusion through aerenchyma versus internal wind (cf. Dacey and Klug 1982) in transporting CH_4 through plants? How important is oxygen transport to the rhizosphere in causing CH_4 oxidation there (cf. Holzapfel-

Pschorn et al. 1985)? Can other metabolic processes in the rhizosphere alter CH_4 oxidation there?

In addition, should other effects of plants on trace gas transport be considered? Certainly organic matter additions to soil by litterfall, root turnover, and rhizodeposition influence the microbial processes which produce and consume trace gases; so does water uptake. Are there more direct effects? Does N_2O move in the transpiration stream of upland plants? Do root channels act as pathways for the rapid movement of gases through soils?

PLANT CANOPIES AS SOURCES AND SINKS OF REACTIVE GASES

Several years ago, the dominant conceptual model for NO_x deposition in plant canopies treated NO as chemically reactive but biologically inert (in the canopy), and NO_2 deposition as being regulated by stomata and hence strongly correlated with water flux. More recently, it has become clear that NO can also be absorbed by foliage, although the rate of uptake is lower than for NO_2. Moreover, flux measurements at sites with small concentrations of NO_x (= NO + NO_2) suggest that deposition process for NO_2 is concentration-dependent, with no net uptake of NO_2 by conifers at concentrations near 0.5–3.0 ppb, a reasonable range for unpolluted regions. It is not clear whether this concentration represents difficulties in measuring net fluxes at low concentrations, a threshold concentration below which uptake does not occur, or even a true compensation point for NO_2 below which net NO_2 flux from leaves occurs. In any case, these results suggest that NO_2 uptake by foliage is best modeled using equations of the general form

$$F = (C_a - C_i)/(r_b + r_s)$$

where F is the flux to (from) a leaf, C_a is the NO_2 concentration in air, C_i is the internal NO_2 concentration, r_b is the boundary layer resistance of the leaf, and r_s is the stomatal resistance.

A major difficulty with this approach is that we know too little about its physiological basis. What mechanisms control internal resistance to NO_2 uptake? NO_2 uptake is known to occur through stomata; the fate of incorporated $^{15}NO_2$ is also known (Rogers et al. 1979). However, the mechanism of uptake is not known, and there is little or no information on how uptake (or the putative compensation point) varies as a function of plant nutrition, condition, or history of exposure to NO_x.

In addition, much of our information on NO_x dynamics in plant canopies is derived from regions with high NO_x concentrations. In such sites it is clear that there is often net deposition of NO_x from the atmosphere, but in areas with small ambient NO_x concentrations we are often uncertain not just of the magnitude but even of the *direction* of net NO_x fluxes.

This set of problems is not unique to NO_x. The suite of chemically and biologically reactive sulfur gases, the NH_3–HNO_3–NH_4NO_3 system, and the isoprene–terpene–ozone–NO system all pose similar or even more complex problems. Add to this complexity the large number of biological species which may be present within some sites (at the extreme, several hundred tree species in a single hectare of certain Amazonian rain forests!) and it is reasonable to ask whether we can hope to determine the processes controlling atmosphere–vegetation exchange of reactive gases within the time span of the "global change" program which is now under development (within 12–15 years). If an adequate understanding requires species-specific information on stomatal uptake, compensation points, or the types of compounds produced, then this task is probably impossible. If, however, most species function similarly, or if we can identify a few functional groups of plants in which species can be classified easily to group, mechanistic models of vegetation–atmosphere exchange are worth pursuing.

Future Research Directions

1. Understanding the physiological mechanisms involved in vegetation–atmosphere exchange. The existence of an apparent compensation point for NO_x uptake by leaves has only been demonstrated for one species (Johansson 1987), and we know very little about the mechanisms causing internal resistance to NO_x uptake. It is particularly important to identify the commonalities and differences in these mechanisms among species and species groups.
2. Extension of this approach to other suites of reactive trace gases (sulfur gases, NH_3–HNO_3–NH_4NO_3, and isoprene–terpene). In some cases (e.g., HNO_3) we understand the mechanisms controlling deposition but are limited primarily by a lack of concentration data; in others (e.g., isoprene–terpene) we need to know considerably more about source strengths.
3. Integrated models for soil and canopy NO and NO_2 exchange need to be developed, parameterized, and validated.
4. Most importantly, multidisciplinary field investigations are fundamental to a timely analysis of vegetation sources and sinks. Progress in this area will be much more rapid if micrometeorological, physiological, and modeling efforts are carried out in an integrated way.

CAN MICROMETEOROLOGICAL MEASUREMENTS PROVIDE FLUXES OF NO, NO_2, AND O_3 AT THE SURFACE?

We answered this question "not now but perhaps soon." The net flux of NO_x between the soil surface and the plant canopy can be measured using micrometeorological techniques. However, the rapid $NO/NO_2/O_3$ chemistry interacts with exchange processes in a way that makes it impossible (at present) to extract individual fluxes of NO, NO_2, and O_3 at the surface. Micrometeorological techniques (Fowler and Duyzer; Desjardins and MacPherson, both this volume and references therein) measure fluxes in a layer of constant vertical flux. This layer exists if storage within the boundary layer, advection, and "internal" sources and sinks are vanishingly small. This last condition is not met in the $NO–NO_2–O_3$ system. The reactions

$$NO_2 + h_\nu \rightarrow NO + O$$

$$O + O_2 \rightarrow O_3$$

$$NO + O_3 \rightarrow NO_2 + O_2$$

can be rapid relative to the rate of vertical mixing. The dynamics of these internal sources and sinks can be described by

$$Q_{O_3} = j[NO_2] - k[O_3][NO] = -Q_{NO_2} \text{ (Fitzjarrald and Lenschow 1983)}$$

where j is the photolysis rate of NO_2 and k is the reaction rate of $NO + O_3 \rightarrow NO_2 + O_2$. The overall rate of these reactions (T_c) may be compared with the characteristic time scale of turbulent mixing. For neutral conditions this can be defined by

$$T_t = Z/kU_*$$

where Z is height, k is von Karman's constant, and U_* is the friction velocity. For a typical mid-day, mid-latitude situation, both T_c and T_t are in the range 30–300 sec at 5 to 10 m above the surface (Lenschow and Delany 1987). Moreover, the deposition velocity of O_3 is generally much larger than that of NO or NO_2.

The measured fluxes of NO, NO_2, and O_3 can be corrected to yield surface fluxes provided that the information necessary to solving the set of coupled partial differential equations which describe the mean budget of NO, NO_2, and O_3 in the planetary boundary layer can be obtained (see Fitzjarrald and Lenschow 1983). *Simultaneous* measurements of concentrations of all three species at two heights, of fluxes and concentrations at one height, or of fluxes alone at two heights are required.

There are a number of difficulties with this approach. As yet, no test of the correction procedure has been published, perhaps due to an inadequate data base. Moreover, the observed magnitude of NO and NO_2 fluxes is small and so the sensitivity and specificity of detectors is crucial. Instruments which allow measurements of $j(NO_2)$ are becoming available, but gradients in $j(NO_2)$ within canopies are very strong. Finally, the high variability in ratios of O_3:NO:NO_2 will need to be considered.

The present state of field measurements leads us to conclude that the answer to the question "can micrometeorological measurements provide fluxes of NO, NO_2, and O_3 at the surface?" is "no." With an increasing number of field campaigns planned, the answer may change to "yes" *as far as simple canopies such as crops* are concerned. Surfaces covered with more complex canopies will be more difficult. The same or similar problems apply to a number of other systems, including sulfur gases, volatile organic carbon (VOC)/NO/O_3, NH_3/HNO_3/NH_4NO_3, and highly soluble trace gases such as HCl and HNO_3 which react with wet aerosols or fog droplets. The existence of strong humidity gradients within canopies and of complex solutions on the surfaces of wet or partially wet canopies presents particular problems for the highly soluble gases.

Discussion within the group pointed out that these analyses illustrate the complexity of the NO–NO_2–O_3 system and other systems very clearly, but what if we are interested primarily in the net NO_x flux? In that case, we are dealing with a conserved quantity, and calculations of net flux at the surface from flux measurements in the atmosphere are indeed possible, at least during the day. During the night, it may be necessary to consider reactions which form nitrate radicals. These represent a sink for NO_x, ultimately leading to the formation of HNO_3 or organic nitrates. The rate of nitrate formation is slow, but it can be important where the rate of NO_x uptake by vegetation is slow.

Even if other reactions are unimportant, measurements of NO_x fluxes by themselves will not provide information on mechanisms of deposition because the dynamics of NO and NO_2 are very different. For the same reason, they cannot be predictive, except in a very empirical way. They are nonetheless possible, and they may be sufficient for many purposes, especially where they can be supplemented by measurements of soil–atmosphere exchange of NO and NO_2 using enclosure methods.

Future Research Directions

1. Application of the corrections for internal sources/sinks to a consistent set of field data.
2. Development of appropriate techniques for work in the background atmosphere (unpolluted regions). Such techniques must include both more sensitive and more specific instrumentation; techniques such as

TDL, laser imaging, detecting, and ranging (LIDAR), Fourier transform infrared (FTIR), and dual optical absorption spectroscopy (DOAS) all have considerable promise.
3. Development of standard aircraft packages for flux measurements.

HOW DOES WITHIN-CANOPY EXCHANGE INFLUENCE TRACE GAS FLUXES?

Our understanding of transport processes within plant canopies has undergone a marked change in the last ten years. The application of fast-response sensors for turbulence measurements has demonstrated that rather than resulting from steady diffusion along the mean concentration gradient, canopy transport is an intermittent process. Gases are transported in and out of the canopy by updrafts and downdrafts occurring at irregular intervals (Denmead and Bradley 1987). The intermittency of transport increases with depth from the canopy top. Typically, near a forest floor, a major transport event may occur once every three minutes on average. Even near the top of the canopy, 50% of the flux may occur during about 5% of the time.

The importance of this understanding of within-canopy exchange processes to the regulation of trace gas flux depends on the mean (+/− variance) residence time of air within the canopy versus the mean (+/− variance) lifetime of a gas. For unreactive gases like N_2O and CH_4, the intermittency of transport and the length of the residence time in the canopy should not affect transport across the soil–air interface. However, where uptake and/or reaction are rapid compared to the time scale of intermittent turbulence, the frequency of mixing itself can control flux. If the mean lifetime (vis-à-vis chemical reaction or uptake) of a soil-derived gas in the air under a forest canopy is 3 min and turbulent mixing is absent, then diffusion should be slow enough that little flux above the canopy will occur. On the other hand, if turbulence vents all the air below the canopy every 3 min, 50% of the gas emitted from soil will escape above the canopy. Such dynamics are on about the right time scale for NO; if we are concerned with NO flux above the canopy, then we need to know a good deal about pathways of air movement under and through the canopy. This requirement is less severe if we require only net NO_x flux; it is much more severe for a number of other trace gases (e.g., NH_3, VOC, reduced sulfur gases).

In order to model the movement of a short-lived gas through a canopy, we need to determine a probability distribution for the typical residence time of a gas molecule in the canopy space (so that we can predict its chemical transformations there), and we need to know the probability distribution that a gas molecule will encounter a vegetation element or the soil surface where uptake or destruction might occur. Explanatory models which make use of measured distributions of sources and sinks and statistical analyses of turbulent transport to predict canopy transport are now being developed. However, ultimately it would be ideal if we could infer sources and sinks of trace gases within canopies through an inverse process based on measured concentrations at a number of points and a knowledge of air-mass trajectories to these points (Raupach 1989). Such an approach would be analogous to inverse modeling of sources and sinks of trace gases on a global scale based on a distributed network of sampling stations (Fung et al. 1987); it would be equally dependent on an adequate number of measurement points to constrain possible source/sink distributions.

The residence time of air under and within the canopy is itself strongly influenced by the biota; canopy structure substantially affects mixing. Heterogeneous features of the canopy such as emergent individuals or treefall gaps cause heterogeneities in the turbulence below the canopy; they help to make the "average residence time" a potentially misleading characterization of trace gas dynamics within canopies.

In summary, the dynamics of air movement within canopies ("micrometeorology") have little influence on overall fluxes of the stable gases N_2O and CH_4; they have rather small effects on the regulation of overall NO_x fluxes. On the other hand, the dynamics of air movement can affect net NO flux above the canopy, and certainly the magnitude of net fluxes of other reactive trace gases can be altered substantially by micrometeorological processes.

GLOBAL CHANGE

This group's primary contribution to the understanding of global-scale biosphere–atmosphere interactions lies in the determination of which processes regulate exchanges of trace gases and in the identification of approaches which will allow an improved understanding of those exchanges. However, it is interesting to speculate whether global-scale changes could themselves cause differences in the relative importance of biological, photochemical, and micrometeorological processes. Such changes have occurred in the past as a consequence of glacial/interglacial cycles (Barnola et al. 1987); they are occurring now as a consequence of human activity.

As this chapter demonstrates, it is difficult to generalize about current vegetation–atmosphere interactions involving reactive gases; many basic questions remain unanswered. We can say still less about those interactions in the past or future. However, it is clear that both past and ongoing changes are of a scope and direction sufficient to alter vegetation–atmosphere interactions. The full-glacial climate had elevated wind velocities and much enhanced aeolian transport relative to the present. The area covered by closed forests was less; open boreal woodlands occurred south of 35°N in North America. It appears that even many areas that are now evergreen tropical rainforest were then savanna woodlands as a consequence of lower CO_2 concentrations (which favor tropical grasses with the C_4 photosynthetic pathway over trees with the C_3 pathway) and increased aridity.

Human-caused land-use change is likely to have similar effects on vegetation–atmosphere interactions in the future. Conversion of forest land to agriculture, deforestation and the spread of derived savannas, and habitat fragmentation in urbanized regions all involve alterations in the openness of canopies; they could affect micrometeorological as well as biological processes.

CONCLUSIONS

We agreed that biological processes regulate above-canopy emissions of the long-lived trace gases N_2O and CH_4; it appears that biological processes, including those in the plant canopy, are also predominant in the regulation of net NO_x fluxes. Micrometeorological and chemical/photochemical processes may interact with biological processes to control atmosphere–surface exchange of a number of other important trace gases, including NO; however, methods which would enable us to determine the relative importance of these processes are not now available.

This surprisingly strong consensus on the importance of biological processes in controlling the trace gases of interest does not imply that we would recommend that all resources in this field should go towards biology. Rather, we concluded that progress in understanding CH_4 and N_2O fluxes, in the development of models and in the determination of global fluxes, is most limited by the lack of field measurements of integrated areal fluxes. Other aspects of the cycles of these long-lived gases are worth pursuing actively; they have and will continue to lead to significant advances. However, the only way that the applicability of these on regional and global scales can be tested is through the development of micrometeorological approaches for measuring integrated fluxes. Such systems are becoming available for CH_4 (above); we hope that they will also be developed for N_2O.

For NO_x, transfer processes and chemical reactions influence fluxes, and a great deal can be gained from further developments of micrometeorological techniques. A particular requirement is the development of sensors which can measure reliably the small net NO_x fluxes in the undisturbed atmosphere. However, our understanding of canopy–atmosphere exchange of NO_x is most strongly limited by a poor understanding of the biological mechanisms involved in NO_x uptake (and possibly release) and a lack of information on how these biological processes vary as a function of species, site, nutrition, condition, and exposure history. Similar problems exist for many of the other reactive trace gases.

OVERALL RECOMMENDATIONS

1. The development of tower- and aircraft-based micrometeorological systems for measuring N_2O and CH_4 fluxes offers the opportunity for a quantum increase in our understanding of the regulation of these globally significant trace gases. It is most important that these techniques be applied in an integrated way as part of a multidisciplinary research program. Such a program could be based on the solid background derived from microbial process studies, flux estimates, model development, and the attempt to construct global budgets. It should include microbial ecologists concerned with mechanisms of production and consumption, soil physicists concerned with transport, terrestrial ecologists concerned with process-model-based extrapolations of chamber flux measurements, remote sensing specialists who can provide forcing functions for the models, and meteorologists with tower- and aircraft-based flux measurement systems. It should be developed within the global context provided by global circulation modelers who can calculate global sources and sinks from seasonal variations in gas concentrations at a network of atmospheric observatories, and carried out in association with a well located set of such observatories which includes continental as well as maritime locations. This integrated program should be focused on the most important and/or least well defined source areas, as described by Schimel et al. (this volume).
2. Multidisciplinary studies of the regulation of the transport of reactive trace gases would be equally rewarding. The "best" way to develop a program to measure, model, and extrapolate such fluxes is not clear; it will undoubtedly vary among the reactive gases. For NO_x, however, such a program should include the capability of making micrometeorological measurements at background concentrations in the unpolluted atmosphere, a set of chamber measurements of NO flux coupled to a model of the mechanisms controlling NO emission (Galbally and Johansson 1989), further developments in modeling turbulent transfer within canopies, and most importantly a better understanding of the biological regulation of NO_x uptake and release.

REFERENCES

Barnola, J.J., D. Raynaud, Y.S. Korotkevich, and C. Lorius. 1987. Vostok ice core provides 160,000-year record of atmospheric CO_2. *Nature* 329:408–414.

Dacey, J.W.H., and M.J. Klug. 1982. Tracer studies of gas circulation in *Nuphar*: $^{18}O_2$ and $^{14}CO_2$ transport. *Physiol. Plant.* 56:361–366.

Denmead, O.T. 1979. Chamber systems for measuring nitrous oxide emissions from soils in the field. *Soil Sci. Soc. Am. J.* 43:89–95.

Denmead, O.T., and E.F. Bradley. 1987. On scalar transport in plant canopies. *Irrig. Sci.* 8:131–149.

Fitzjarrald, D.R., and D.H. Lenschow. 1983. Mean concentration and flux profiles for chemically reactive species in the atmospheric surface layer. *Atmos. Envir.* 17:2505–2512.

Fung, I.Y., C.J. Tucker, and K.C. Prentice. 1987. Application of Advanced Very High Resolution Radiometer vegetation index to study atmosphere-biosphere exchange of CO_2. *J. Geophys. Res.* 92:2999–3105.

Galbally, I.E., and C. Johansson. 1989. A model relating laboratory measurements of rates of nitric oxide production and field measurements of nitric oxide emission from soils. *J. Geophys. Res.*, in press.

Holzapfel-Pschorn, A., R. Conrad, and W. Seiler. 1985. Production, oxidation, and emission of methane in rice paddies. *FEMS Microb. Ecol.* 31:343–351.

Johansson, C. 1987. Pine forest: a negligible sink for atmospheric NO_x in rural Sweden. *Tellus* 39B:426–438.

Jury, W.A., J. Letey, and T. Collins. 1982. Analysis of chamber methods used for measuring nitrous oxide production in the field. *Soil Sci. Soc. Am. J.* 46:250–256.

Lenschow, D.H., and A.C. Delany. 1987. The analytical formulation for NO and NO_2 flux profiles in the atmospheric surface layer. *J. Atmos. Chem.* 5:301–309.

Mosier, A.R., S.K. Mohanty, A. Bhadraschalam, and S.P. Chakravorti. 1989. Effect of rice plants on N_2 and N_2O emissions from flooded soil. In: Proceedings of International Denitrification Workshop, Giessen, F. R. Germany, in press.

Raupach, M.R. 1989. Applying Lagrangian fluid mechanics to infer scalar source distributions from concentration profiles in plant canopies. *Agric. For. Meteor.*, in press.

Rogers, H.H., J.C. Campbell, and R.J. Volk. 1979. Nitrogen-15 dioxide uptake by *Phaseolus vulgaris* L. *Science* 206:333–335.

Sachse, G., et al. 1989. Atmospheric methane and CO over arctic tundra. AGU Abstr. Baltimore: AGU.

Sellers, P.J., F.G. Hall, G.S. Asrar, D.E. Strebel, and R.E. Murphy. 1988. The first ISLSCP field experiment (FIFE). *Bull. Am. Meteor. Soc.* 69:22–27.

Exchange of Trace Gases between Terrestrial Ecosystems and the Atmosphere
eds. M.O. Andreae and D.S. Schimel, pp. 263–280
John Wiley & Sons Ltd

Atmospheric Deposition and Nutrient Cycling

J.M. Melillo[1], P.A. Steudler[1], J.D. Aber[2], R.D. Bowden[1]

[1] *The Ecosystems Center*
Marine Biological Laboratory
Woods Hole, MA 02543, U.S.A.

[2] *Institute for the Study of Earth, Oceans, and Space*
Center for Complex Systems
University of New Hampshire
Durham, NH 03824, U.S.A.

Abstract. The magnitude of nitrogen deposition in the temperate and boreal zones of the Northern Hemisphere is estimated to be about 18 Tg yr^{-1} (6 Tg yr^{-1} in North America and 12 Tg yr^{-1} in Europe). The linkages between nitrogen inputs to forest ecosystems and fluxes of trace gases between the atmosphere and forest soils are discussed. Experimental evidence is presented that shows (*a*) increases in N_2O, COS, and CS_2 emissions from forest soils with increases of nitrogen inputs and (*b*) decreases in CH_4 consumption by forest soils with increases of nitrogen inputs. Future research on the linkages between nitrogen inputs and trace gas fluxes between the atmosphere and forest soils should include tests of the generality of the couplings described here and a consideration of the relationship between nitrogen inputs and CO fluxes. In addition, work must be done on the couplings among nitrogen inputs to soils, the nutrient status of plants, and the exchange of trace gases between plants and the atmosphere. Gases of interest would certainly include COS and isoprene. Finally, there is a need for a great deal of geographically referenced information on atmospheric deposition patterns, soil and plant distribution, etc. These data should be coupled with dynamic process models so that regional- and global-scale estimates of trace gas exchanges between the land and the atmosphere can be made.

OVERVIEW

We are powerful agents of change on planet Earth. Because the Earth is a coupled system of land, atmosphere, and ocean, changes we cause in any one component of this megasystem affect the others. Some of the couplings between system components are obvious and well understood. For example, the conversion of forests and grasslands to agricultural use has caused a net flux of carbon from the land to the atmosphere of at least 100×10^{15} g C over the past few centuries.

Other couplings are more subtle and less well known, such as the effects of atmospheric deposition of nitrogen on ecosystem structure and function and on the flux of trace gases between the "fertilized" systems and the atmosphere. In this chapter we consider several aspects of this coupling.

The paper is divided into four sections. In the first we briefly review the "anthropogenic" emissions of nitrogen oxides and ammoniacal nitrogen to the atmosphere and the deposition patterns of nitrogen in the temperate and boreal regions of the Northern Hemisphere. We estimate that about 18 Tg of nitrogen are deposited in the 22,800,000 km^2 area of these regions.

We consider the effects on ecosystem structure and function of chronic additions of high levels of nitrogen to forests in the second section. Here we present data that link chronic nitrogen additions to increases in nitrous oxide fluxes from forest soils to the atmosphere.

In the third section we explore the linkage between chronic nitrogen additions to forest ecosystems and methane fluxes. We discuss the results of field studies in which we have shown that the ability of upland soils to consume methane is reduced by nitrogen additions. We also discuss the biological mechanisms that may be responsible for this behavior.

Finally, we discuss the linkage between chronic nitrogen additions to forests and sulfur gas fluxes between forests and the atmosphere. We focus on field studies that have shown a linkage between nitrogen additions and organic sulfur gas emissions. In these studies we have found that fluxes of carbonyl sulfide and carbon disulfide from the soil to the atmosphere increase when nitrogen is added to the soil. Here again we discuss the biological mechanisms that may be responsible for this phenomenon.

LARGE-SCALE NITROGEN EMISSION AND DEPOSITION PATTERNS

Emission Estimates

Human activities have dramatically influenced the emissions of nitrogen compounds from the land to the atmosphere. Two classes of compounds are important in this regard: the nitrogen oxides (NO_x) and ammoniacal nitrogen (NH_x). The major source of nitrogen oxides is fossil fuel combustion, while the major source of ammoniacal nitrogen is related to agricultural activities (i.e., animal husbandry and fertilizer production and use).

TABLE 1. Estimates of annual "anthropogenic" emissions of nitrogen to the atmosphere by continent ca. 1979; units are Tg N per year (after Rodhe et al. 1988).

Continent	Emission Rate
North America	7.0
South America	0.8
Europe	8.0
Africa	0.5
Asia	5.0
Australia	0.5

Rodhe et al. (1988) have estimated that 92% of all "anthropogenic" NO_x emitted to the atmosphere around 1979 came from the Northern Hemisphere, and about 70% was emitted from the industrial regions of Europe and North America (Table 1). A detailed analysis of "anthropogenic" emissions of NO_x from Europe in about 1975 has been made by Bónis et al. (1980). They identified "the combustion of fossil fuels and automobile traffic" as accounting for 97% of "anthropogenic" NO_x emissions to the atmosphere from the European landscape (Table 2).

Bónis et al. (1980) also studied NH_x emissions from Europe at about the same time (ca. 1975). Livestock was found to be the greatest "anthropogenic" source of NH_x. The combined stocks of cattle, pigs, sheep, and other farm

TABLE 2. NO_x and NH_x anthropogenic sources to the atmosphere and emission rates in Europe ca. 1975; units are Tg N per year (modified from Bónis et al. 1980).

	Sources	Emission Rates
NO_x	Fossil Fuels	2.3–6.2
	Automobiles	1.1–1.5
	Nitric acid production	<0.1–0.1
	Fertilizer production	0.1
	Aviation	<0.1
	Burning of waste materials	<0.1
	Σ	3.5–7.9
NH_x		
	Domestic animals	3.0–4.5
	Fertilizer production	0.5–0.9
	Coal combustion	<0.1
	Automobiles	<0.1
	Σ	3.5–5.4

animals were estimated to account for between 83% and 86% of the "anthropogenic" NH_x emissions (Table 2).

Using the data of Rodhe et al. (1988), Bónis et al. (1980), and data on livestock populations in North America (WRI 1988), we have developed annual "anthropogenic" emission estimates for NO_x and NH_x for North America and Europe (Table 3). These estimates of between 10.3 and 12.1 Tg N for North America and 11.0 and 12.5 Tg N for Europe serve as upper bounds on annual nitrogen deposition rates on these continents if we assume relatively short atmospheric residence times for these compounds and no other major sources of them (Bónis et al. 1980).

TABLE 3. Estimates of annual "anthropogenic" emission rates of nitrogen oxides and ammoniacal nitrogen for North America and Europe ca. 1975–1980; units are Tg N per year.

	NO_x-N	NH_x-N	Σ N
North America	7.0	3.3–5.1	10.3–12.1
Europe	8.0	3.0–4.5	11.0–12.5

Deposition Estimates

Networks established in North America and Europe to monitor acid rain deposition have provided a large amount of data on nitrogen inputs to terrestrial ecosystems in the temperate and boreal zones. We have used the summaries of nitrogen deposition prepared by Barrie and Hales (1984), Buijsman and Erisman (1988), and Bónis et al. (1980), in conjunction with the global vegetation map of Olson et al. (1983), to produce estimates of the patterns of wet plus dry nitrogen deposition in the temperate and boreal zones of North America and Europe. Dry deposition is assumed to be equivalent in magnitude to wet deposition. Europe is broadly defined to include all of the temperate and boreal areas of the Soviet Union; that is, the vast forested landscape east of the Ural Mountains as well as "European Russia" west of the Urals.

Our estimate of the atmospheric nitrogen deposition for the temperate and boreal zones of the Northern Hemisphere is 17.8 Tg N per year for the period between 1975 and 1980 (Table 4), with North America receiving 5.8 Tg N annually and Europe receiving 12.0 Tg N annually. For the temperate and boreal zones of the Northern Hemisphere, 80% of the nitrogen deposited during that period fell on about half of the area and about 35% of the nitrogen was deposited on 15% of the area. The "hot-spot" nature of nitrogen deposition is a well-recognized phenomenon.

The Federal Republic of Germany is one of the hot spots for nitrogen deposition in Europe. On the Solling Plateau in central Germany, high levels of nitrogen deposition have been measured for more than a decade (Fig. 1). The deposition rates for Solling are about twice our estimated mean high deposition value for Europe (Table 4). The consequences of "chronic" inputs of large amounts of nitrogen for the function of forest ecosystems is the topic of the remainder of this paper. Special attention is given to the coupling of nitrogen inputs and trace gas fluxes.

CHRONIC NITROGEN ADDITIONS: EFFECTS ON PLANT PROCESSES AND NITROGEN CYCLING IN FOREST ECOSYSTEMS

Forest ecosystems' responses to chronic, elevated nitrogen additions involve a complex set of interactions among the processes affecting nitrogen cycling. These include plant uptake and allocation, litter production, immobilization and mineralization during litter and soil organic matter decay, nitrification, ion leaching, and trace gas emissions.

We have recently developed an integrated set of hypotheses about the time course of changes in major plant and soil processes in response to chronic nitrogen deposition (Aber et al. 1989; Fig. 2). We predict a progression of changes in both plant and soil processes as the period of chronic fertilization lengthens. This progression is set out in Fig. 2, where the time dimension is divided into four stages of ecosystem response relative to the duration of elevated nitrogen inputs. For the purposes of this discussion, the nitrogen inputs can be either "inadvertent fertilization" associated with atmospheric deposition or scheduled fertilizer applications. To simplify the graphic, nitrogen inputs are described in Fig. 2 as a step increase commonly associated with scheduled fertilizer applications.

Stage O. Stage O is the pre-treatment condition where nitrogen inputs are at a background level that we estimate to be about 2 kg N ha^{-1} yr^{-1} in the temperate and boreal regions (Table 4). At this level of nitrogen deposition, forest systems of these regions are generally nitrogen limited, and so they show significant increases in tree growth in response to nitrogen additions (e.g., Mitchell and Chandler 1939).

The rates of soil processes can also be limited in systems with suboptimal nitrogen availability. For example, the addition of nitrogen will both speed the litter decay process and increase the amount of nitrogen immobilized during the early stages of that process (Melillo et al. 1984). As another example, nitrification (the transformation of ammonium to nitrate) is absent or is a relatively minor soil process in many forest soils of the temperate and boreal zones where nitrogen availability is suboptimal. In general,

TABLE 4. Estimates of wet plus dry nitrogen deposition in the temperate and boreal regions of North America and Europe; H = high deposition, M = medium deposition, and L = low deposition.

Continent	Region	Deposition level	Area km^2	Annual deposition kg/km^2	Annual deposition by region 10^9 kg	Annual deposition by continent (Tg)
North America	Temperate	H	595,000	1700	1.01	
		M	3,197,000	1130	3.61	
		L	464,000	200	0.09	
	Boreal	M	510,000	1130	0.58	
		L	2,819,000	200	0.56	
	Temperate + Boreal					5.85 Tg
Europe	Temperate	H	2,745,000	2000	5.49	
		M	5,021,000	900	4.52	
		L	635,000	200	0.13	
	Boreal	M	1,031,000	900	0.93	
		L	4,385,000	200	0.88	
	Temperate + Boreal					11.95 Tg

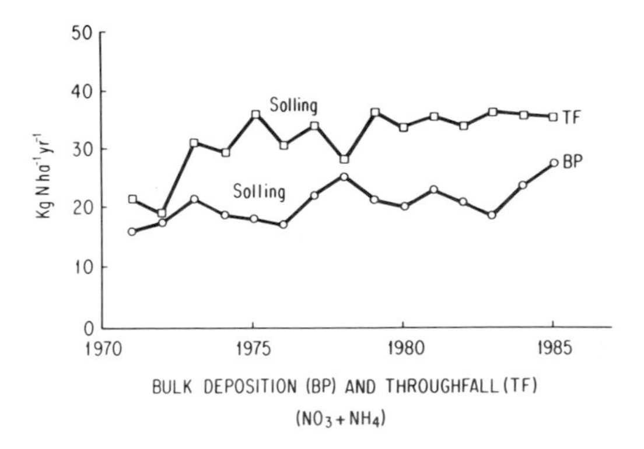

Fig. 1—The sum of NO_3-N and NH_4-N in bulk precipitation (BP) and throughfall (TF) at Solling between 1971 and 1984 (Hans, pers. comm.). We consider bulk precipitation to represent wet deposition plus some fraction of dry deposition. We view throughfall as being an upper bound or wet deposition plus all of dry deposition.

limited ammonium availability is thought to reduce nitrification in forest ecosystems (Vitousek and Matson 1985; Robertson 1982). This fact is particularly relevant to the production of nitrous oxide (N_2O) since it is linked to the nitrification process either directly or indirectly; that is, N_2O can be a direct product of nitrification, or it can be produced from the primary end product of nitrification, nitrate, which is the substrate that denitrifying microorganisms use to produce N_2O and N_2.

Stage 1. In stage 1, increased nitrogen deposition is occurring and for a limiting nutrient such as nitrogen, this should result in increased plant production. Any changes in plant production are likely to be preceded by changes in other plant characteristics such as foliar nitrogen concentration and foliar biomass as shown in Fig. 2.

The rate of the key soil process of nitrogen mineralization, the transformation of organically bound nitrogen to ammonium, should also increase. As ammonium becomes more abundant, we hypothesize further that the rate of nitrification will begin to increase, although not enough to result in either substantial nitrate leaching or large increases in N_2O fluxes from the system. We suggest that most of the nitrate produced through nitrification, as well as nitrate entering the system in precipitation during stage 1, will be assimilated by the biotic components of the system and remain part of the system's internal nitrogen cycle.

Stage 2. Upland forest ecosystems have a finite capacity to assimilate and retain nitrogen. With the exception of the cation exchange property, there

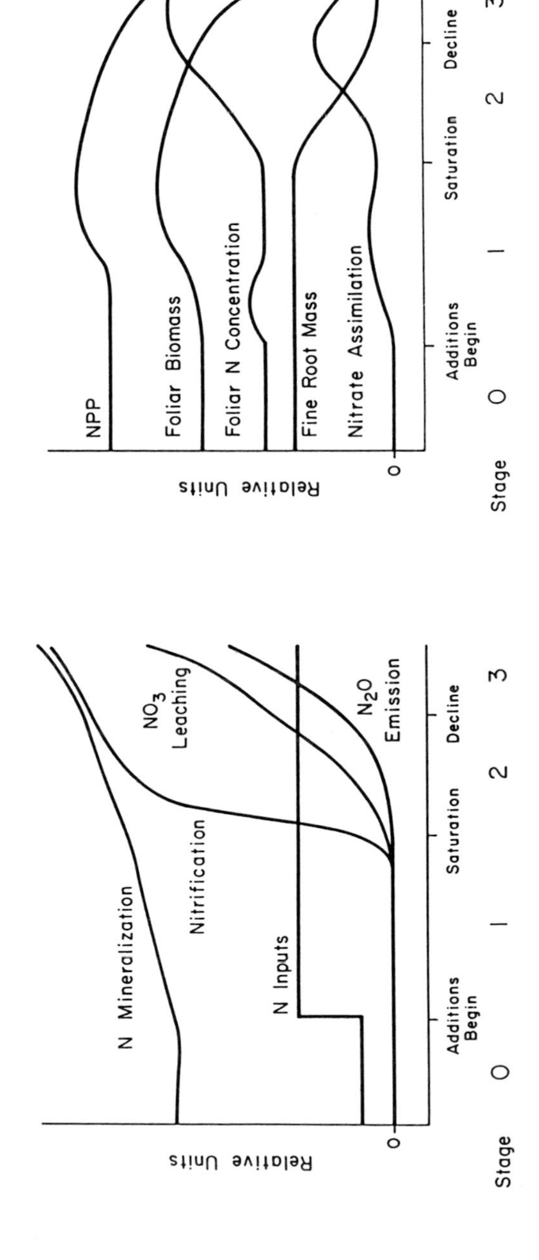

Fig. 2—Hypothesized responses of a forest ecosystem to chronic N additions; axis scales in relative units (Aber et al. 1989).

are no important chemical retention mechanisms in the soil for large amounts of ammonium. Nitrogen retention in these systems occurs largely through plant uptake of inorganic nitrogen or incorporation into decaying organic matter. As the amount of inorganic nitrogen in a soil increases, biotic uptake (either plant or microbial) becomes limited by the availability of other resources essential for growth.

Nitrogen saturation is reached when nitrogen availability, both through current atmospheric deposition and mineralization of previously accumulated soil nitrogen, exceeds the biotic capacity of the system. Stated another way, it occurs when other essential resources limit plant and microbial growth. In temperate and boreal systems, the "secondary limiting factors" include water and phosphorus for the plants and carbon availability for the microbes.

Prolonged nitrogen saturation can decrease plant production in a number of ways (Aber et al. 1989). Soil processes can also be changed under conditions of nitrogen saturation, and perhaps the most important change conjectured by us is a dramatic increase in nitrification rate. The production of large amounts of nitrate in the soil in excess of plant demand can lead to elevated nitrate leaching from the soil and increased emissions of N_2O. Also, nitrate entering forest ecosystems directly in precipitation may be denitrified.

Stage 3. This is the period along the chronic nitrogen fertilization progression when we hypothesize that forest decline will become very obvious. We also predict that this is the period when nitrate leaching and N_2O emissions will reach their peaks.

Nitrous oxide production. Nitrous oxide is a very long-lived (about 180-year lifetime) nitrogen gas and its concentration in the atmosphere is increasing (Bolle et al. 1986). Nitrous oxide emissions from undisturbed forest ecosystems are generally very low. From an extensive review of the literature, Bowden (1986) lists average rates of no more than 1 kg N_2O-N ha^{-1} yr^{-1} for temperate deciduous forests and only slightly higher for temperate coniferous and boreal forests. He also notes, however, that the N_2O losses from forests can be much higher and that, at least in agricultural systems, fertilization can increase losses of nitrogen gases.

Currently we are conducting a study of the ecosystem-level effects of chronic nitrogen additions in temperate forest ecosystems in Massachusetts, U.S.A. Deciduous and coniferous forest plots are receiving chronic nitrogen additions at one of two rates: 50 kg N ha^{-1} yr^{-1} (low N) or 150 kg N ha^{-1} yr^{-1} (high N) above the ambient deposition rate of about 10 kg N ha^{-1} yr^{-1}. Preliminary results from these studies support a number of the hypotheses outlined in Fig. 2. The results most germane to this workshop are our measurements of trace gas fluxes, including N_2O fluxes. Using the

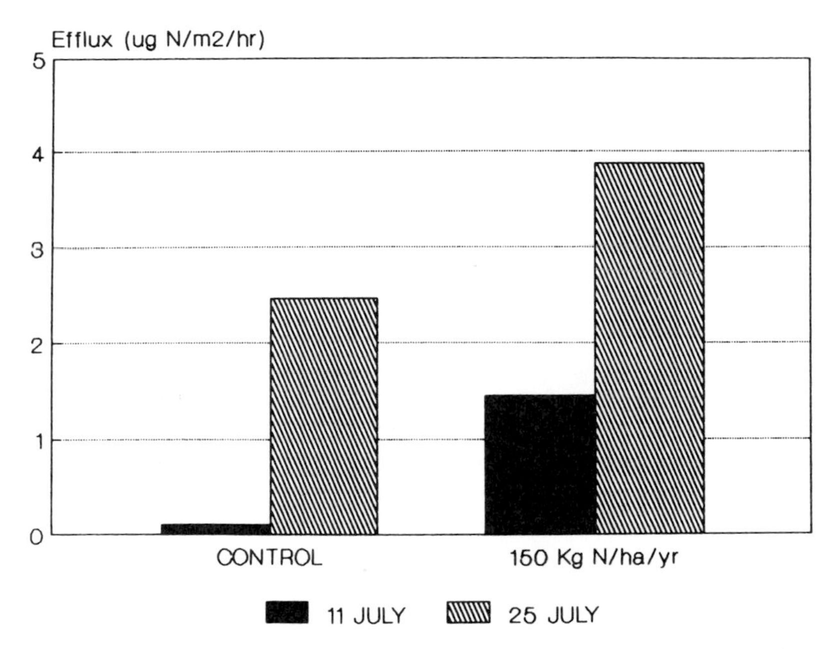

Fig. 3—Nitrous oxide efflux from control and nitrogen-fertilized conifer forest plots at Harvard Forest, Massachusetts. Measurements made before (11 July 1988) and immediately after (25 July 1988) a 6 cm rain event (Bowden et al., pers. comm.).

small, closed chamber technique, we have measured significant differences between N_2O fluxes from control plots where there have been no chronic nitrogen additions above ambient levels and plots receiving chronic applications of nitrogen.

These differences are illustrated by results from the control and high-N plots in a coniferous forest stand for two dates in July 1988 (Fig. 3). The 11 July sampling took place in the middle of a long dry period and the 25 July sampling occurred immediately after a 6 cm rain event. In both instances, the N_2O emissions from the high-N plots were significantly greater than those from the control plots. As expected, the moisture status of the soil also had an influence on the rates of N_2O emissions. Clearly, the relative differences between N_2O emissions from control and high-N plots were greatest during dry soil moisture conditions.

We have also seen a similar pattern for the relationship between chronic nitrogen additions and N_2O emissions in deciduous stands in the same area of Massachusetts. These chronic addition plots (both coniferous and deciduous) are, however, only a year old and so we have no basis to judge how prolonged additions of nitrogen to these systems will further affect N_2O emissions.

CHRONIC NITROGEN ADDITIONS AND THE METHANE CYCLE

Methane (CH_4) is another long-lived trace gas (about 8 years), the tropospheric abundance of which has been increasing at about 1.1% annually over the last decade (Bolle et al. 1986). The recently measured increases in tropospheric CH_4 may be due to either increased production or decreased consumption of this trace gas (Khalil and Rasmussen 1983; Seiler 1984; Seiler and Conrad 1987). Major sources of CH_4 include rice cultivation, ruminants, termites, wetlands, and biomass burning (about 400 Tg CH_4 yr^{-1}). Methane sinks can be either chemical or biological. The largest chemical sink of CH_4 (about 300 Tg yr^{-1}) is the reaction with hydroxyl radical in the troposphere. If human activities have decreased tropospheric OH abundances (Khalil and Rasmussen 1987), for example by increasing the atmospheric burden of carbon monoxide, then CH_4 depletion rates would be slower and atmospheric concentrations would increase.

The largest biological sink for CH_4 (currently estimated at about 32 Tg yr^{-1}) is microorganisms in aerobic soils. Some evidence from the literature indicates that aerobic tropical soils are CH_4 sinks (Keller et al. 1983; Keller et al. 1986). The role of aerobic soils of the temperate zone in the global CH_4 budget is not well studied. Data from a New Hampshire forest (Keller et al. 1983) and preliminary data from our research on trace gases in Massachusetts (Steudler et al. 1989; Fig. 4) suggest that forest soils of the temperate zone are also CH_4 sinks.

Working on the control and fertilized deciduous and coniferous sites in Massachusetts, described in the previous section on N_2O fluxes, we have observed that both moisture conditions and nitrogen status of the sites affect

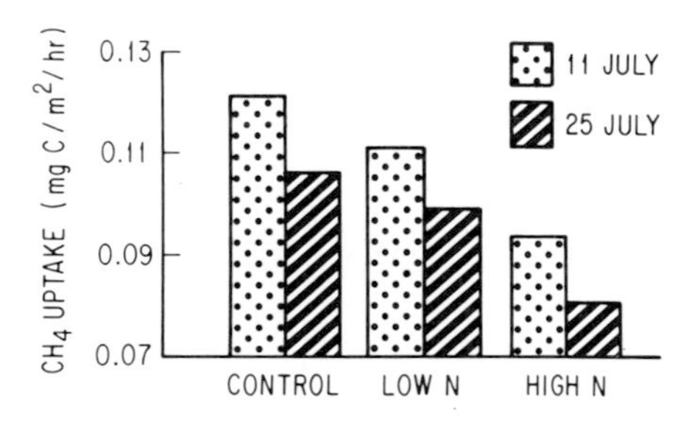

Fig. 4—Methane uptake by control and nitrogen-fertilized conifer forest soils at Harvard Forest, Massachusetts. Measurements made before (11 July 1988) and immediately after (25 July 1988) a 6 cm rain event (Steudler et al. 1989).

the magnitude of the CH_4 sink. We have found that when sites are dry they are greater sinks for CH_4 than when they are wet. Both CH_4 oxidizers and CH_4 producers are thought to be widespread; in fact, they often exist in the same areas, with the methane producers occupying the deeper, wetter soil horizons, while the methane oxidizers occupy the surface, aerobic horizons (Keller et al. 1983). Changes in soil moisture result in changes in the net balance of CH_4 production and consumption, with wetter conditions tipping the balance towards production.

A new finding we have made in the Massachusetts study is that the higher the nitrogen content of the soil at a site, the lower the magnitude of the CH_4 sink. After only four months of fertilization the CH_4 consumption rates were reduced by 14% in the high-N fertilization plots (annual application rate of 150 kg N ha^{-1}) relative to the control plots (Fig. 4). A preliminary analysis of CH_4 consumption data taken after six months of fertilization indicates that the difference became more pronounced, such that CH_4 consumption rates were reduced by about 33% in the high-N plots relative to the controls. Clearly, we do not know the long-term trajectory of this relationship.

The mechanism responsible for the effect of nitrogen on CH_4 is not well understood. Laboratory studies indicate that the oxidation of CH_4 by a variety of organisms is competitively inhibited by nitrogen, especially ammonium. For example, the ammonium ion was found to inhibit CH_4 oxidation by *Methylosinus trichosporium* (Whittenbury et al. 1970) and by *Pseudomonas methanica* (Lineweaver and Burk 1934).

Interestingly, nitrifying bacteria, including the dominant terrestrial species, *Nitrosomonas europaea*, have the ability to oxidize methane, even at low atmospheric concentrations (Jones and Morita 1983b; Hyman and Wood 1983). Of further interest is the fact that at high ammonium levels (>15 ppm), CH_4 oxidation by *N. europaea* is dramatically reduced (Jones and Morita 1983b).

The results of these laboratory studies are consistent with CH_4 oxidation in soils being suppressed in systems which receive high inputs of nitrogen. The global-scale consequences of the reduced CH_4 sink with nitrogen fertilization are uncertain because we lack data on the geographic generality of the nitrogen–methane interaction.

CHRONIC NITROGEN ADDITIONS AND SULFUR GAS FLUXES

Carbonyl sulfide (COS) and one of its important tropospheric precursors (Jones et al. 1983), carbon disulfide (CS_2), gases produced mainly by biological processes (Khalil and Rasmussen 1984; Steudler and Peterson 1984), play important roles in the energy budget of the Earth and the chemistry of the stratosphere. COS is generally the dominant sulfur source for the stratospheric aerosol layer (Crutzen 1976; Servant 1986) which affects the Earth's heat budget by reducing the amount of solar energy entering the troposphere (Turco et al. 1980). Increases in the aerosol layer would increase the temperature of the stratosphere and this may affect the reactivity and composition of other stratospheric gases (Servant 1986). The sulfuric acid aerosol layer recently has been implicated in the depletion of stratospheric ozone by acting as a catalytic surface for the production of reactive chlorine species (Oppenheimer 1987; Tolbert et al. 1988). Indirect evidence suggests that the flux of COS and CS_2 from the land and sea to the atmosphere may be increasing due to human activities (Hofmann and Rosen 1981; Kerr 1988).

The role of forest ecosystems in global budgets of COS and CS_2 is not well known. We propose two scenarios in which we relate the magnitudes of nitrogen inputs to forest ecosystems and the forms and magnitudes of gaseous sulfur outputs from these systems. The first scenario considers nitrogen-limited forest ecosystems and can be summarized as follows. In nitrogen-limited systems, plants accumulate excess SO_4 in their leaves (Kelley and Lambert 1972). During photosynthesis, SO_4 is reduced, via a light-dependent reaction, to H_2S that is then emitted from the plants through their stomates (Rennenberg 1984). These H_2S emissions have very little effect on global atmospheric processes because H_2S has a very short lifetime in the atmosphere before it is oxidized to SO_4, which is then returned to the land or sea in precipitation.

The second scenario focuses on nitrogen-saturated systems. In these systems, where nitrogen is available in excess of plant demand, the plants transform a large fraction of the SO_4 in their leaves into carbon-bonded sulfur such as the sulfur-containing amino acids (Kelley and Lambert 1972). While some of this carbon-bonded sulfur may be transformed into H_2S in a nonlight-dependent reaction (Rennenberg et al. 1982), most of it will enter the soil as plant litter. In the soil, microbes will transform the sulfur-containing amino acids into a number of compounds including COS and CS_2.

To test the hypotheses associated with these scenarios, we began to study the effects of chronic nitrogen additions to forest ecosystems on the COS and CS_2 exchanges between the soils of these systems and the atmosphere (Melillo and Steudler 1989). Nitrogen fertilization increased the total of the combined COS plus CS_2 emissions from the soils in both stands (Table 5). The total emissions were increased by nearly a factor of three in the deciduous stand and were more than doubled in the coniferous stand. Interestingly, nitrogen fertilization did not affect COS and CS_2 emissions in the same way in the two stands. The added nitrogen caused a dramatic increase in COS emissions from the deciduous stand (a factor of three increase), while CS_2 emissions from this site were not affected. We observed

TABLE 5. Mean carbonyl sulfide and carbon disulfide emissions from nitrogen fertilized and nonfertilized forest soils. Flux units are micrograms sulfur per m^2 per day and the numbers in parentheses are the standard errors.

Stand	Treatment	COS	CS_2	$COS + CS_2$
Hardwood	Control	8.66 (6.92)	2.04 (0.53)	10.71* (6.64)
	Fertilized	27.40 (7.34)	1.71 (0.23)	29.11 (2.06)
Pine	Control	19.86 (8.98)	2.36* (0.17)	22.21 (10.18)
	Fertilized	26.63 (10.33)	22.01* (3.55)	48.64 (13.88)

* Difference between control and fertilized treatments was significant at 0.05 level, n = 4 chambers.

the opposite response in the coniferous stand; that is, the nitrogen fertilization had no affect on COS emissions but did stimulate CS_2 emissions (a factor of more than nine increase).

The mechanisms responsible for the observed linkage between sulfur and nitrogen in forest ecosystems are not yet understood. We do know that many forests in New England, including the ones studied, are nitrogen limited (Aber et al. 1983). We also know that sulfur-containing amino acids, methionine, cystine, and cysteine are precursors of COS and CS_2 (Bremner and Steele 1978; Minami and Fukushi 1981a, 1981b). Perhaps nitrogen additions to forest ecosystems increase the pool of sulfur-containing amino acids in soils through microbial synthesis of these compounds. Alternatively, the added nitrogen could be taken up by the trees, the sulfur-containing amino acids could be synthesized by the vegetation, and eventually the plant-produced amino acids could enter the soil system in plant litter or in leaf or needle leachate. There is support in the literature for both alternatives (see Melillo and Steudler 1989).

The linkage between nitrogen inputs and sulfur gas outputs may have important implications for the global budget of COS and CS_2 and postulated changes in the budget. This topic certainly deserves further consideration.

FUTURE RESEARCH NEEDS

We have discussed a number of linkages between nitrogen deposition and fluxes of trace gases between the atmosphere and soils in forest ecosystems of the temperate and boreal zones. While we have not focused on agricultural systems of these regions, these systems must exhibit at least some of the same linkages. In fact, we already know that there is a coupling between nitrogen fertilization and N_2O emissions from agricultural soils. We feel that the potential link between nitrogen fertilization and CH_4 consumption is certainly worth investigating in agricultural systems.

The linkage between the nitrogen status of soils and trace gas fluxes is also important in tropical ecosystems. Since many tropical ecosystems are not limited by nitrogen, these systems may exhibit some "extreme" behaviors relative to the couplings we have recognized.

In our work to date we have not considered carbon monoxide (CO) fluxes between the soil and the atmosphere. There is reason to believe that CO may behave like CH_4 in response to nitrogen fertilization since the microbial enzyme systems involved in CO oxidation are very similar to those involved in CH_4 and NH_4 oxidation (Ferenci et al. 1975; Suzuki et al. 1976; Jones and Morita 1983a).

Moving the focus from soils to plants, we have a number of questions about how chronic nitrogen deposition affects the nutrient status of plants and how this in turn affects trace gas exchanges between plants and the atmosphere. Work certainly needs to be done in this area for a number of trace gases including COS and isoprene. There is some evidence that the nitrogen status of plants can affect the terpenoids (Gershenzon 1984), but there is no specific information about the compounds of interest such as isoprene.

Finally, we need a great deal of geographically referenced information to extrapolate our site-specific findings to regional- and global-scale estimates of trace gas exchanges between the land and the atmosphere as influenced by nitrogen deposition. The information needed includes detailed nitrogen deposition data (inorganic and organic N, wet plus dry) and data on a variety of ecosystem characteristics. These data should be organized in a geographic information system in such a way as to allow the coupling of "the background information" with processes models that would "predict" the magnitude of trace gas fluxes between the land and the atmosphere as the environment changes. In our opinion, the development of a geographically referenced "ecological" data base should be among the highest priorities of the upcoming IGBP activity.

Acknowledgements. This research was supported by the Ecosystems Studies and the Atmospheric Chemistry Programs of the National Science Foundation and by the Exxon Corporation. We thank the staff of the Harvard Forest for their cooperation.

REFERENCES

Aber, J.D., J.M. Melillo, C.A. McClaugherty, and K.N. Eshleman. 1983. Potential sinks for mineralized nitrogen following disturbance in forest ecosystems. In: Environmental Biogeochemistry, ed. R.O. Hallberg, vol. 5. *Ecol. Bull.* 35:179–192.

Aber, J.D, J.K. Nadelhoffer, P.A. Steudler, and J.M. Melillo. 1989. Nitrogen saturation in northern forest ecosystems—hypotheses and implications. *Bioscience*, in press.

Barrie, L.A., and J.M. Hales. 1984. The spatial distributions of precipitation acidity and major ion wet deposition in North America during 1980. *Tellus* 36B: 333–355.

Bolle, H.-J., W. Seiler, and B. Bolin. 1986. Other greenhouse gases and aerosols: assessing their role for atmospheric radiative transfer. In: The Greenhouse Effect, Climatic Change, and Ecosystems, ed. B. Bolin, B.R. Doos, J. Jager, and R.A. Warrick, pp. 157–203. SCOPE 29. Chichester: Wiley.

Bónis, K., E. Mészáros, and M. Putsay. 1980. On the atmospheric budget of nitrogen compounds over Europe. *Időjárás* (periodical of the Hungarian Meteorological Service) 84:57–68.

Bowden, W.B. 1986. Gaseous nitrogen emissions from undisturbed terrestrial ecosystems: an assessment of their impacts on local and global nitrogen budgets. *Biogeochem.* 2:249–279.

Bremner, J.M., and C.G. Steele. 1978. Role of microorganisms in the atmospheric sulfur cycle. *Adv. Microb. Ecol.* 2:155–201.

Buijsman, E., and J.-W. Erisman. 1988. Wet deposition of ammonium in Europe. *J. Atmos. Chem.* 6:265–280.

Crutzen, P.J. 1976. The possible importance of CSO for the sulfate layer of the stratosphere. *Geophys. Res. Lett.* 3(2):73–76.

Ferenci, T., T. Strom, and J.R. Quayle. 1975. Oxidation of carbon monoxide and methane by *Pseudomonas methanica. J. Gen. Microbiol.* 91:79–91.

Gershenzon, J. 1984. Changes in the levels of plant secondary metabolites under water and nutrient stress. In: Phytochemical Adaptations to Stress, ed. B.N. Timmermann, C. Steelink, and F.A. Loewys, pp. 273–320. New York: Plenum.

Hofmann, D.J., and J.M. Rosen. 1981. On the background stratospheric aerosol layer. *J. Atmos. Soc.* 38:168–181.

Hyman, M., and P.M. Wood. 1983. Methane oxidation by *Nitrosomonas europaea. Biochem. J.* 212:31–37.

Jones, B.M.R., R.A. Cox, and S.A. Penkett. 1983. Atmospheric chemistry of carbon disulfide. *J. Atmos. Chem.* 1:65–86.

Jones, R.D., and R.Y. Morita. 1983a. Carbon monoxide oxidation by chemolithotrophic ammonium oxidizers. *Can. J. Microbiol.* 29:1545–1551.

Jones, R.D., and R.Y. Morita. 1983b. Methane oxidation by *Nitrosococcus oceanus* and *Nitrosomonas europaea. Appl. Env. Microbiol.* 45:401–410.

Keller, M., T.J. Goreau, S.C. Wofsy, W.A. Kaplan, and M.B. McElroy. 1983. Production of nitrous oxide and consumption of methane by forest soils. *Geophys. Res. Lett.* 10:1156–1159.

Keller, M., W.A. Kaplan, and S.C. Wofsy. 1986. Emissions of N_2O, CH_4 and CO_2 from tropical forest soils. *J. Geophys. Res.* 91:11791–11802.

Kelley, J., and M.J. Lambert. 1972. The relationship between sulfur and nitrogen in the foliage of *Pinus radiata. Plant Soil* 37:395–407.

Kerr, R.A. 1988. Ozone hole bodes ill for the globe. *Science* 241:785–786.

Khalil, M.A.K., and R.A. Rasmussen. 1983. Sources, sinks, and seasonal cycles of atmospheric methane. *J. Geophys. Res.* 88:5131–5144.

Khalil, M.A.K., and R.A. Rasmussen. 1984. Global sources, lifetimes, and mass balances of carbonyl sulfide (OCS) and carbon disulfide (CS_2) in the earth's atmosphere. *Atmos. Envir.* 18:1805–1813.

Khalil, M.A.K., and R.A. Rasmussen. 1987. Atmospheric methane: trends over the last 10,000 years. *Atmos. Envir.* 21:2445–2452.

Lineweaver, H., and D. Burk. 1934. The determination of enzyme dissociation constants. *J. Am. Chem. Soc.* 56:658–666.

Melillo, J.M., R.J. Naiman, J.D. Aber, and A.E. Linkins. 1984. Factors controlling mass loss and nitrogen dynamics of plant litter decaying in northern streams. *Bull. Mar. Sci.* 35:341–356.

Melillo, J.M., and P.A. Steudler. 1989. The effect of nitrogen fertilization on the COS and CS_2 emission from temperate forest soils. *J. Atmos. Chem.*, in press.

Minami, K., and S. Fukushi. 1981a. Detection of carbonyl sulfide among gases produced by the decomposition of cystine in paddy soils. *Soil Sci. Pl. Nutr.* 27(1):105–109.

Minami, K., and K.S. Fukushi. 1981b. Volatilization of carbonyl sulfide from paddy soils treated with sulfur-containing substances. *Soil Sci. Pl. Nutr.* 27(3):339–345.

Mitchell, H.L., and R.F. Chandler. 1939. The nitrogen nutrition and growth of certain deciduous trees of the northeastern United States. Black Rock Forest Bull. 11.

Olson, J.S., J.A. Watts, and L.J. Allison. 1983. Carbon in live vegetation of major world ecosystems. Environmental Sciences Division Publ. 1997. Oak Ridge National Laboratory. Oak Ridge, TN: NTIS, US Dept. of Commerce.

Oppenheimer, M. 1987. Stratospheric sulfate production and the photochemistry of the Antarctic circumpolar vortex. *Nature* 328:702–704.

Rennenberg, H. 1984. The fate of excess sulfur in higher plants. *Ann. Rev. Plant Physiol.* 35:121–153.

Rennenberg, H., J. Sekijo, L.G. Wilson, and P. Filner. 1982. Evidence for an intracellular sulfur cycle in cucumber leaves. *Planta* 154:516–524.

Robertson, G.P. 1982. Nitrification in forested ecosystems. *Phil. Trans. R. Soc. Lon. B* 296:445–457.

Rodhe, H., E. Cowling, I.E. Galbally, J.N. Galloway, and R. Herrera. 1988. Acidification and regional air pollution in the tropics. In: Acidification in Tropical Countries, ed. H. Rodhe and R. Herrera, pp. 1–42. SCOPE 36. Chichester: Wiley.

Seiler, W. 1984. Contribution of biological processes to the global budget of CH_4 in the atmosphere. In: Current Perspectives in Microbial Ecology, ed. M.J. Klug and C.A. Reddy, pp. 468–477. Washington, D.C.: Am. Soc. Microbiol.

Seiler, W., and R. Conrad. 1987. Contribution of tropical ecosystems to the global budgets of trace gases, especially CH_4, H_2, CO, and N_2O. In: The Geophysiology of Amazonia: Vegetation and Climate Interactions, ed. R.E. Dickinson, pp. 133–162. New York: Wiley.

Servant, J. 1986. The burden of the sulfate layer of the stratosphere during volcanic "quiescent" periods. *Tellus* 38B:74–79.

Steudler, P.A., R.D. Bowden, J.M. Melillo, and J.D. Aber. 1989. Influence of nitrogen fertilization on methane uptake in temperate forest soils. *Nature*, in press.

Steudler, P.A., and B.J. Peterson. 1984. Contribution of gaseous sulfur from salt marshes to the global sulfur cycle. *Nature* 311:455–457.

Suzuki, I., S.-C. Kwok, and I. Dular. 1976. Competitive inhibition of ammonium oxidation in *Nitrosomonas europaea* by methane, carbon monoxide or methanol. *FEBS Lett.* 72:117–120.

Tolbert, M.A., J.J. Rossi, and D.M. Golden. 1988. Heterogeneous interactions of chlorine nitrate, hydrogen chloride, and nitric acid with sulfuric acid surfaces at stratospheric temperatures. *Geophys. Res. Lett.* 15(8):847–850.

Turco, R.P., R.C. Whitten, O.B. Toon, J.B. Pollack, and P. Hamill. 1980. COS, stratospheric aerosols and climate. *Nature* 283:283–296.

Vitousek, P.M., and P.A. Matson. 1985. Causes of delayed nitrate production in two Indiana forests. *Forest Sci.* 31:122–131.

Whittenbury, R., K.C. Phillips, and J.F. Wilkinson. 1970. Enrichment, isolation and some properties of methane-utilizing bacteria. *J. Gen. Microbiol.* 61:205–218.

WRI. 1988. World Resources 1988–1989. The World Resources Institute and the International Institute for Environment and Development. New York: Basic Books.

Exchange of Trace Gases between Terrestrial Ecosystems and the Atmosphere
eds. M.O. Andreae and D.S. Schimel, pp. 281–290
John Wiley & Sons Ltd

Global Climate and Trace Gas Composition: From Atmospheric History to the Next Century

S.H. Schneider[1]

National Center for Atmospheric Research[2]
Boulder, CO 80307–3000, U.S.A.

Abstract. The sawtooth signature of glacial/interglacial cycles over the past half-million years has been attributed to external factors (e.g., orbital variation), internal factors (e.g., trace gas changes or bedrock depression), or both. Since future climate changes from increasing greenhouse gases such as CO_2 or CH_4 can alter climate sufficiently to warm the surface several degrees C or alter soil moisture by tens of percent, these climate changes could alter microbial production or consumption of these gases, with substantial potential feedback on climate change. It is argued that interdisciplinary research on the interactions among climate change and trace gas composition be accelerated.

Over the past few million years, the Earth has experienced more than a dozen oscillations in climate in what are known as ice age/interglacial cycles (Fig. 1). In the most recent of these, temperatures varied by about 5°C globally (about 3°C over the oceans and 8°C or so on the continents, with maximum changes in the high latitudes and relatively little temperature change in the tropics [CLIMAP 1981]. The glacial/interglacial transitions make somewhat of a sawtooth pattern (see Fig. 1), with about a 70,000- to 80,000-year decline in temperature and sea level (sea levels vary about 100 meters from glacial to interglacial) followed by an abrupt termination that typically takes 10,000 years and then about a 10,000-year interglacial period before the cooling part of the cycle begins again. The transitions are not

[1] Any opinions, findings, conclusions or recommendations expressed in this article are those of the author and do not necessarily reflect the views of the National Science Foundation.
[2] The National Center for Atmospheric Research is sponsored by the National Science Foundation.

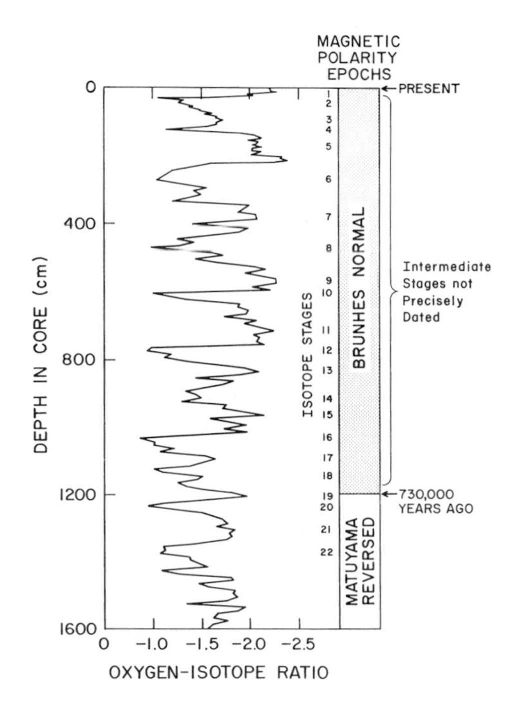

Fig. 1—Variation in the oxygen isotope ratios in the shells of fossil forams living near the ocean floor taken from a deep-sea core in the Pacific Ocean. If all other factors are constant, less negative values of this oxygen isotope ratio index (δO^{18}) indicate colder climates corresponding to increased ice volumes. However, a decrease in bottom-water temperature would also cause a less negative δO^{18}. Each major change of direction in the oxygen isotope ratio curve is called a stage, as indicated on the figure. Inasmuch as similar stages are found from deep-sea cores taken all over the world, many paleoclimatologists believe that these major shifts in oxygen isotope ratio index indicate a record of global climatic change over the past million years or so. The only certain globally synchronous dates, however, are the present and the time when the Earth's magnetic polarity reversed roughly 730,000 years ago. Source: Shackleton and Opdyke (1973).

always smooth; there is some evidence of abrupt changes in at least regional climates (e.g., the Younger Dryas).

Although no two of these events look identical, the major "cycle" occurs with about 100,000-year frequency over the past three quarter of a million years. There also is substantial variation occurring with 41,000- and about 20,000-year periods (these coincide with the frequencies of variations in characteristics of the Earth's orbit around the sun caused by the gravitational tugs of other planets). These orbital parameters primarily affect the latitudinal and seasonal cycle of solar radiation, changing the winter-to-summer radiation difference by as much as 5 to 10%. The causes of the ice ages are still debated. Proposed mechanisms range from the astronomical theory, to volcanic eruptions, to fully internal explanations. The latter include the interactions among atmosphere, oceans, ice fields, bedrock changes, and biological connections—especially those that affect the composition of greenhouse gases such as carbon dioxide and methane. It is, of course, possible—even likely—that all of these mechanisms may be operating simultaneously, with the most popular current thinking suggesting that the 20,000- and 40,000-year cycles in the geological record almost certainly were externally forced by orbital changes (with perhaps some feedback on cloud albedo from DMS production in the oceans or CO_2 and methane from the biosphere). However, the rate at which temperature dropped off at the end of the previous interglacial is not identical in different proxies, e.g., compare Fig. 1 to Fig. 2 about 120,000 years ago. Both records show some recovery after the initial glacial expansion, followed by a noisy drift toward the last glacial maximum at 18,000 YBP. However, the 100,000-year cycle in ice volume and bottom water temperature in the geologic record (which is what is represented by the δO^{18} curve on Fig. 1) and the 100,000-year cycle in the Earth's eccentricity may be fortuitous; at least no clear mechanism has been demonstrated. (For a discussion of the climatic interpretation of Fig. 1 see Schneider and Londer [1984].) This 100,000-year periodicity is often attributed to internal nonlinear dynamics of large ice sheets and bedrock (e.g., Pollard 1983).

Regardless of the actual cause of these events, a remarkable historic record covering the past 150,000 years has been uncovered at Vostok in Antarctica. Gas bubbles trapped in ice are preserved over these periods of time and have shown that carbon dioxide and methane levels were both substantially reduced during the glacial periods relative to interglacial periods. The Vostok records show that CO_2 (Fig. 2) was about 190 to 200 ppm at the height of the last ice age (about 18,000 years ago) and the ice age before that (some 150,000 years ago), and that CO_2 changed in very close association with the change in the oxygen or deuterium isotope ratio in the ice in Antarctica. The δO^{18} and D/H ratios are generally believed to be good proxies of the local temperature over that continent during snowfall.

Debate takes place as to whether the CO_2 or temperature signal is a lead or lag; in other words, a cause and effect or effect and cause. Correlation of the ice record with ocean cores (such as Fig. 2 compared to Fig. 1) may help to provide insight, except for the complication that precise dating in the glaciers and oceans is still an issue of some controversy. For example, the ice sheet record is dated by modeling; also the δO^{18} signal in the oceans represents an ice volume change that is not necessarily time synchronous with the actual change in ice volume because the δO^{18} signature in ice sheets varies internally since large ice sheets have characteristic flow times

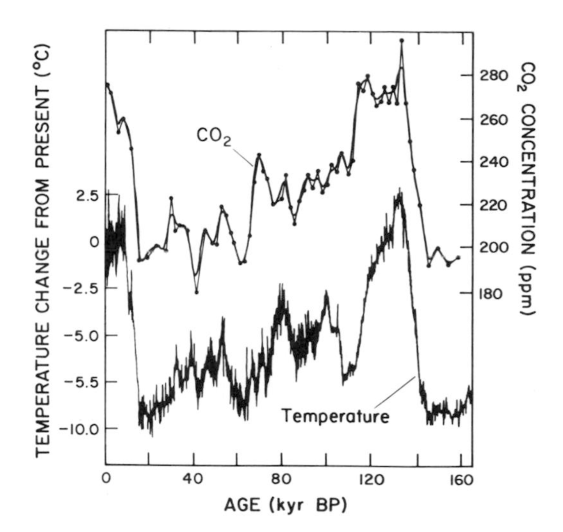

Fig. 2—CO_2 concentrations ("best estimates") and smoothed values (spline function) in ppmv. plotted against age in the Vostok record (upper curves) and local atmospheric temperature change derived from the isotopic profile (lower curve). Source: Barnola et al. (1987).

of tens of thousands of years (e.g., see Covey and Schneider 1984). Also, the marine δO^{18} signal as a proxy for ice volume is confounded by any changes in bottom water temperature. Nevertheless, the ice core records show that not only has CO_2 closely tracked the δO^{18} signal in ice (that is, the proxy for local temperature signal) in the region of the ice cores in both Greenland and Antarctica, but so has methane. The one major exception is that local temperature proxies in Antarctica seem to plunge at the end of the previous interglacial a few thousand years earlier than CO_2. Whether, for example, this is a local phenomenon related to a change in the δO^{18} source region for evaporation that later ends up as snow, or a surge of the Antarctic ice sheet (East or West), or a truly global phenomenon is not yet clear. However, the ocean proxy data on Fig. 1 resembles the CO_2 curve on Fig. 2. Explanation of the CO_2 association with temperatures over the bulk of the record varies. Some people have argued that changes in nutrient runoff associated with the covering or uncovering of shelves with changing sea level could be a biologically modulated feedback (through altered phytoplankton productivity) that makes CO_2 vary closely with temperature. Other explanations (see Sundquist and Broecker 1985 for references) include changes in deep ocean circulation or biological productivity (perhaps even driven by insolation changes associated with changing orbital elements). These largely internal mechanisms typically rely upon ocean chemistry and biological productivity.

Another internal mechanism is actually mechanical. David Erickson at Scripps Institution of Oceanography (submitted) has suggested that gas exchange rates between atmosphere and ocean may have varied between ice ages and interglacial times because of the substantially different pattern of surface winds over the ocean. These surface winds, of course, modulate gas exchange rates given a vertical profile of temperature and gaseous concentration in both the atmosphere and oceanic components. Using the general circulation model results of Kutzbach and Guetter (1986), which produce three-dimensional wind fields characteristic of 18,000 years ago and five other times right up to the present, Erickson speculated that increased wind speeds in areas of CO_2 undersaturation in the oceans might have led to increased CO_2 flux into the oceans. This occurred in the GCM at the last glacial maximum relative to the present and is a competitive internal explanation of how whatever caused the ice ages to come and go and thereby modify the wind field led to a change in the atmospheric concentration of the greenhouse gas CO_2, which may have been an amplifying feedback.

Lee Klinger (1988) has also suggested that the CO_2 cycle was internally generated, not only by ocean biota but by land biota. He argues that bog formation in both high latitudes and tropical areas was very large during the ice buildup phase just before the last glacial maximum and that large carbon storage on land could then account for some of the observed draw down in the atmospheric CO_2. Further, during glacial termination, the growth of forests and the reduction of the stored organic matter on land could explain some of the buildup of CO_2.

Also related is the methane record. This is potentially important to climate because methane is about 20 to 30 times more effective per molecule than CO_2 as an infrared-absorbing greenhouse gas. Presently, a great deal of methane is stored under permafrost and off continental shelves in a form known as clathrates. Below certain temperatures and above certain pressures, this matrix locks methane in. Gordon MacDonald (submitted) has suggested that as ice ages are waning, the downward penetration of thermal warming into permafrost would have allowed CH_4 trapped in some of the clathrates to escape. Such warming could have released potentially vast quantities of methane. That could help to explain the approximate doubling of methane observed between 1800 A.D. and the last ice age. Also, since methane today chemically converts to CO_2 in the atmosphere in about 10 years, the relation between CH_4, CO_2, and other chemical constituents over geological time needs to be carefully studied. It is important that the Dahlem Workshop participants examine what the causes are of the methane stored in the clathrates as well as the larger questions of carbon transformation, storage, and release, as these issues are significant not only for biogeochemical cycling per se but for climate change as well.

Also in need of study are variations of sulfate in ice cores (perhaps driven

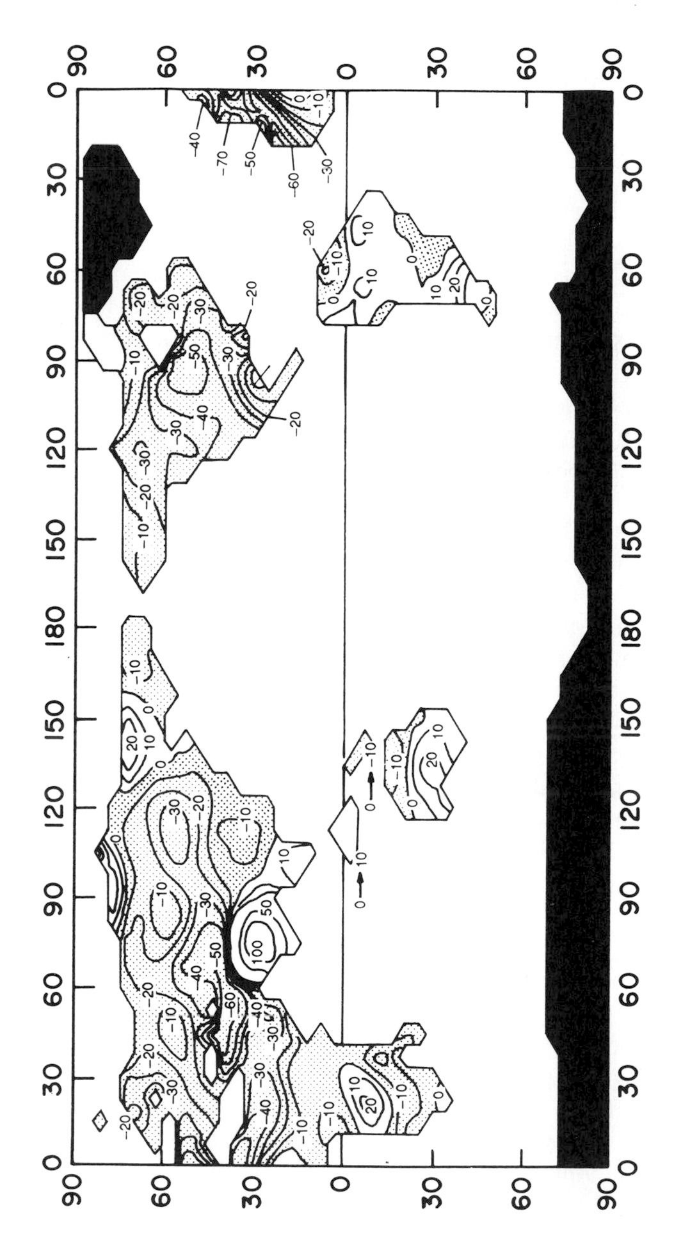

Fig. 3—CO_2-induced change in soil moisture expressed as a percentage of soil moisture obtained from a computer model with doubled CO_2 compared to a control run with normal CO_2 amounts. Note the nonuniform response of this ecologically important variable to the uniform change in CO_2. Source: Manabe and Wetherald (1986).

by orbitally induced changes in DMS production) as well as ice and ocean core records of mineral dust, whose changing composition over time needs more elaboration.

Concerning the future of climate change, a major debate has broken out over the relative role of the biosphere as a climate feedback mechanism given accelerating human-induced injections of trace greenhouse gases or CO_2 from fossil fuels or deforestation. For example, it has long been argued that an increase in CO_2 will cause a photosynthetic response in the biosphere. This would remove some of that added CO_2 providing a negative feedback mechanism (e.g., Goudriaan and Ketner 1984). On the other hand, George Woodwell (pers. comm.) at Woods Hole has argued that there is roughly twice as much carbon stored in the form of dead organic matter in the soil (necromass) than there is stored in live form in forests (which is comparable to that in the atmosphere as CO_2). Moreover, Woodwell notes that the decomposition rate of that necromass into CO_2 and methane depends upon the microbial activity in the soil. This, in turn, depends on soil moisture and temperature. If all other factors were equal, Woodwell has argued, increasing temperature will increase the rate of microbial activity, thereby increasing the rate of the decomposition of necromass, which is a potentially large methane and (some decades later) carbon dioxide source. Thus, Woodwell has suggested that this would be a positive feedback, which on the time frame of a century could substantially accelerate the greenhouse effect due to anthropogenic injections of CO_2, methane, and other trace gases. However, soil moisture could also change substantially if greenhouse gases continued to increase and the climate is changed (see Fig. 3). How temperature and soil moisture changes could affect microbial action (e.g., altering the balance between methane-producing and consuming species) needs more careful attention.

Dan Lashof (1989) of the U.S. Environmental Protection Agency has tried to analyze a number of climatic feedback mechanisms and has produced a table in which he adds these processes together (reproduced as Table 1). These appear as linear "gains," in which the sensitivity of the climate to greenhouse forcing is inversely proportional to 1 minus the total gain factor. While such simple addition of feedbacks ignores the interactions among them, Lashof concludes that synergism of all gains is more likely to be an amplifying positive feedback (i.e., positive global gain) than a negative feedback. In any case, it seems appropriate that the global implications of changing greenhouse gas composition remain a principal focus of the Dahlem Workshop and the issues of methane and CO_2 exchanges between soils, plants, atmosphere, and oceans especially be considered, given the obvious importance of these exchanges both in explaining past changes and in predicting future ones.

TABLE 1. Estimated gain from climate and biogeochemical feedbacks (source: Lashof 1989).

Feedback	Gain		Source
Geophysical			
Water Vapor[a]	0.39	(0.28–0.52)	
Ice and Snow	0.12	(0.03–0.21)	Dickinson (1986)
Clouds	0.09	(−0.12–0.29)	
Sub-Total[b]	0.64	(0.17–0.77)	
Biogeochemical			
Methane Hydrates	0.1	(0.01–0.2)	after Revelle (1983)
Tropospheric Chemistry	−0.04	−(0.01–0.06)	Hameed and Cess (1983)
Ocean Chemistry	0.008		$\partial \ln(pCO_2)/\partial T = 4\%$
Ocean Eddy-Diffusion	0.02		$1/K \propto (\partial T/\partial Z)^2$
Ocean Biology and Circulation	0.06	(0.0–0.1)	after Sarmiento et al. (1984)
Vegetation Albedo	0.05	(0.0–0.09)	after Hansen et al. (1984); Dickinson and Hanson (1984)
Vegetation Respiration	0.01	(0.0–0.03)	Flux = 0.5 Pg $y^{-1} {}^\circ C^{-1}$
CO_2 Fertilization	−0.02	−(0.01–0.04)	15% biomass increase for 2 × CO_2
Methane from Wetlands	0.01	(0.003–0.015)	Lashof and Fung, unpublished
Methane from Rice	0.006	(0.0–0.01)	after Holzapfel-Pschorn and Seiler (1986)
Electricity Demand	0.001	(0.0–0.004)	after Linder and Inglis (1988)
Sub-Total[c]	0.16	(0.05–0.29)[d]	
TOTAL	0.80	(0.32–0.98)[d]	

NOTES:
[a] Includes the lapse rate feedback and other geophysical climate feedbacks not included elsewhere.
[b] Based on 1.5–5.5 K for doubling CO_2. The individual values do not sum to these values—see Dickinson (1986) for details.
[c] Based on selected biogeochemical feedbacks, considering which could occur together during the next century—see text for details.
[d] Ranges are combined using a least-squares approach. That is, by letting $C - L = [\Sigma(c_i - l_i)^2]^{0.5}$, where C is the central estimate for the total, c_i is the central estimate for feedback i, L is the lower uncertainty bound for the total, and l_i is the lower uncertainty bound for feedback i. A similar calculation is performed for the upper uncertainty bounds.

REFERENCES

Barnola, J.M., D. Raynaud, Y.S. Korotkevich, and C. Lorius. 1987. Vostok ice core provides 160,000-year record of atmospheric CO_2. *Nature* 329:408–414.

CLIMAP Project Members. 1981. Sea surface temperature anomaly maps for August and February in the modern and last glacial maximum. Geological Society of America map and chart series M-36, maps 5A and 5B.

Covey, C., and S.H. Schneider. 1984. Models for reconstructing temperature and ice volume from oxygen isotope data. In: Milankovitch and Climate Change, ed. A. Berger, J. Imbrie, J. Hays, G. Kukla, and B. Saltzman, pp. 699–705. Dordrecht: Reidel.

Dickinson, R. 1986. The climate system and modelling of future climate. In: The Greenhouse Effect, Climate Change and Ecosystems, ed. B. Bolin, B. Doos, J. Jager, and R. Warrick, pp. 207–270. Chichester: Wiley.

Dickinson, R., and B. Hanson. 1984. Vegetation-albedo feedbacks. In: Climate Processes and Climate Sensitivity, ed. J. Hansen and T. Takahashi. *Geophys. Monog.* 29:180–186. Washington, D.C.: Am. Geophys. Union.

Goudriaan, J., and P. Ketner. 1984. A simulation study for the global carbon cycle, including man's impact on the biosphere. *Clim. Change* 6:167–192.

Hameed, S., and R. Cess. 1983. Impact of a global warming on biospheric sources of methane and its climatic consequences. *Tellus* 35B:1–7.

Hansen, J., A. Lacis, D. Rind, G. Russell, P. Stone, I. Fung, R. Ruedy, and J. Lerner. 1984. Climate sensitivity: analysis of feedback mechanisms. In: Climate Processes and Climate Sensitivity, ed. J. Hansen and T. Takahashi. *Geophys. Monog.* 29:130–163. Washington, D.C.: Am. Geophys. Union.

Holzapfel-Pschorn, A., and W. Seiler. 1986. Methane emission during a cultivation period from an Italian rice paddy. *J. Geophys. Res.* 91:11803–11814.

Klinger, L. 1988. Successional change in vegetation and soils of southeast Alaska. Thesis submitted to the Faculty of the Graduate School, Dept. of Geography and Institute of Arctic and Alpine Research, Univ. Colorado, Boulder, CO.

Kutzbach, J.E., and P.J. Guetter. 1986. The influence of changing orbital parameters and surface boundary conditions on climate simulations for the past 18,000 years. *J. Atmos. Sci.* 43:1726–1759.

Lashof, D. 1989. The dynamic greenhouse: feedback processes that may influence future concentrations of atmospheric trace gases and climatic change. *Clim. Change* 14:213–242.

Linder, K., and M. Inglis. 1989. The Potential Impacts of Climate Change on Electric Utilities: Regional and National Estimates. Report of ICF Inc. to the U.S. Environmental Protection Agency, in press.

Manabe, S., and R. Wetherald. 1986. Reduction in summer soil wetness induced by an increase in atmospheric carbon dioxide. *Science* 232:626–628.

Pollard, D. 1983. Ice-age simulations with a calving ice-sheet model. *Quatern. Res.* 20:30–48.

Revelle, R. 1983. Methane hydrates in continental slope sediments and increasing atmospheric carbon dioxide. In: Report of the Carbon Dioxide Assessment Committee, National Research Council, Changing Climate, pp. 252–261. Washington, D.C.: National Academy Press.

Sarmiento, J., and J. Toggweiler. 1984. A new model for the role of the oceans in determining atmospheric pCO_2. *Nature* 308:621–624.

Schneider, S., and R. Londer. 1984. The Coevolution of Climate and Life, pp. 64–71. San Francisco: Sierra Club.

Shackleton, N.J., and N.D. Opdyke. 1973. Oxygen isotopes and paleomagnetic stratigraphy of equatorial Pacific core v28-238: oxygen isotope temperatures and ice volumes on a 10^5 and 10^6 year scale. *Quatern. Res.* 3:39–55.

Sundquist, E.T., and W.S. Broecker, eds. 1985. The Carbon Cycle and Atmospheric CO_2: Natural Variations Archean to Present. *Geophys. Monog.* 32. Washington, D.C.: Am. Geophys. Union.

Exchange of Trace Gases between Terrestrial Ecosystems and the Atmosphere
eds. M.O. Andreae and D.S. Schimel, pp. 291–301
John Wiley & Sons Ltd

Experimental Design for Studying Atmosphere–Biosphere Interactions

R.C. Harriss

Institute for the Study of Earth, Oceans, and Space
University of New Hampshire
Durham, NH 03824, U.S.A.

Abstract. The study of trace gas exchange between terrestrial ecosystems and the atmosphere at regional to global scales requires simultaneous, integrated biophysical and biogeochemical measurements across a wide range of space (m^2 to 10^6 km^2) and time (10^{-1} to 10 yr) scales. Intensive field campaigns with simultaneous ground-based, aircraft, and satellite measurements can be used to determine which ecosystem components are major sources and sinks of specific atmospheric gases. Ground-based observatories can be best used to provide networks of continuous biophysical and biogeochemical measurements along important climatic or disturbance gradients within the high priority source/sink regions identified by intensive field campaigns. This combination of spatial and temporal data will elucidate feedback processes which determine how ecosystems respond to large-scale disturbances produced by climate variability and/or human activities. The highest priority ecosystems for comprehensive studies are northern high-latitude boreal and tundra regions (which are especially sensitive to climatic change), tropical moist forest and savanna regions in Africa and South America (which are subject to intensive land disturbance), and regions downwind of rapidly developing industrial regions of Asia and Latin America.

INTRODUCTION

This book, like many other recently published documents, recommends a "geophysiological" (Lovelock 1987) approach to understanding Earth systems. This change in primary emphasis, from individual investigators probing the detailed aspects of the geology or ecology of a local area to studies of regional and global coupling of biosphere–atmosphere–ocean–lithosphere components of the Earth, has resulted in a need to explore new approaches to field experimentation. In geophysiological research the emphasis is placed on measuring and understanding large-scale interactions which influence the structure and function of ecosystems and the composition of the global atmosphere.

It is extremely probable that human activities will continue to produce greenhouse gases in the next 50–100 years at rates which will result in continuing increases in global ambient air concentrations (e.g., for a review see Dickinson and Cicerone 1986). Scenarios of changes in globally averaged temperatures that might occur in response to continued emissions of atmospheric greenhouse gases are shown in Fig. 1, along with observed temperature data for the past 120 years. The middle scenario assumed a linear extrapolation of recent atmospheric concentration trends (except for chlorofluorocarbons) and a moderate climate sensitivity. In the professional judgement of a group of climate experts, there was a consensus for a 90 percent probability that the actual future pattern of greenhouse gas-induced climatic change will lie within the bounds set by the upper and lower curves (see Jager 1988 for a comprehensive discussion of the 1987 workshop held in Villach, Austria, which produced Fig. 1). Unfortunately, the scientific data base on sources certain of greenhouse gases (e.g., methane, nitrous oxide) is fragmentary and very inadequate for supporting such policy studies. Studies on the response of atmospheric photochemical processes to increasing

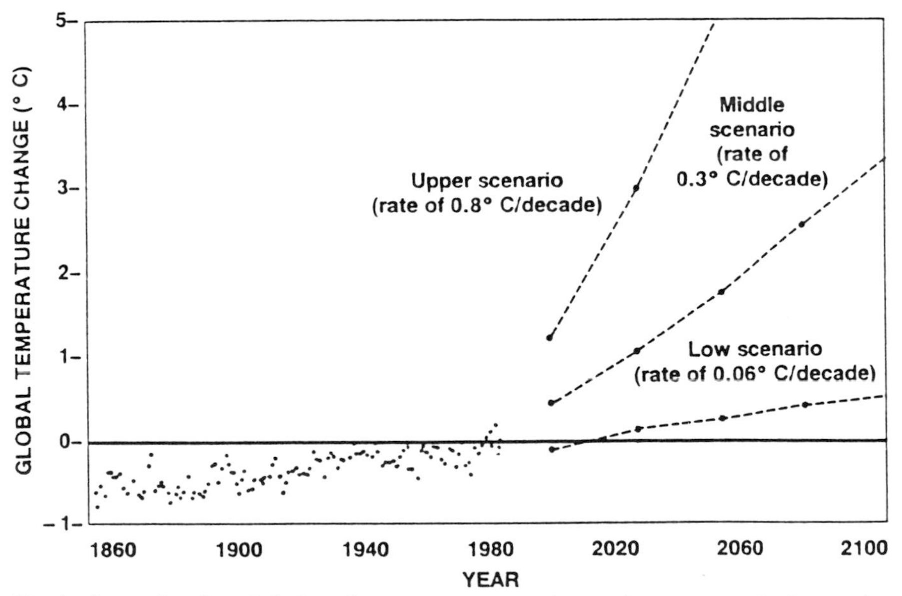

Fig. 1—Scenarios for global surface temperature change in response to increasing atmospheric greenhouse gas concentrations. Actual temperature data for 1860 to 1980 are also plotted. (From Jäger 1988).

concentrations of pollutant gases and/or changing climate are also just beginning (e.g., Isaksen and Hov 1987; Thompson and Cicerone 1986).

The premise of this paper is that our understanding of how the biosphere will respond to future climatic variability and/or disturbance by human activities in terms of large-scale exchange of trace gases with the atmosphere cannot progress much further without major improvements in the quantification of trace gas exchange rates and their controlling processes. The entire discussion of the "greenhouse problem" is theory-rich and data-poor. The following sections of this paper suggest elements of a strategy for a focused, internationally coordinated program of field expeditions and experiments, aimed specifically at understanding the effects of landscape change induced by either human activities or climate change on trace gas emissions. This is a subtle, albeit important, change in approach away from the more traditional "steady state" assumption used in designing field measurements to obtain data for geochemical mass balance modeling. In our approach a field measurement program is designed to study the effects of gradients in variables such as soil moisture, nutrient and/or pollutant flux, and landscape disturbance on regional and global scale biosphere–atmosphere trace gas interactions. It is assumed that the task of accurately determining the absolute magnitudes of many global trace gas sources is presently an unachievable objective; both fiscal and human resources are too limited. However, it should be possible in the next decade to understand some of the major factors which could enhance rates of global environmental change. It is the potential increase in rates of environmental change which is most likely to be the focus of social concern.

EXPERIMENTAL DESIGN FOR MEASURING LARGE-SCALE TRACE GAS EXCHANGE

The research discussed in this volume imposes a requirement for understanding trace gas exchange with terrestrial ecosystems at spatial scales compatible with grid scales used in global climate models. Pioneering efforts to develop terrestrial vegetation and soil data sets for global modeling purposes used 1° latitude by 1° longitude resolution (Matthews 1983; Wilson and Henderson-Sellers 1985). Current field programs aimed specifically at global understanding, such as the First ISLSCP Field Experiment (FIFE) and the Hydrologic Atmospheric Pilot Experiment (HAPEX), have conducted intensive measurement campaigns at scales of a General Circulation Model (GCM) grid square, i.e., of the order 10^4 km^2 (Andre et al. 1986; Hall et al. 1989).

The U.S. National Aeronautics and Space Administration (NASA), Global Tropospheric Experiment (GTE), Atmospheric Boundary Layer Expedition (ABLE) program attempted a regional-scale study of atmospheric trace gas and aerosol distributions and exchanges in the Brazilian Amazon (Harriss et al. 1988). European and African-based researchers have conducted similar studies in the equatorial rainforest region of the North Congo (Fontan and Çros 1988). The DECAFE (Dynamique et Chimie Atmosphérique en Forêt Equatoriale) project and GTE/ABLE provide the first comprehensive data base on biosphere–atmosphere interactions over tropical wet forests. All of these field campaigns have included some consideration of temporal variability in gas exchange due to seasonality in rainfall and consequent variations in soil moisture and vegetation physiological status.

Future research on biosphere–atmosphere trace gas exchange can combine elements of traditional, small-scale, process-oriented biogeochemical research with the "scaling-up" techniques being developed in current programs such as FIFE, HAPEX, DECAFE, and GTE/ABLE. Such experiments will require major resources and will only be conducted in a selected number of ecosystems. The following sections of this paper briefly describe components for this type of large-scale measurement program.

Characterizing Temporal and Spatial Variability

The initial step in understanding the significance of an ecosystem or region to atmospheric chemistry is to characterize both the absolute magnitude and the spatial and temporal scales of variability in patterns of trace gas exchange. There are also associated variables which may regulate the biological production and/or consumption of gases. The goal of such an intensive initial study is to identify specific trace gas sources and/or sinks which can potentially impact the regional or global atmosphere. Once the important source and sink environments have been identified, a focused, less resource-intensive, ground-based observatory program can be put in place to characterize impacts on trace gas exchange of variability resulting from long-term climate change, unpredictable episodic geophysical events (e.g., fire, drought), long-term effects of changes in land-use practices, etc.

An initial field experiment in a poorly understood region requires measurements across a range of spatial scales. Such a campaign will generally involve enclosure gas flux measurements, micrometeorological gas flux measurements, aircraft flux measurements, and satellite remote sensing of habitat distribution (Fig. 2). These simultaneous, integrated measurement programs address issues such as the relative importance of distributed, low intensity gas exchange processes versus localized, intense "hot spots." Following the definition of the general spatial and temporal characteristics of trace gas sources and sinks in a regional landscape, a longer-term, smaller-scale, focused investigation can be put in place to investigate effects of perturbations and/or natural variability.

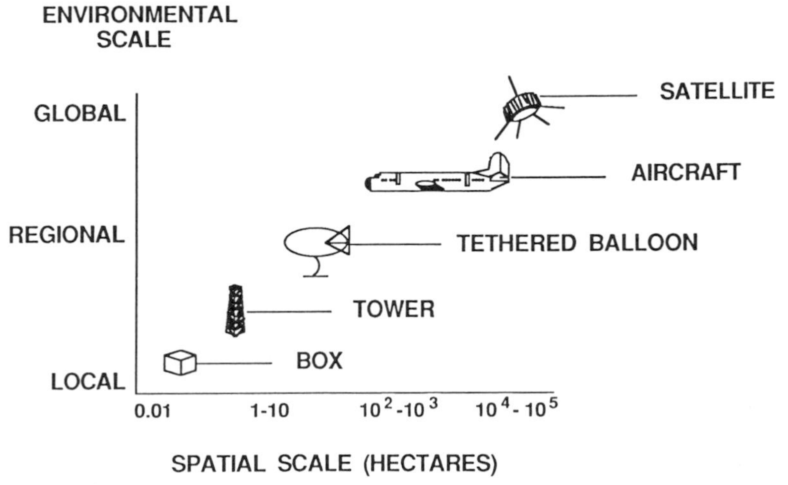

Fig. 2—A schematic illustration of the relationship between measurement techniques and platforms and the typical spatial coverage provided by their observations.

Recent studies on methane sources in subtropical and tropical wetland and forest environments illustrate the value of an initial intensive survey approach to identify critical environments for long-term research. In the Okefenokee Swamp, located in the southeastern United States, a complex variety of habitats was studied with remote sensing techniques to determine their spatial distribution and extent; seasonal measurements of methane flux were conducted in each habitat to derive an estimate of annual methane flux from each habitat. The combination of these two data sets provided a regional-scale estimate of methane source strengths (Table 1). Contrary to the expectations of the investigators, one habitat (shrub vegetation) was responsible for approximately 86 percent of the total methane emissions from the entire region. Future research on methane in this region could be focused on understanding feedbacks to environmental change in the shrub habitat. In this example, an improved understanding of methane emissions at the regional scale provided the rationale for a longer-term detailed study at local scales.

In a study of methane sources in the Amazon Basin, initial large-scale surveys of wetland habitats produced the expected conclusion that tropical floodplain environments are an important component of the global source from natural wetlands (Bartlett et al. 1988; Matthews and Fung 1987). However, methane studies from a tower in the upland forest showed a surprising result, pointing to a potential ecosystem response to climatic change. Even though enclosure measurements indicated that upland soils are a weak sink for atmospheric methane, there was a nocturnal accumulation of ambient methane under the forest canopy (S.C. Wofsy, pers. comm.).

TABLE 1. Methane flux from habitats in and adjacent to the Okefenokee Swamp, Georgia, U.S.A.

Vegetation Type	Area (hectares)	CH_4 Flux ($gCH_4/m^2/yr$)	Total Annual Flux (10^{10} gCH_4/yr)
Shrub	110,200	98	10.8
Prairie	20,500	48	1.0
Cypress	19,020	23	0.4
Gum/Bay	15,630	19	0.3
Open Water	1,600	96	0.1
Dry Forest	28,560	0	0

The source of this methane in the upland forest environment may be highly dispersed, waterlogged soils in topographic depressions, termites, or some other dispersed source which was not detected with soil enclosure techniques. Follow-up studies of globally significant methane sources in the tropics will need to investigate the origin of dispersed sources of methane in the forest. These sources may be particularly sensitive to human disturbance (e.g., deforestation) and climate variability, which would alter soil moisture and the extent of soil flooding.

In summary, initial efforts toward understanding ecosystem-scale biosphere–atmosphere interactions have made progress toward an important first step in measuring the large-scale exchange of gases between the biosphere and atmosphere. These experiments demonstrated the need for very comprehensive surveys of biosphere–atmosphere interaction across a wide range of spatial scales. Simultaneous trace gas measurements and remote sensing of ecological and biophysical characteristics of a landscape provide the basis for scaling the experimental design to meet the requirements for regional and global modeling. The primary weakness in the large-scale field experiments conducted to date has been the lack of integration of biophysical objectives (e.g., FIFE) with biogeochemical objectives (e.g., GTE/ABLE). The factors determining the moisture and heat budgets in soil should be studied concurrently with trace gas exchange.

Assessing Trace Gas Response to Climatic Change

Because trace gas exchange is sensitive to variations in temperature and moisture, biogeochemical measurements may contribute to early detection of climate change and to understanding the potential response of major ecosystems to changes in climate. Detecting climate change trends in ambient air temperature records and precipitation records has proved to be extremely difficult and controversial. An alternative, and perhaps more sensitive, method for detecting climate change trends is to use integrating measures such as trends in the near surface geothermal gradient (Lachenbruch and Marshall 1986) and trends in the emission of trace gases from ecosystems. For example, it is generally accepted that microbial processes in soils are sensitive to temperature. Methane emissions from northern boreal and tundra wetland soils have been shown to be exceptionally sensitive to soil temperature changes (Crill et al. 1988; Fig. 3). A ground-based observatory

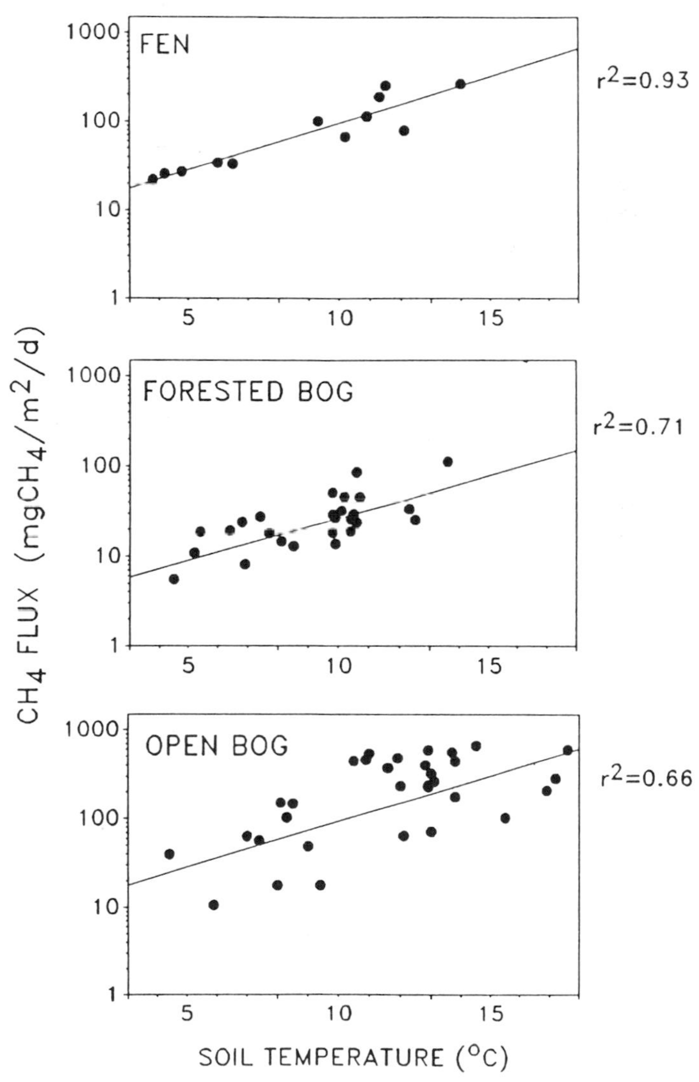

Fig. 3—Influence of soil temperature on methane emissions from peatlands in Marcell Experimental Forest, Minnesota.

system studying trends in soil geothermal characteristics and annual methane emissions along a boreal to polar continental wetlands transect could be an important future experiment for detecting climate change and for understanding feedbacks induced by changes in soil temperature and/or moisture regimes.

An alternate and complementary approach to investigating ecosystem and atmospheric chemical response to climate change is to design future experiments for regions characterized by high natural climate variability. The philosophy in this approach is to use natural extremes where they occur typically on time scales of less than once per decade (e.g., El Niño events) as large-scale experiments on the effects of climate change on the emission of trace gases from ecosystems. An experiment conducted from a ground-based observatory system in the southwestern Pacific region, ranging from northern Australia to southeast Asia, is an example of this approach. This region includes extensive areas of tropical rainforest and wetlands (Gore 1983; Matthews and Fung 1987). It is also a region with highly variable interannual rainfall and therefore frequently experiences large-scale drought. This unique coincidence of tropical ecosystems (which are important sources of methane, nitrous oxide, nitric oxide, and other trace gases to the atmosphere with highly variable moisture input) provides an excellent basis for a study of climate–ecosystem–atmospheric chemistry interactions.

Several studies have linked the 1982–1983 El Niño related climate extremes (e.g., extensive tropical droughts) to unexpected variability in global atmospheric carbon dioxide (Keeling and Revelle 1985; Gaudry et al. 1987). These studies concluded that observed changes in the pattern of global atmospheric carbon dioxide accumulation and seasonal variability associated with, and following, this El Niño event were not simply the result of changes in atmospheric circulation, but were likely to involve significant changes in global biospheric sources and sinks for carbon dioxide. Drought associated with El Niño also lowers the water table in large areas of tropical wetlands, which could reduce emissions of methane to the atmosphere. My preliminary model calculation indicated that a reduction of up to 28 Tg CH_4 yr^{-1} could have resulted from the drying of wetlands during the 1982–1983 drought period. However, these calculations do not take into account the possibility of enhanced methane emissions from the above normal occurrence of fires that accompanied the drought period.

Methods for forecasting El Niño events are improving rapidly. It is not unreasonable to expect that a combination of continuous, ground-based observations and intensive field campaigns could, over the period of about a decade, document the influence of climate variability associated with El Niños on global trace gas emissions from terrestrial ecosystems in the Tropics.

Assessing Ecosystem and Atmospheric Chemical Response to Human Activities

The spread of forest decline in Europe and eastern North America has been hypothesized to be related to complex interactions of possible changes in climate, atmospheric chemistry, and other factors which influence growth processes. This problem may be ideally suited to a research approach with integrated measurements of biological, biogeochemical, and physical processes across a wide range of space and time scales. In fact, many of the necessary components of such a study are already in place, but operating independently so that data integration is difficult. Initial studies indicate that acid deposition has a significant influence on trace gas exchange from some temperate zone forest ecosystems (Melillo et al., this volume).

The impact of population growth and land-use change in the tropics is a widely recognized problem with environmental implications (e.g., see papers in Dickinson 1987). Studies of the coupling of tropical ecosystems, both natural and disturbed, to global biogeochemical processes are urgently needed because the high rates of change in these environments may soon preclude critical baseline experiments in undisturbed forests. The most important regions for study are the regions of South America and Africa with rapidly growing rural populations. The moist forest and woodland–savanna regions of west Africa, southern Africa, and moist forest environments along the southern edge of Amazonia in South America are excellent candidates for future geophysiological research.

The projected economic development of countries like China and India will result in significant changes in both the total fluxes and patterns of pollutant emissions to the atmosphere. If these populous countries follow the industrial development paths of Japan, South Korea, and Taiwan, a major enhancement of pollutant gases can be expected over poorly studied regions downwind. Studies of sources of trace gas emissions associated with intensive agricultural regions (e.g., rice paddies) of Asia have also been identified as a high priority for future research (NAS 1984).

CONCLUSIONS

A regional or ecosystem approach to understanding the exchange of trace gases between terrestrial ecosystems and the atmosphere places priority on the following:

1. *Measurement of trace gas exchange at spatial scales similar to the resolution of global climate models* (e.g., 1°–5° grid squares). Scaling up to the global climate model grid scale, or to regional ecosystem

scales, requires an integration of remote sensing and *in situ* measurement techniques which characterize ecosystem biophysical properties and trace gas source/sink processes at scales ranging from m^2 to 10^6 km^2.

2. *Measurement of long-term trends in trace gas exchange in relation to potential perturbations caused by human activities or climate change.* Temporal variability of dominant trace gas sources and sinks can be determined with a network of ground-based observatories. These observatories should be networked along gradients of temperature, moisture, and land disturbance. The focus of such measurements is on determining feedbacks to changes in both external and internal forcing factors.
3. *Experiments in regions characterized by maximum potential for changes in climate or land disturbance.* High priority regions for such experiments include polar and boreal ecosystems which are expected to be particularly sensitive to climate change, tropical moist forest and savanna regions (where both natural variability in climate and land disturbance rates are high), and regions downwind of the rapidly developing industrial regions in Asia.

REFERENCES

Andre, J.-C., J.P. Goutorbe, and A. Perrier. 1986. HAPEX-MOBILHY, a hydrologic atmospheric pilot experiment for the study of water budget and evaporation at the climatic scale. *Bull. Am. Meteor. Soc.* 67:138–144.

Bartlett, K.B., P.M. Crill, D.I. Sebacher, R.C. Harriss, J.O. Wilson, and J.M. Melack. 1988. Methane flux from the central Amazonia floodplain. *J. Geophys. Res.* 93:1571–1582.

Crill, P.M., K.B. Bartlett, R.C. Harriss, E. Gorham, E.S. Verry, D.I. Sebacher, L. Modzar, and W. Sanner. 1988. Methane flux from Minnesota peatlands. *Glob. Biogeochem. Cyc.* 2:371–384.

Dickinson, R.E., ed. 1987. The Geophysiology of Amazonia. New York: Wiley.

Dickinson, R.E., and R.J. Cicerone. 1986. Future global warming from atmospheric trace gases. *Nature* 319:109–115.

Fontan, J., and B. Cros. 1988. The DECAFE 88 experiment: overview and large-scale meteorology. *EOS* 69:1065.

Gaudry, A., P. Monfray, G. Poliar, and G. Lambert. 1987. The 1982–1983 El Niño: a 6 billion ton CO_2 release. *Tellus* 39B:357–360.

Gore, A.J.P., ed. 1983. Ecosystems of the World. Parts 4a and 4b: Mires, Swamp, Bog, Fen, and Moor. New York: Elsevier.

Hall, F.G., P.J. Sellers, I. MacPherson, R.D. Kelly, S. Verma, B. Markham, B. Blad, J. Wang, and D.E. Strebel. 1989. FIFE: analysis and results—a review. *Adv. Space Res.*, in press.

Harriss, R.C., et al. 1988. The Amazon boundary layer experiment (ABLE 2A): dry season 1985. *J. Geophys. Res.* 93:1351–1360.

Isaksen, I.S.A., and O. Hov. 1987. Calculation of trends in the tropospheric concentration of O_3, OH, CH_4, and NO_2. *Tellus* 39B:271–285.

Jager, J. 1988. Developing policies for responding to climatic change. WCIP-1, WMO/TD-No. 225. Geneva: WMO.

Keeling, C.D., and R. Revelle. 1985. Effects of El Niño southern oscillation on the atmospheric content of carbon dioxide. *Meteoritics* 20:437–451.

Lachenbruch, A.H., and B.V. Marshall. 1986. Changing climate: geothermal evidence from permafrost in the Alaskan Arctic. *Science* 234:689–696.

Lovelock, J. 1987. In: Geophysiology of Amazonia, ed. R. Dickinson, pp. 11–24. New York: Wiley.

Matthews, E. 1983. Global vegetation and land use: new high-resolution data bases for climate studies. *J. Clim. Appl. Meteor.* 22:474–487.

Matthews, E., and I. Fung. 1987. Methane emission from natural wetlands: global distribution, area, and environmental characteristics of sources. *Glob. Biogeochem. Cyc.* 1:61–86.

National Academy of Sciences. 1984. Global Tropospheric Chemistry: A Plan for Action. Washington, D.C.: National Academy.

Thompson, A.M., and R.J. Cicerone. 1986. Possible perturbations to atmospheric CO, CH_4, and OH. *J. Geophys. Res.* 91:10853–10864.

Wilson, M.F., and A. Henderson-Sellers. 1985. A global archive of land cover and soils data for use in general circulation climate models. *J. Climatol.* 5:119–143.

Standing, left to right:
George Zavarzin, Michael Keller, Robert Delmas, Maria Kanakidou, Inez Fung, Heinz Bingemer, Jan Duyzer

Seated, left to right:
Paul Crutzen, Robert Harriss, Phil Robertson, Jerry Melillo, Andi Andreae

Exchange of Trace Gases between Terrestrial Ecosystems and the Atmosphere
eds. M.O. Andreae and D.S. Schimel, pp. 303–320
John Wiley & Sons Ltd

Group Report
Trace Gas Exchange and the Chemical and Physical Climate: Critical Interactions

G.P. Robertson, Rapporteur
M.O. Andreae
H.G. Bingemer
P.J. Crutzen
R.A. Delmas
J.H. Duyzer
I. Fung
R.C. Harriss
M. Kanakidou
M. Keller
J.M. Melillo
G.A. Zavarzin

INTRODUCTION

Of the major factors that regulate trace gas production and consumption in terrestrial ecosystems, perhaps none are more important than those that constitute the physical and chemical climate of the system. Directly or indirectly, both the physical climate (temperature, rainfall) and the chemical climate (the chemical composition of the atmosphere) affect virtually every aspect of ecosystem function, and thus can profoundly influence the net exchange of trace gases between terrestrial ecosystems and the atmosphere.

A unique feature of trace gas fluxes in terrestrial ecosystem functioning is their potential for participating in regional- and global-scale climate feedbacks. At local scales, trace gas exchange can affect within-canopy microclimates as well as individual plant-carbon balances, with concomitant effects on soil microbial communities. Such feedbacks make the relationships between climate and trace gas exchange an extremely important scientific topic. Past (glacial–interglacial), recent, and expected future changes in the global chemical climate provide further impetus for closely examining climate–gas flux interactions. The effects of increasing CO_2 levels on trace gas exchange, for example, is totally unknown. Global CO_2 fertilization may be leading to higher rates of carbon fixation in some environments and possibly to higher leaf C:N ratios (Lemon 1983), leading to changes in both the quantity and quality of leaf litter in these environments and subsequently to changes in the rates of soil organic matter turnover and N_2O, NO_x, and CH_4 fluxes.

A further reason to examine trace gas–climate interactions is to help balance existing atmospheric trace gas budgets. The N_2O budget is a particular case in point (Cicerone 1987). Atmospheric N_2O is long-lived enough to be relatively well mixed. Its concentration is about 306 ppb in the Northern Hemisphere and about 305 ppb south of the Equator. Measurements show that these concentrations have increased by 0.7 ($\pm$0.1) ppb per year since 1976, and other data show that this trend began before 1965. According to analysis of air from dated ice cores (Pearman et al. 1986), N_2O was present at concentrations of 280 to 285 ppb about 100 years ago. This contemporary increase of 0.25% per year represents a net annual addition to the global atmosphere of 3.5 Tg N_2O–N. The only sinks for the 1500 Tg N_2O–N now in the atmosphere are photodissociation by short wavelength ultraviolet light and reaction with O('D) in the stratosphere, which occur at a rate of about 12 Tg yr^{-1}. This implies an atmospheric loading today that is almost 30% above the likely steady-state preindustrial rate of 12 Tg yr^{-1}.

While the sizes of the preindustrial N_2O source and of the contemporary N_2O increase are clear from atmospheric knowledge, the identities of the individual components of these sources are not. It is not at all clear whether recognized candidate sources such as agricultural soils, native temperate and tropical soils, open oceans, and combustion can account for the total. At least one atmospheric model incorporating observed interhemispheric ratios suggests large tropical sources. This model (Prinn, pers. comm.) is based on empirically derived atmospheric transport rates and on time series measurements of N_2O at background stations in both hemispheres. Our present ignorance concerning N_2O sources and the recent discovery that an experimental artifact has led us to severely overestimate industrial combustion as an N_2O source (Muzio and Kramlich 1988) lead us to major questions about these sources: Where are they and what are their identities? What biological processes are most responsible for natural (preindustrial) and contemporary fluxes? What is the role of nitrogen-enriched ecosystems in Europe and North America, and of disturbed ecosystems in the tropics? And to what extent can the observed increases be attributed to trace gas interactions with the chemical or physical climate?

In the pages that follow we address four research issues related to our current understanding of interactions between trace gas exchange processes and the chemical and physical climate: (*a*) trace gas exchange at high latitudes, (*b*) changes in trace gas emissions resulting from increased N and S deposition, (*c*) effects of global CO_2 fertilization on trace gas exchange, and (*d*) trace gas exchange as affected by major land use changes and in particular by tropical deforestation and the widespread implementation of agricultural technologies. The need to understand issues (*b*) and (*d*) is especially urgent, as significant alterations in biospheric dynamics have already occurred as a result of accelerating human intervention. The need to understand issues (*a*) and (*c*) may be equally urgent if postulated atmosphere–biosphere feedbacks are as important as they appear to be. We conclude with specific research recommendations.

TRACE GAS EXCHANGE AT HIGH (> 50° N) NORTHERN LATITUDES

Northern Wetlands as a Source of Tropospheric Methane

Atmospheric CH_4 concentrations are chiefly a function of three processes: (*a*) the activity of microbial methanogenic communities, (*b*) the activity of methanotrophs (CH_4 consumers) at aerobic–anaerobic interfaces in wetlands and in a wide range of upland soils, and (*c*) photochemical CH_4 destruction in the atmosphere, which occurs mainly at low latitudes. Northern wetland ecosystems (those above 50° N) may well be a globally significant source of atmospheric methane (CH_4) (Aselmann, this volume), although emission rates are uncertain. This region is expected to be especially sensitive to climate change. Model predictions indicate that climate warming related to a doubling of atmospheric carbon dioxide (CO_2) will be most pronounced (6–8° C) in subarctic and polar regions. The production and emission of CH_4 from wetland soils is related, in part at least, to variations in soil temperature and moisture (Reeburgh and Whalen 1989; Crill et al. 1988; Svensson 1980; Baker-Blocker et al. 1977). Thus, we can expect the northern wetland CH_4 source to be potentially sensitive to regional variations in climate parameters such as precipitation and temperature. Because CH_4 is also a greenhouse gas it has a potential biospheric feedback: enhanced emissions of CH_4 to the atmosphere may further accelerate rates of regional and global climate warming.

Coupled photochemical reactions involving CH_4 also determine the oxidizing capacity of the atmosphere. The photochemistry of most of the background troposphere is to a large degree determined by the action of solar ultraviolet radiation on O_3, H_2O, CH_4, CO, and NO_x. At the low concentrations of NO_x that occur over much of the marine troposphere and in the middle troposphere over the continents, the photochemistry of this gas system has a substantial element of instability due to positive feedbacks (e.g., Crutzen 1986):

1. Increased emissions of CH_4 lead to less OH as OH is consumed in reactions with CH_4 to form ultimately CO_2 and H_2O, with CO as an intermediate product.

2. Because CH_4 is removed from the atmosphere mainly by reaction with OH, atmospheric CH_4 concentrations increase.

3. Because a substantial fraction of atmospheric CO is produced from CH_4 oxidation, and because CO is lost from the atmosphere by reaction with OH, CO also increases in the atmosphere.

4. As OH is lost by reaction with CO, atmospheric OH concentrations are lowered even further, allowing even more CH_4 and CO to build up.

Conversely, with lower concentrations of CH_4 in the atmosphere during glacial periods, global coverage concentration of OH radicals may have been much larger, perhaps at least partially explaining low CH_4 concentrations in glacial ice. In any case, it is clear that changing CH_4 sources can strongly affect the chemical climate as well as the physical climate of the Earth.

A number of important questions are raised by these issues. Is the northern wetland CH_4 source sufficiently well understood to allow us to predict how a regional-scale change in air temperature would influence the contribution of this flux to the global source on annual or longer time scales? Could trends in CH_4 flux at the regional scale or trends in global ambient air methane be used as an indication of the large-scale biospheric response to climate change? Will the magnitude of the feedback from changes in a wetland CH_4 source be sufficient to influence rates of climate change and/or the global atmospheric oxidizing capacity?

Changing Climate as a Factor in Methane Fluxes

Climate scenarios from modeling studies such as NOAA/GFDL (Manabe and Wetherald 1987) and NASA/GISS (Hansen et al. 1988) predict significant variability in regional climatic patterns over the next 10–100 years in response to the accumulation of greenhouse gases in the atmosphere. Because there is considerable uncertainty with regards to changes in precipitation patterns as a result of global warming, consideration must be given to two climate change scenarios. In the first, climate warming leads to lower precipitation in boreal and tundra areas and the water table drops. In this scenario the landscape changes from a wetland landscape to one dominated by small lakes; exposed peat becomes aerobic and the decomposer community switches from dominance by bacterial anaerobes to dominance by fungal aerobes with an associated switch in trace gas production from CH_4 to CO_2 and perhaps also to N_2O and NO_x. In the second scenario precipitation increases and the areal extent of existing wetlands increases, with subsequent increases in CH_4 production.

Critical Gaps in Knowledge and Potential Research Approaches

Critical gaps in our knowledge of trace gas exchange at high latitudes include both our lack of data on trace gas sources (fluxes and spatial extents) and on interactions between rates of change in macroclimatic and microclimatic factors and trace gas production/consumption processes. For CH_4, for example, there are no data on fluxes from the two largest northern wetland regions in the world: the Siberian lowlands and the Canadian Hudson–James Bay lowlands. Even the spatial extent and seasonal variations in soil microclimatic characteristics (e.g., soil moisture) of these regions is poorly known. Consideration must also be given to the possibility of extensive drying with a lowering of the water table and subsequent reduction of anaerobic habitat.

In addition to ecosystem-level field measurements, a continued program of regional and global modeling studies is required to integrate the flux measurements with concomitant observations of global and regional concentrations of CH_4 in the atmosphere. A critical data gap for these studies is the lack of continental research sites for monitoring variations in atmospheric CH_4 concentrations over wetland ecosystems.

We can envision two specific approaches to fulfilling these research needs. The first is a network of regional research sites in relatively remote regions of the tundra and boreal zones to monitor long-term changes in the physical/chemical climate and resulting biospheric responses. A network of sites is needed along primary environmental gradients such as vegetation type and associated controls on methane emission (e.g., permafrost depth, precipitation, slope, aspect). In the tundra (permafrost) regions, three sites in North America and a similar number in the Soviet Union and Europe might be reasonable. At these sites ambient air measurements of CO_2, CH_4, N_2O, NO_x, O_3, and climate variables together with measurements of vegetation, hydrology, and soil should be carried out. The second approach could be a comparative study of trace gas fluxes from sites with manipulated temperatures and water tables and from nearby sites held under existing conditions. A strong focus would be placed on the consequences of shifting between anaerobic and aerobic soil environments for the exchange rates of CH_4, CO_2, N_2O, and NO_x in these sites. Both deposition gradients and the locations of other study sites could be suggested by climate data analyses and mesoscale and global-scale atmospheric transport models in which the regional variations of deposit rates in response to known emission distributions can be simulated (e.g., Levy and Moxim 1987). The interaction between the models and measurements will elucidate knowledge gaps and enable identification of measurement strategies.

CHANGES IN TRACE GAS EXCHANGE RESULTING FROM INCREASES IN NITROGEN AND SULFUR DEPOSITION

Rate, Extent, and Composition of Deposition Inputs

Human activities have already significantly influenced rates of sulfur and nitrogen cycling in many globally important, relatively undisturbed ecosystems, and particularly dramatically in the temperate and boreal zones of the Northern Hemisphere. Excess rates of annual inorganic nitrogen deposition in these zones (ca. 23×10^6 km^2), for example, may be as large as 18 Tg N yr^{-1}. About two-thirds of this inorganic N is in the nitrate form. While our current knowledge of the absolute amounts of this input is incomplete—in particular the relative amounts of NH_4^+ −N and NO_3^- −N, the ratio of wet to dry deposition, and the amounts of organic-N inputs—it is clear that overall inputs have been greatly increased.

System-level Consequences of Deposition

Melillo et al. (this volume) describe a number of likely effects of enhanced N inputs on ecosystem processes. These include changes in plant tissue chemistry, in decomposition rates, and on soil mineral N pools. Possible changes in leaf tissue chemistry include reduced C:N ratios, changes in the form of amino-N compounds in leaves, and changes in cellulose:hemicellulose ratios. As a function of such changes, leaf decomposition rates may accelerate and soil organic matter turnover may also increase. Over the long term, evidence suggests that N may become sufficiently available that nitrification rates may increase substantially in areas where rates are now low. On the other hand, N inputs may be sufficient to inhibit existing levels of N_2-fixation, lending to no net change in system-wide N status.

The significance of potential changes for trace gas fluxes can be considered at several levels. First, N_2O and NO_x fluxes may increase significantly, especially if nitrification increases substantially. N-gas fluxes may also increase in the absence of nitrification as a result of direct additions of deposition NO_3^-–N to the soil pool. Second, elevated NH_4^+ levels in these soils may inhibit the oxidation of CH_4 (Melillo et al.; Rosswall et al., both this volume), leading to increases in regional CH_4 fluxes due to a reduction of soil sink strengths. Third, added N may affect plant production of terpenes, isoprenes, and other nonmethane hydrocarbons, as well as the strength of the carbonyl sulfide (COS) sink in plants and the strength of the COS source in soils.

Critical Gaps in Knowledge and Potential Research Approaches

The rate and extent of total deposition fluxes of S and N species are relatively well known for some parts of the Northern Hemisphere. At this stage, the

Southern Hemisphere information is very scarce. The establishment of a well-distributed network of research sites in areas of both hemispheres to measure wet deposition of N and S will therefore be especially useful. Most of these fluxes can be measured easily with sufficient accuracy using wet-only rainfall samplers. For dry deposition the situation is more complicated since very little monitoring of dry deposition has been attempted. Moreover, in very remote sites concentrations and fluxes of gases such as NO_x and SO_2 are likely to be very low. This may very well exclude the possible use of conventional micrometeorological methods to measure dry deposition fluxes over an entire network in the near future. It will probably be best, then, to perform dry deposition studies only in those areas where they expect to contribute significantly to S and N cycling. For this purpose, the dry deposition fluxes can be inferred from measurements of trace gas concentrations and meteorological variables using micrometeorological models that have already been developed (Hicks and Matt 1988).

We also need experimental, ecosystem-level studies to document the effects of chronic N input disturbances on plant communities, and in particular, effects on leaf litter quality, subsequent litter decomposition rates, fine root turnover, and rates of nitrification in soils. Simultaneous measurements of gas flux changes (especially changes in N_2O, NO_x, and CH_4) are equally important.

It is worth noting that many of these potential effects will interact significantly with CO_2 fertilization, which may have quite opposite effects on plant tissue chemistry (see next section). It thus will be important to conduct experimental studies of N and S deposition in concert with CO_2 fertilization experiments.

Experimental studies could be conducted along existing N:S deposition gradients to investigate the influence of N and S deposition of plant tissue chemistry. Gradients for precipitation of N:S ratios exist across Europe, North America, China, and parts of the Southern Hemisphere, and studies along such gradients could provide substantial insight into likely long-term effects of N and S deposition in a given location. It is also important to recognize that the location of sampling sites with respect to major source regions is crucial.

GLOBAL CO_2 ADDITIONS AND TRACE GAS EXCHANGE

Plant Responses to CO_2 Increases

The concentration of atmospheric CO_2 has increased by about 25% since the beginning of the industrial era. The effects of this increase on *in situ* plant communities are essentially unknown but potentially of critical global importance. Greenhouse and phytotron studies show that elevated levels of

CO_2 positively affect the stomatal exchange of CO_2, and hence water vapor in C3 plants, and consequently can increase rates of photosynthesis and/or water use efficiency (Goudriaan and Ajtay 1979; Lemon 1983; Strain 1985). Several isolated studies of ring width and $\delta^{13}C$ values in tree rings have suggested that CO_2 fertilization may be already occurring in some locations (LaMarche et al. 1984; Francey and Farquhar 1982). The only deliberate field experiment, a small enclosure of tundra ecosystem in a 2x CO_2 atmosphere, showed that after a short period of equilibration, ecosystem net primary production (NPP) did not change as a result of elevated CO_2 levels because of nutrient limitations on the ecosystem, though species distributions did change (Tissue and Oechel 1987).

Recent CO_2 enrichment experiments have also shown that in addition to increased production and improved water use efficiency, CO_2 fertilization may also increase the C:N ratio in plant biomass (Norby et al. 1986) and the allocation of C to recalcitrant compounds (Strain 1985). Other studies suggest that CO_2 fertilization can alter the density of stomatal openings on leaf surfaces (Woodward 1987). Although the occurrence and magnitude of CO_2 fertilization on ecosystems is far from being established, such changes may have impacts on substrate quality and rates of litter decomposition, with subsequent impact on fluxes of trace gases to the atmosphere.

Consequent Effects on Trace Gas Exchange

The assessment of potential consequences of CO_2 fertilization on the exchange of trace gases between terrestrial biomass and the atmosphere is based on assumed changes in rates of photosynthesis, elemental composition of leaves, and consequent changes in amounts and quality of litterfall and in species distributions. Changes in soil microbial communities may result and may themselves modify the turnover rates of elements and composition of gases emitted.

For example, a higher C:N ratio in plant leaves due to CO_2 fertilization may, in N-limited systems, increase N_2 fixation and reduce the rates of soil N turnover and subsequent N_2O and NO_x emissions. In fertilized systems, on the other hand, N-gas fluxes from denitrification may be larger as C-limited denitrifiers are stimulated at the end of the growing season by carbon from root death (Mosier, pers. comm.) and carbon-rich litter inputs. Methane fluxes may also be affected by litter quality, in that the strength of CH_4 sinks in upland soils may depend on the availability of labile C in these soils relative to the amount of available NH_4^+ (Melillo et al.; Rosswall et al., both this volume). As the C:N ratio increases, NH_4^+ will become more scarce, and CH_4 oxidation could decrease. Fluxes of trace gases to and from plants, particularly those under stomatal control (e.g., terpenes, COS, NH_3), may also be altered as stomata are opened less frequently (Johansson, this volume; Goldan et al. 1988).

Critical Gaps in Knowledge and Potential Research Approaches

We can identify two major areas in which a lack of knowledge inhibits our ability to predict the effects of CO_2 fertilization on trace gas exchange:

1. the effects of changing amounts of leaf and root litter and the changing quality of this material on decomposition, nitrification, denitrification, and related soil microbial processes with subsequent effects on N_2O, NO_x, and CH_4 fluxes; and
2. the effects of changing leaf stomatal resistance and photorespiration rates on plant hydrocarbon emissions and the exchange of other gases such as COS.

Assessment of these effects can best be addressed using a two-tiered approach. First, we need to establish process-level studies to determine (*a*) the response of plant communities under *in situ* conditions to enhanced CO_2 levels, particularly with respect to NPP, leaf and root tissue quality, and hydrocarbon emissions; and (*b*) the response of microbial communities to changes in substrate inputs and quality, particularly with respect to CO_2, N_2O, NO_x, and CH_4 fluxes. Second, we need to use this information to parameterize existing system-level C and N models to predict the effects of CO_2 fertilization on global trace gas exchanges from terrestrial systems in general.

Because changes in CO_2 concentrations will be accompanied by changes in temperature and precipitation, taken together outside the range of current experience, it will be necessary to manipulate experimentally all of these factors simultaneously in order to parameterize existing system-level C and N models. These experiments must be done on a whole-ecosystem basis in order to allow us to understand system-level couplings at a level sufficient to model the systems accurately. Such experiments could be approached with open-air, CO_2-amended plant communities or perhaps better with *in situ* whole-system growth chambers.

EFFECTS OF MAJOR LAND-USE CHANGES ON TRACE GAS EXCHANGE

Tropical Deforestation

Effects on climate and system-level N cycling. In equatorial rainforests 80% of the rainwater is recycled within the region. Forest conversion to pasture and continuous cultivation decreases the rate of evapotranspiration while runoff increases. This can lead to reduced precipitation in downwind areas. Resulting changes in these downwind regions in NPP, together with associated changes in biological processes such as N-mineralization and changes in physical processes such as leaching rates, may substantially affect trace gas emissions. Effects of similar magnitude may occur at microclimatic scales following forest conversion. Average soil temperatures commonly increase, soil moisture can decrease, and wind velocity at the soil surface can be high following clearing. All of these factors can affect trace gas emissions via both direct physical effects and effects on vegetation and the soil biota.

Major changes also occur in system-level N cycling following vegetation removal. Immediately following clearing, soil nutrients (including N) are generally more available due to the addition of organic matter and ash to the soil (N from incompletely burned slash and roots). Consequently, rates of N mineralization, nitrification, and denitrification increase (Robertson 1984; Matson et al. 1987; Robertson and Tiedje 1988). This increased N cycling, however, will persist only so long as labile soil organic matter remains available or some other source of N is added to the system. External N sources might include fertilizer N in the case of cultivation or enhanced rates of N_2-fixation, where vegetation management includes legumes or C_4 pasture grasses with associative N_2-fixing bacteria.

Direct burn effects on trace gas exchange. Tropical biomass burning, especially frequent burning of savanna systems, is a potent source for atmospheric trace gases including NO_x, N_2O, CO, CH_4, and nonmethane hydrocarbons (Crutzen et al. 1979, 1985). The combination of these emissions with high UV flux in the tropics leads to the efficient production of ozone from tropical burning emissions.

We currently lack knowledge of the annual area burned and the emission factors for various trace gases relative to CO_2 for burns of different types. In the case of forest clearing burns, even the proportion of material volatilized has never been quantified. Remote sensing approaches to quantify burning are hampered by resolution too low to identify individual fires. Fire surveys on a regional basis, whether ground based or based on aerial surveys of fire scars, would greatly improve our accounting of annual areas burned. Ground-based studies are also needed in forest clearing burns to quantify the proportion of material volatilized and that left on the ground unburned and as charcoal. Aircraft-based experiments are needed to determine emission factors for trace gases for fires. The critical quantity to measure is the NO_x emitted since NO_x is the fulcrum upon which tropospheric production or consumption of O_3 balances. Several determinations of fire emission factors exist, including surveys of the Congo, the Brazilian *cerrado*, and the Amazon area of Brazil. Reported average or typical emission factors from these studies greatly underestimate the variability in fires. Attention should be paid to all stages in burning including the long period of smouldering that can follow forest clearing fires. Variability in fuel quality should also be determined. Fuel mass burned is also uncertain. Ecosystem (and agricultural) modeling may help to supplement knowledge of fuel mass when production and decomposition rates are coupled to climate and timing of burns.

Post-burn effects on trace gas exchange. The process of biomass burning represents a vast reallocation of nutrients in cleared tropical forest and savanna systems. Large proportions of system C, N, and S are volatilized. Soils are affected by changes in nutrient levels and pH, with associated changes in microbial communities. These effects on soils may result in large post-clearing fluxes of N_2O and NO_x. Research must identify the cause of these fluxes (physicochemical versus biological, and the mechanisms involved), and the duration of the post-clearing enhancement ought to be carefully traced to identify other possibilities for significant trace gas releases.

Critical knowledge gaps and potential research approaches. We have very little information about how trace gas fluxes change following forest clearing in rainforest and savanna woodland areas. What little evidence we have suggests that changes in N_2O, NO, and CH_4 fluxes can be dramatic immediately following clearing and that subsequent fluxes can be high or low depending on subsequent land use. For most major areas of the tropics we do not know the direction, magnitude, or duration of these changes.

Clearly the first step for evaluating the global importance of trace gas fluxes from deforested areas of the tropics is to expand our data base of flux estimates from tropical regions undergoing major land-use change. At the primary level this data base should include sites on major tropical soil types replicated across continents, each subjected to a combination of post-clearing management practices. Suggested post-clearing management practices examined should include pasture establishment (grazed vs. ungrazed), cultivation (low chemical input vs. high chemical input; continuous vs. fallowed), and reforestation. A nearby site under pre-burn natural vegetation (whether mature or successional) should be included in the comparison. At each site we suggest measuring for a period of 3–5 years N_2O, NO_x, and CH_4 fluxes, together with important C and N cycle characteristics (e.g., mineralization, nitrification, denitrification, soil organic matter turnover). Trace gas measurements must also include those gases lost during the burn itself.

These studies should be accompanied by continental ground-based stations in tropical savanna and forest regions for measurement of key burning-related species such as O_3, NO_x, and CO. Remote sensing may also be useful for such measurements. The rapid rate of land-use change in tropical ecosystems makes the need for measurements in these areas especially acute. Certain habitats, especially dry forests, have already been converted to other uses. In the savanna zone, overgrazing and fuel wood removal are leading to accelerated deforestation. Biogeochemical studies of baseline properties of these biomes are required now while relatively undisturbed examples of these ecosystems still exist.

Agricultural Technologies

Changing agricultural technologies. Both the global extent of cropped areas and crop management practices have changed remarkably in recent decades. In the tropics, native vegetation has been converted to cropland, and in temperate regions, fertilizer use has become intense. The net effect is a potential for much greater quantities of N cycling through systems with equivalent or lower quantities of plant biomass, resulting in substantially more mineral N available to microbial N_2O and NO_x producers, and potential changes in the strengths of soil CH_4 sinks.

The extent of the increase in N_2O and especially NO_x emissions from most cropped areas is poorly known, existing studies in wheat, corn, rice, and irrigated vegetable production areas notwithstanding (Sahawarat and Keeney 1986). System-level factors known to affect N-gas fluxes from cropped areas include crop phenology and planting densities, fertilization, water regimes (including irrigation and, in the case of rice, flooding cycles), and organic matter management.

Knowledge gaps and potential research approaches. Areas of greatest research needs include knowledge about emission rates from existing cropping systems in both tropical and temperate regions. We also need information on likely rates from the various conservation tillage systems now being implemented on a widespread basis and from the low chemical input, high productivity cropping systems now under development. If the latter systems are as nutrient efficient as current projections suggest they could be, the widespread adoption of these technologies may attenuate current global fluxes from agricultural areas directly. Fluxes from coastal areas may also be reduced if current cropping systems that are now loading these areas with N are modified, consequently affecting coastal nitrogen oxide emissions. We also see a strong need for the development of process-specific N_2O and NO_x models linked to crop growth models (such as CERES) now in widespread use.

The most straightforward way to assess the current importance of N_2O, NO_x, and CH_4 fluxes in cropped systems would likely be to take advantage of current worldwide networks of agricultural field stations. At these stations, trace gases from a combination of different cropping systems could be monitored for 2–3 year periods. Concurrent measurements of important crop and C and N cycle variates could be used to parameterize the process-specific N_2O and NO_x models mentioned above.

SUMMARY RECOMMENDATIONS

General Approach

To make efficient and rapid progress on evaluating the consequences of global change phenomena for the trace gas fluxes, it is clear that the research community will have to integrate several research approaches, including long-term measurement activities, manipulative experimentation, and modeling.

Long-term measurements should be made at a number of sites across the globe, with special emphasis on those areas undergoing large land-use change or known to be especially sensitive to climate change. The measurements should include not only atmospheric trace gas concentrations and trace gas fluxes, but also measurements of factors that control these fluxes such as changes in substrate availability and soil moisture. These measurements should be made over the long term so that we can develop a reliable, well-documented record of change.

The data from these long-term measurements can be used in several ways in modeling efforts. First, some subset of the data can be used to build models that parameterize processes of trace gas fluxes; that is, the correlative relationships between controlling factors and fluxes can guide model development. Second, the data not used in model development can be used in model validation—an independent check on model results.

It is likely that early generation models will identify gaps in our understanding of the way system processes are linked, especially in the new physical and chemical environments we are now beginning to encounter. Understanding these linkages and new conditions will require manipulative experiments at various scales. Examples of manipulations include laboratory studies of the effects of changes in substrate quality derived from plants grown in high CO_2 environments on the rate of CH_4 production or the ratio of N_2O to NO production. In addition, there may be a need for whole-system manipulations that test our understanding of complex ecosystem responses to environmental change. We may want to alter, for example, the CO_2 composition and climate of ecosystems at sub-hectare scales so that we can observe system-level responses such as plant carbon allocation or net ecosystem production to environmental change, and so that we can determine how these responses affect trace gas fluxes. Experiments of this type will have to be long term to avoid confusing short-term transient responses with decade-to-century responses. The data can be used either to further model development or to validate existing models.

The ultimate goal of this research is to develop predictive models of how regional and global environmental changes will affect trace gas fluxes and to develop methods to counteract the enhanced release rates of greenhouse and chemically active trace gases. Scaling the results of small scale (cm^2 to ha) studies to the region (10^6 km^2) or globe is problematic. Clearly one of the tools needed in this scaling effort is a set of global data sets on the physical, chemical, and biological factors that control ecosystem processes that affect trace gas emissions. Geographically referenced data on soil texture, vegetation, land-use history, and so on will be essential to the scaling task. This geographically referenced data will be used in conjunction with mechanistic models of ecosystem processes and trace gas fluxes to produce regional and global maps of trace gas exchanges between the land and the atmosphere.

Validation of these regional estimates of trace gas fluxes will be very difficult but absolutely necessary. It may be done in one of several ways. One alternative is to compare flux estimates derived from process-model-based predictions with flux estimates derived from inverse model predictions in turn derived from calculations based on the regional or global distribution of trace gas concentrations. A second alternative is to link process-model-based predictions for an area with mesoscale meteorological models to predict atmospheric concentrations and fluxes of trace gases, and then check these predictions with aircraft measurements.

Clearly a variety of approaches will be needed to evaluate the consequences of global change phenomena for trace gas fluxes. These approaches must be integrated into a stratified plan for attacking the problem. Included in this plan must be (*a*) long-term field measurements, including both atmosphere-sampling stations and stations at which ecosystem fluxes are measured, (*b*) manipulative experiments at both the ecosystem and soil–plant microcosm level, and (*c*) the development of models and the creation of geographically referenced data bases and modeling efforts at ecosystem, regional, and global scales to integrate our understanding of trace gas flux interactions and to allow us to predict the effects of future changes in the physical and chemical environment on these interactions.

Specific Recommendations

We propose six specific activities to attain these objectives. Three activities are aimed toward quantifying contemporary and historical trace gas fluxes in globally important biomes, two are aimed toward understanding the mechanisms that regulate trace gas exchange at ecosystem levels, and the final activity is directed towards integrating this information into models that can be used to predict the effects of anthropic activities on future gas fluxes.

1. Biospheric sample archives should be established to document future change, and careful syntheses should be made of existing historical records for evidence of past climate change coupled to factors suspected to affect gas fluxes. Ice cores, sediment cores, and tree ring samples are examples of historical records that may yield useful information using stable isotope, gas fractionation, or other techniques; such syntheses should be concentrated in regions where high climate variability is expected from modeling studies.
2. A modest global network of long-term sampling stations should be established to collect integrated, self-consistent sets of information about (*a*) atmospheric concentrations of trace gases and the chemical composition of precipitation, (*b*) trace gas fluxes from surrounding habitats, and (*c*) climatological and ecosystem-level parameters needed for model development (e.g., air temperature, soil organic matter concentrations, nitrogen mineralization rates, soil moisture, etc.).
3. Short-term regional campaigns should supplement the global sampling noted above (point 2) for regions either particularly sensitive to change with respect to gas fluxes or rapidly undergoing large-scale, land-use change. These 12–24 month campaigns would measure the same fluxes and parameters as at the global stations but would do so at a much more geographically intense scale; two candidate regions that come immediately to mind are (*a*) humid tropical rainforest regions and (*b*) high latitude wetlands.
4. At the ecosystem level, we propose ecosystem manipulations that simulate expected or extreme climate changes in order to examine process-level interactions. Three types of manipulations appear particularly important with respect to trace gas–climate interactions:
   (*a*) the manipulation of northern wetland temperatures and water tables to gauge the effects of precipitation changes on CH_4 fluxes in particular;
   (*b*) ecosystem-level CO_2-fertilization experiments to gauge the effects of increased CO_2 levels on system-level processes such as primary production and organic matter quality, and resulting effects on trace gas fluxes; and
   (*c*) ecosystem-level climate manipulations, e.g., *in situ* sub-hectare greenhouses that enclose portions of intact ecosystems, and in which temperature, precipitation, and CO_2 levels can be manipulated to simulate expected climate changes; such systems are needed to test hypotheses related to the interactions of climatic factors with trace gas fluxes.

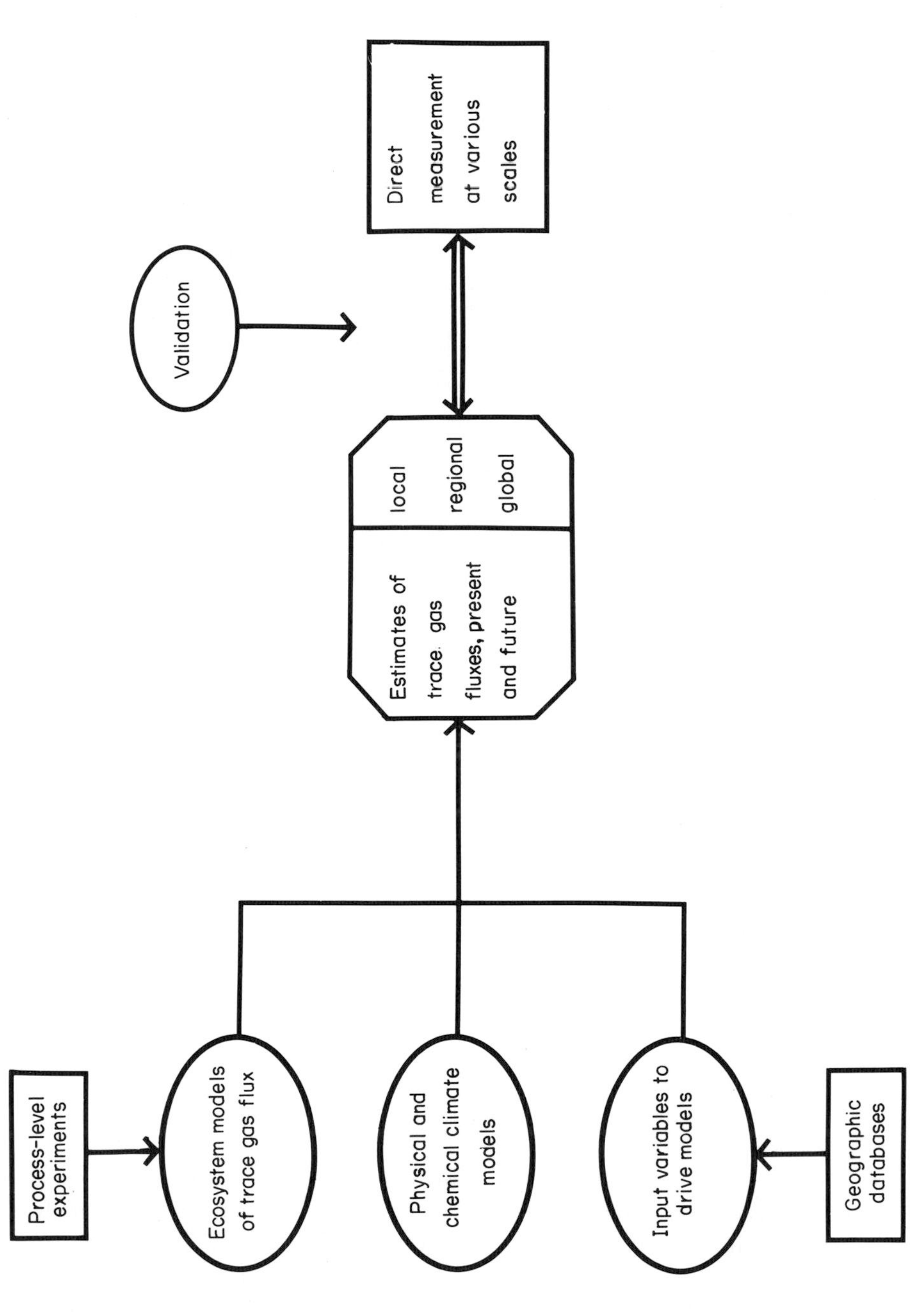

Fig. 1—Conceptual model of a strategy for assessing the impact of interactions between changes in the physical and chemical climate and changes in trace gas fluxes.

5. Also at the ecosystem level, process-level experiments are proposed to delineate the specific mechanisms responsible for changes in trace gas exchange resulting from climate change. Such experiments should be conducted as part of the experiments noted above but they can also be conducted in existing field sites or in greenhouses or growth chambers. They might include, for example, greenhouse pot experiments to assess the effects of enhanced CO_2 on leaf tissue chemistry, or soil microcosm or field experiments to examine changes in soil microbial communities and associated trace gas fluxes in response to altered litter quality.
6. Finally, we propose to integrate this information with models that will allow us first to test our understanding of how various ecosystem-level processes (including trace gas fluxes) interact with the physical and chemical climate at various geographic scales, and second to predict the effects of changes in climate on trace gas exchanges across different geographic and temporal scales. Two specific activities are needed for such syntheses: (*a*) the continued development of models to link research efforts synthetically, and (*b*) the continued development of geographical, perhaps remotely sensed data bases in order to provide a basis for regional- and global-level extrapolations.

Figure 1 presents a conceptual model for this research plan. In it we link ecosystem models based on process-level experiments, general circulation models of physical and chemical climate, and climatic and ecosystem-specific variates derived from geographically referenced data bases to develop estimates of trace gas fluxes at local, regional, and global scales. These estimates are validated—and models are appropriately amended—using direct measurements of gas fluxes and atmospheric concentrations also at local, regional, and global scales.

Only in such an integrated manner, we believe, can a more reliable assessment of the interactions of climate change and trace gas fluxes be made. Without such an assessment it will continue to be extremely difficult to predict the ultimate effects of climate change on the biosphere.

REFERENCES

Baker-Blocker, A., T.M. Donahue, and K.H. Mancy. 1977. Methane flux from wetlands. *Tellus* 29:245–250.

Cicerone, R.J. 1987. Changes in stratospheric ozone. *Science* 237:35–42.

Crill, P.M., K.B. Bartlett, R.C. Harriss, E. Gorham, E.S. Verry, D.I. Sebacher, L. Madzar, and W. Sanner. 1988. Methane flux from Minnesota peatlands. *Glob. Biogeochem. Cyc.* 2:371–384.

Crutzen, P.J. 1986. The role of the tropics in atmospheric chemistry. In: Geophysiology of Amazonia, ed. R.E. Dickinson, pp. 107–130. New York: Wiley.

Crutzen, P.J., A.C. Delany, J. Greenberg, P. Haagenson, L. Heidt, R. Lueb, W. Pollock, W. Seiler, A. Wartburg, and P. Zimmerman. 1985. Tropospheric chemical composition measurements in Brazil during the dry season. *J. Atmos. Chem.* 2:233–256.

Crutzen, P.J., L.E. Heidt, J.P. Krasnec, W.H. Pollock, and W. Seiler. 1979. Biomass burning as a source of atmospheric gases CO, H_2, N_2O, NO, CH_3Cl and COS. *Nature* 282:253–256.

Francey, R.J., and G.D. Farquhar. 1982. An explanation of $^{13}C/^{12}C$ variations in tree rings. *Nature* 297:28–31.

Goldan, P.D., R. Fall, W.C. Kuster, and F.C. Fehsenfeld. 1988. Uptake of COS by growing vegetation: a major tropospheric sink. *J. Geophys. Res.* 93:14186–14192.

Goudriaan, J., and G.L. Ajtay. 1979. The possible effects of increased CO_2 on photosynthesis. In: The Global Carbon Cycle, ed. B. Bolin, E. Degens, J. Kempe, and P. Ketner, pp. 237–249. New York: Wiley.

Hansen, J., I. Fung, A. Lacis, D. Rind, S. Lebedeff, R. Ruedy, and G. Russell. 1988. Global climate changes as forecast by Goddard Institute for Space Studies three-dimensional model. *J. Geophys. Res.* 93:9341–9364.

Hicks, B.B., and D.R. Matt. 1988. Combining biology, chemistry, and meteorology in modelling and measuring dry deposition. *J. Atmos. Chem.* 6:117.

LaMarche, V.C., Jr., D.A. Graybill, H.C. Fritts, and M.R. Rose. 1984. Increasing atmospheric carbon dioxide: tree ring evidence for growth enhancement in natural vegetation. *Science* 225:1019–1021.

Lemon, E.R., ed. 1983. CO_2 and Plants: The Response of Plants to Rising Levels of Atmospheric Carbon Dioxide. Boulder, CO: Westview.

Levy, H., II, and W.J. Moxim. 1987. Fate of U.S. and Canadian combustion nitrogen emissions. *Nature* 328:414–416.

Manabe, S., and R.T. Wetherald. 1987. Large-scale changes of soil wetness induced by an increase in atmospheric carbon dioxide. *J. Atmos. Sci.* 44:1211–1235.

Matson, P.A., P.M. Vitousek, J.J. Ewel, M.J. Mazzarino, and G.P. Robertson. 1987. Nitrogen transformations following tropical forest felling and burning on a volcanic soil. *Ecology* 68:491–502.

Muzio, L.J., and J.C. Kramlich. 1988. An artifact in the measurement of nitrous oxide from combustion sources. *Geophys. Res. Lett.* 15:1369–1372.

Norby, R.J., J. Pastor, and J. Melillo. 1986. Carbon-nitrogen interactions in CO_2-enriched white oak: physiological and long-term perspectives. *Tree Physiol.* 2:233–241.

Pearman, G.I., D. Etheridge, F. DeSilva, and P.J. Fraser. 1986. Evidence of changing concentrations of atmospheric carbon dioxide, nitrous oxide, and methane from air bubbles in antarctic ice. *Nature* 320:248–250.

Robertson, G.P. 1984. Nitrification and nitrogen mineralization in a lowland rainforest succession in Costa Rica, Central America. *Oecologia* 61:99–104.

Robertson, G.P., and J.M. Tiedje. 1988. Deforestation alters denitrification in a lowland tropical rainforest. *Nature* 336:756–759.

Sahawarat, K.L., and D.R. Keeney. 1986. Nitrous oxide emission from soils. *Adv. Soil Sci.* 4:103–148.

Strain, B.R. 1985. Physiological and ecological controls on carbon sequestering in terrestrial ecosystems. *Biogeochem.* 1:219–232.

Svensson, B.H. 1980. Carbon dioxide and methane fluxes from the ombrotrophic parts of a subarctic mire. *Ecol. Bull.* 30:235–250.

Tissue, D.T., and W.C. Oechel. 1987. Response of *Eriophorum vaginatum* to elevated CO_2 and temperature in the Alaskan tussock tundra. *Ecology* 68:401–410.

Whalen, S.C., and W.S. Reeburgh. 1989. A methane flux time-series for tundra environments. *Glob. Biogeochem. Cyc.*, in press.

Woodward, F.I. 1987. Stomatal numbers are sensitive to increases in CO_2 from pre-industrial levels. *Nature* 327:617–618.

Exchange of Trace Gases between Terrestrial Ecosystems and the Atmosphere
eds. M.O. Andreae and D.S. Schimel, pp. 321–331
John Wiley & Sons Ltd

Priorities for an International Research Program on Trace Gas Exchange

D.S. Schimel
M.O. Andreae
D. Fowler
I.E. Galbally
R.C. Harriss
D. Ojima
H. Rodhe
T. Rosswall
B.H. Svensson
G.A. Zavarzin

OVERVIEW

The participants of the Dahlem Workshop on "Exchange of Trace Gases between Terrestrial Ecosystems and the Atmosphere" identified a number of exciting and important areas for future research. These areas range from biological process studies to investigations on the consequences of changing land use and climate for trace gas fluxes. While the reports from the four discussion groups contain specific research recommendations, it was agreed that there should be a chapter in which these suggestions would be synthesized into the outline of a research plan for the international, multidisciplinary investigations which are to be conducted during the upcoming decade by the International Geosphere-Biosphere Program (IGBP). This chapter, therefore, discusses those recommended research activities which require the development of large-scale, sustained, international programs for their execution.

PROCESS AND METHODOLOGICAL ISSUES

Three biological issues appear most critical. First, factors controlling the partitioning of nitrogen gas production between N_2O, N_2, and NO are not well understood. Second, the control and importance of CH_4 oxidation require much more study since a less efficient sink could contribute to globally increasing concentrations. Finally, the controls over nonmethane hydrocarbon (NMHC) exchange between plants and the atmosphere are a key new topic. NMHC emissions should be studied in ecosystems with both polluted and clean atmospheres.

The development of methodology for measuring averaged fluxes of CH_4, N_2O, NO, and NO_x must continue using micrometeorological techniques. There is an urgent need for field studies in which the biological processes regulating gas production and their environmental controls are investigated in combination with flux measurements. The flux measurements should be made at multiple scales (chamber, tower, aircraft) to test our ability to predict fluxes using process-based models. Combining such studies with remote sensing will provide the basis for further extrapolation of fluxes to biome and global scales. This type of research must be conducted for both long- and short-lived chemical species. Isotopic budgets provide an additional constraint on global budgets of trace gases, an approach which is currently of most interest for methane. Accurate estimates of the isotopic composition of sources to the atmosphere are essential for accurate isotope-based models. Integration of isotopic measurements with coordinated measurement campaigns will strengthen the linkage between local studies and global estimates.

Several issues were raised that are important, but may not yet require coordinated international research. First, CH_4 flux from landfills may be significant and must be better known. Second, the unprecedented rates of climate change suggest that some ecosystems may be unable to adapt, raising the possibility that regions without stable vegetation will develop. The impact of such changes on trace gas exchange is not known. We suggest that pilot measurements from drastically disturbed areas may provide models for unstable ecosystems. Such disturbed areas could include heavily polluted soils, urban soils, landfills, areas of forest decline, and forest areas affected by pathogens. We encourage pilot initiatives to better define these problems prior to the development of integrated programs.

RECOMMENDED RESEARCH PROGRAMS

Uncertainties in the estimates of present fluxes of CH_4 and N_2O are very serious. In the case of CH_4, critical research areas to reduce these uncertainties are northern wetland ecosystems and rice agriculture regions. Effects of changing land use in the tropics include consequences for soil nutrient and carbon status, aeration, porosity, and water storage. These effects, or others not presently understood, are likely causes of observed high N_2O fluxes, and this is a priority research area. Biomass burning in tropical areas is also a very important source of gases, including tropospheric O_3, and should receive intensive study. Measurements of NO fluxes are also critical, since the global data base for this species is particularly sparse.

Changes in physical and chemical climate impact the biota at a whole-system level. Experimental manipulations of ecosystems are required for analysis of ecological and biogeochemical consequences of changes in climate, atmospheric deposition, and CO_2 concentration. Several system manipulation experiments are proposed. These include hydrological manipulations in northern wetlands to evaluate the consequences of climate change for CH_4 fluxes. It is not clear whether global climate change will cause northern wetlands to become wetter due to increased rainfall or drier because of permafrost melting and water table depression; both possibilities should be simulated experimentally. Large greenhouses, enclosing entire ecosystems, are a way of simulating effects of climate change and increased CO_2 concentration. These are advocated as a means to study system-level effects of temperature and moisture change. Consequences of CO_2 fertilization could also be evaluated. Finally, increased deposition of N and S species is a chronic problem in the midlatitudes and is increasing in the tropics. Preliminary studies suggest that experiments to evaluate N-saturation effects on C, N, and S trace gas fluxes are essential. Interactions of ozone with vegetation and with N and C gas exchange must also be investigated in this context.

NORTHERN WETLAND STUDIES

Northern circumpolar wetlands over permafrost are an important source of atmospheric methane and a unique type of ecosystem in that they give a two-month long peak of CH_4 emission. CH_4 production depends on physical climate parameters and is suspected to be very sensitive to changes in growing season, temperature, and hydrologic regime. CH_4 emission is further assumed to be dependent on the type of vegetation and its net primary production (NPP), on the structure and chemistry of decomposing organic matter, on the microbial decomposition rate, and on the balance between microbial CH_4 production and oxidation. The influence of changing UV flux on the photochemistry of CH_4 oxidation in polar regions remains unclear.

There is an urgent need to identify positive and negative feedbacks in plant-microbial communities resulting from anticipated climate change. Climate warming and changes in precipitation patterns might express themselves differently depending on geographical location, either as increasingly wet conditions with expansion of bog areas or as drying with transition to landscape patterns dominated by small lakes. Both types of change should be modeled experimentally to predict future patterns.

At present, only very broad model analyses can be made based on observed seasonal variations in atmospheric CH_4 concentration; these suggest globally important CH_4 emissions in northwestern Siberia and in the Hudson

Bay lowlands. A critical problem for modeling and predictive studies is the lack of continuous measurements of ambient concentrations and of fluxes of CH_4 and other trace gases in these globally important regions. The establishment of an observational network is a necessary prerequisite for forecasting future changes.

The following questions should be answered by the proposed research program:

1. What are the current CH_4 fluxes from northern regions and what determines their magnitude and variability in space and time?
2. How will northern ecosystems react to the expected climatic change?

Research addressing the first objective must bring together information on flux rates at different time and space scales in order to be able to extrapolate regionally and globally. A thorough understanding of the appropriate driving factors at each scale is therefore also required. The technical means to obtain net flux rates at different scales (from chambers to airborne measurements) are available today, while the interlinkage between these levels has to be developed and tested. The most important factors determining the production rate for methane are likely to be moisture and temperature, although these two factors may be overruled by substrate and nutrient availability. The influence of the two latter variables should be emphasized since they, together with soil content, may provide a process-based modeling approach to extrapolation. Within this context, the importance of CH_4 oxidation as a regulator of emission fluxes should be stressed. Any change in this ubiquitous sink may be at least as important as possible changes in the CH_4 production rates.

^{13}C analysis is a key tool in the understanding of the global methane cycle. Many sources have distinctive isotopic compositions, and oxidation in the atmosphere produces an isotopic signal as well. The balance between production and oxidation of methane causes its carbon and deuterium/hydrogen isotopic abundance to vary seasonally, and this is poorly understood. Measurements of isotopic composition aid in both process studies and in studies of global sources and sinks. Studies of methane emissions should include measurements of $^{13}C/^{14}C$ and D/H of methane emitted over time.

Sites for the proposed program of flux measurements have to be selected carefully to represent the most important regions of methane emissions. The sites should be situated along a latitudinal gradient reflecting climate variability. Five stations along a regional transect would be appropriate; at the ends and in the middle, detailed studies should be conducted on the effects of moisture, substrate, and nutrient availability. Microbial and plant studies should be made at selected sites within this larger design, to aid in the process-level understanding required for mechanistic models. Measurements of transport via ebullition and through plants should be included, in addition to diffusive flux. Measurements should include plant production, decomposition rates, acetate availability, O_2, and redox potential. Isotopic or inhibitor measurements of CH_4 production and consumption in the sediment are desirable.

The vast northern ecosystems in North America (e.g., the Hudson Bay area) and northern USSR (around the Ob River) are priority areas for study. These two areas are large and provide rates from relatively undisturbed environments. Within these areas, sites should be identified for long-term observation programs. In order to investigate a somewhat different climatic regime, a similar study could also be undertaken in Scandinavia, where the oceanic influences dominate climate and air chemistry. Within this area, perturbation from acid precipitation including sulfur deposition could easily be included as a superimposed gradient.

Since CH_4 is a relatively stable gas and the concentration gradient between soils and the atmosphere is large, appropriate use of chambers makes it possible to obtain net fluxes on a local basis. Micrometeorological and airborne measurements are also now available. This gives an excellent possibility to achieve integration of fluxes at different scales, which is essential to verify the outcome of the ecosystem gradient extrapolation method. Such a combination of flux measurement techniques will also provide the means to obtain variability measures at different scales. However, in order to utilize these tools extensively they should be used and compared under a diverse set of conditions. Thus, simultaneous measurements with chambers, and micrometeorological and aircraft measurements over the same areas, should be a part of this program. Experimental manipulations at the ecosystem level should be based at sites with long meteorological/botanical records (e.g., arctic botanical gardens) and with well-defined air pollution and soil dust transport patterns. The patchiness of the representative sites determines the minimum critical area. Two types of manipulated experiments should be established in (*a*) greenhouses and (*b*) experimental fields, including sites with the water table lowered by drainage and flooded sites. The minimal set of measurements should include:

—soil temperature profiles,
—water content,
—soil porosity, texture measurements,
—ambient concentrations of CO_2, CH_4, N_2O, CO, O_3,
—flux measurements for the gases listed above, using chamber and micrometeorological techniques, and
—organic matter decomposition rates.

TROPICAL LAND USE AND NITROGEN CYCLING

Land-use changes modify the fluxes of NO_x and N_2O from the soil into the atmosphere. Atmospheric N_2O, an important greenhouse gas, is changing at an alarming rate; this is likely the result of increased tropical emissions. Also, the production of ozone in the lower atmosphere is regulated by the ambient NO concentration. In the tropics, soil emissions of NO dominate the sources of NO_x to the lower atmosphere outside of the burning season. Long-term studies of land use, biological nitrogen cycling, and emissions to the atmosphere of N_2O and NO_x in tropical regions are urgently needed. This work needs to combine modeling and measurement programs. The modeling should initially include mechanistic models which couple soil microbial activity and available substrate with transport processes and micrometeorology to predict local trace gas fluxes. The models' results will serve to design field studies of emission rates with chambers and micrometeorological methods and to provide a mechanistic basis for larger-scale ecosystem models. The ecosystem models should be coupled to data obtained from major observatories located in key representative areas of the tropics. It is envisaged that a set of long-term observatories be maintained at sites on major tropical soil types replicated across the continents. These observatories should have sufficient land available so that sites representative of the natural system plus successional areas and other land-use types can be compared. At each site, flux measurements of N_2O, NO_x, and NO together with important C and N cycle characteristics (e.g., mineralization, nitrification, denitrification, soil organic matter turnover) should be measured initially for 3–5 years. Later, these trace gas exchange studies should be repeated at each observatory when the interactions between the observations and the ecosystem models indicate a further need to understand the changing environment and trace gas emissions at that observatory.

Associated with the long-term studies and survey investigations in tropical regions, a need exists for a set of studies to identify how to extrapolate across various scales. The effects of land cover and land use on N fluxes from the soil and biota into the atmosphere provide a basis for biome- and global-scale extrapolation. This set of studies will incorporate plot-scale studies and eddy flux measurements (both tower and aircraft) to scale up to the landscape level (e.g., integrations over several to hundreds of kilometers) and to relate spatial variability to remote sensing data bases. Spatial variability in those land cover variables that regulate trace gas flux must be associated with variables detectable by remote sensing to allow regional flux estimates to be made. The results of these studies will be vital in order to develop the appropriate methodology for extrapolating local measurements (observations) from the long-term studies and process studies to the regional scale; to determine the relative importance of various ecosystem properties to regional fluxes across a heterogeneous landscape;

and to assess land cover and land-use effects on N trace gas fluxes. An important part of the development of these extrapolation techniques will be the development of small-scale mechanistic models and the incorporation of the relevant processes into ecosystem models. This, coupled with extensive model validation studies using experimental data, will provide a timely solution to the extrapolation problem. The study sites to be used should encompass areas in the order of 100 × 100 km. These regions should have ongoing observations or measurement studies and have adequate remote sensing coverage. The extrapolation studies will be short-term, intensive studies lasting approximately 3–5 years. These studies are essential for tropical studies but are also generally needed in trace gas research.

THE ROLE OF PADDY RICE CROPPING SYSTEMS AS A SOURCE OF ATMOSPHERIC METHANE AND NITROGEN TRACE GASES

Preliminary data available for methane fluxes from rice paddies indicate that this system is a significant source of global tropospheric CH_4. It is also important to consider the entire rice production process for potential sources and/or sinks of CH_4, including organic fertilizer production in digestors, flooded soils not in production, and dry paddy soils which may function as CH_4 sinks. It has also been noted that ammonia (NH_3) emissions are significant from rice agricultural systems, especially during periods of fertilizer application. The emission of N_2O and presumably NO has been observed during periods of wetting and drying of rice paddies. Several factors make rice a unique and particularly important target for international collaborative research: the observed trace gas fluxes, the importance and large area of rice production as a food crop in regions of high population growth, and potential opportunities for technological innovations which could mitigate CH_4 and/or nitrogen gas emissions from the rice ecosystem.

Experimental Design

Primary issues of concern in experimental design include:

1. the relationships between agricultural practice (e.g., irrigation pattern, fertilization methods, cropping frequency, etc.) and trace gas production, transport, and emission to the atmosphere,
2. determining the sampling scales (temporal and spatial) necessary to quantify the fluxes of CH_4 and nitrogen gases from the world's major rice-growing regions, and
3. investigating how emerging agricultural technologies might influence CH_4 production/oxidation processes in regional rice production systems.

Approaches to the experimental design will require a definition phase study to identify the range of agricultural practices used in each major rice-growing region, (e.g., India, China). For each region and each rice ecosystem with unique agricultural techniques, a multidisciplinary team of microbiologists, soil scientists, and biogeochemists should design process-oriented studies to quantify the mechanisms and pathways of CH_4, N_2O, NO, and NH_3 production, transport, and *in situ* consumption or emission to the atmosphere. Following definition of priority regions for study, field campaigns must be designed that involve exploratory sampling of potential sources/sinks. It is likely that different combinations of enclosure, tower-based eddy correlation, and aircraft eddy correlation techniques will be required in different regions. As noted above, stable isotopes of C and H in methane are valuable in understanding the global methane cycle. Measurements of isotopic abundance in methane should be made in a variety of paddy systems.

If mitigation of CH_4 loss from rice production systems is identified as a primary objective, particular attention should be placed on CH_4 oxidation processes and on engineering approaches to CH_4 recovery for use as a fuel. The development of a geographic information system appropriate for use in modeling the global source of trace gas emissions from rice agriculture will require a collaboration of agricultural geographers and members of the global atmospheric modeling community.

INTERACTIONS OF THE TROPICAL BIOSPHERE WITH A CHANGING PHYSICAL AND CHEMICAL CLIMATE: DIRECT AND INDIRECT EFFECTS OF BIOMASS BURNING

The tropical regions of the Earth are subject to major perturbations, caused by processes active both within the tropics (e.g., biomass burning and land-use change) and by the influence of global change resulting from processes like the global greenhouse warming and the depletion of the ozone layer. Recent studies on the chemical composition of the tropical troposphere show the existence of regions with strongly elevated levels of trace gases, including O_3, CO, and NO_x, that have been attributed to biomass burning emissions and subsequent photochemical reactions in the emission plumes. Airborne and satellite-based measurements have shown these regions of perturbed atmospheric composition to extend to hundreds, and in some cases even thousands, of kilometers from the source areas. Burning during conversion of tropical forests is significant, but huge areas of natural and derived savanna are also burned and may currently be the dominant source for combustion inputs to the tropical atmosphere.

The investigations proposed here are based on a three-pronged approach, linking together ground-based measurements of burning characteristics, airborne measurements of atmospheric distributions and reactions of trace species, and remote sensing techniques to make large-scale extrapolation possible. The ground-based measurements should focus on characterization of the pre-burn and post-burn biomass, linked with determinations of emission factors for key species, e.g., CO_2, CH_4, NO_x, N_2O, CO, hydrocarbons, SO_2, COS, and aerosols. Airborne measurements should emphasize determination of emission ratios ($\Delta X/\Delta CO_2$, where ΔX and ΔCO_2 stand for the concentration difference between burning plume and background for species X and CO_2, respectively) and the possibility of relating aircraft measurements to parameters amenable to remote sensing, e.g., haze, O_3, and CO distribution. Satellites should contribute global and regional distributions of these parameters; they also should play an essential role in determining the spatial and temporal distribution of biomass burning. Ground-based emission studies should be conducted at a variety of sites in different fire-managed ecosystems. Linkage with studies at ground research sites appears particularly appropriate. Wherever possible, isotope ratio measurements should be made; combustion inputs to the global methane pool are quite uncertain, in part due to uncertainty about the isotopic composition of combustion products. Airborne measurements should focus on the tropical belt in Africa and South America where currently significant increases in tropospheric ozone are seen.

Potential effects of the perturbation of tropical atmosphere result from (*a*) the impact of the enhanced deposition of O_3, N and S compounds, organic acids, etc., (*b*) changes in the radiation field by tropical haze, and (*c*) modification of the hydrological cycle. Studies on these issues should be conducted at long-term tropical research sites. Long-term ecosystem studies should be combined with modification experiments. Short-term changes in exposure, e.g., variability due to seasonal cycles, should also be used as a tool to investigate biospheric response. The temporal and spatial variability of the distribution of O_3, nitrogen species, organic acids, sulfate, and acidity should be determined from a tropical measurement network; wet deposition studies are an important first step. Ground-based measurements should be complemented by regular determination of vertical profiles of key chemical species, especially tropospheric ozone.

Perturbations of the radiation field are expected to result from changes in UV-B radiation due to depletion of the ozone layer and from increased diffuse vs. direct radiation due to aerosols derived from biomass burning. We lack stations monitoring radiation parameters at continental sites, and the response of tropical ecosystems to current and predicted changes of these parameters remains unknown. Again, manipulative experiments at selected sites appear the most promising approach.

Changes in the hydrological cycle in the tropics are most difficult to predict in both extent and character. We recommend long-term studies of ecosystem characteristics, especially trace gas emissions, combined with a remote-sensing-based assessment of the change in hydrological cycling and ecosystem distribution. Further refinement of climate models, in order to make it possible to predict changes in the regional hydrological cycle resulting from interactions of global climate with land use and vegetation cover, will be essential.

THE RESPONSE OF TRACE GAS EXCHANGE IN MIDLATITUDE FORESTS TO CHANGES IN THE CHEMICAL AND PHYSICAL CLIMATE

The major changes in chemical climate considered are (*a*) regional-scale modification of nitrogen and sulfur deposition and (*b*) the global increase in CO_2 concentrations in temperate and boreal zones of the Northern Hemisphere. This research area is also considered relevant for those regions of the globe in which N and S inputs may become a problem, particularly in rapidly developing countries. The major physical climate variables considered are temperature, amount and temporal pattern of precipitation, and UV radiation. The work is most conveniently divided between comparative field studies and experimental manipulations to determine the sensitivity of trace gas regulating processes (e.g., to changes in temperature, CO_2, etc.).

For field studies it is necessary to define regions with contrasting atmospheric inputs of sulfur and nitrogen, particularly areas with very different ratios of N and S in the deposition. There are areas with pronounced gradients in these quantities, for example Europe, where the N/S ratio is much smaller in Eastern than in Western Europe. The wet deposition fluxes are reasonably well defined by current monitoring networks whereas for dry deposition, major uncertainties include the concentrations of HNO_3 and NH_3 and the dry deposition rates for aerosols (especially nitrate, ammonium, and sulfate).

Considering a range of sites with contrasting N/S inputs, the following trace gas fluxes are required:

1. NO and N_2O from soil into the plant canopy,
2. NO_x exchange between the plant canopy and the atmosphere,
3. CH_4 and N_2O emission from soil and atmosphere,
4. H_2S, DMS, and COS exchange between soil, canopy, and atmosphere,
5. NMHC exchange with the plant canopy.

The soil NO and CH_4 emissions into the forest canopy space are best measured using chamber methods, while the net exchange of NO_y and CH_4 between the forest canopy and the atmosphere require micrometeorological techniques. As the chamber and micrometeorological approaches average over very different scales and are not directly comparable, the problems of scaling require the development of models.

Experimental manipulations of temperature, CO_2, and UV to study the response of trace gas fluxes require enclosure methods. The effects of the enclosure on the physical and chemical environment are of such importance that studies on the influence of the enclosure on environmental properties and on biological processes must form a component of these studies. Enclosures for the study of CO_2 effects must be large enough to contain entire plant communities. This is because of the complex feedbacks between plant physiological responses, plant chemistry, decomposition, and thus nutrient availability to plants. Problems measuring physiological and ecosystem response in chambers are serious; given the importance of manipulations for a “preview of global change,” work should begin on designing experiments to address problems associated with this methodology. CO_2 response experiments are of course of interest for overall ecosystem response, as well as for the trace gas responses.

SUMMARY

Several areas of research are common to all of the above topics. First, manipulations of whole systems are urgently required to “preview” effects of global climate change. Second, integrated field and modeling experiments are necessary to develop the techniques for estimating regional exchange rates of gases and predictive models for gas exchange rates. These field experiments should include studies of production, micrometeorological measurements of flux, and stratified sampling to link landscape variability to variation in trace gas exchange. Third, coordinated measurements and modeling of reactive gases (e.g., NMHCs, O_3, NO) should be emphasized. Simultaneous measurements of interactive gas species are not common because of the small number of available instruments and investigators capable of these sophisticated measurements. Coordinated campaigns must be encouraged. Fourth, measurements of isotope ratios of gases are valuable in developing and testing global budgets. Measurements of the isotopic composition of gases should be included in as many field studies as possible. Finally, it has become clear that vegetation plays important roles in trace gas exchange as source, sink, conduit, and controller of microbial activity. Plant biologists will have to be centrally involved in the next phase of trace gas research.

Exchange of Trace Gases between Terrestrial Ecosystems and the Atmosphere
eds. M.O. Andreae and D.S. Schimel, pp. 333–337
John Wiley & Sons Ltd

List of Participants with Fields of Research

M.O. ANDREAE
Abt. Biogeochemie
Max-Planck-Institut für Chemie
Postfach 3060
6500 Mainz, F.R. Germany

Biogeochemistry, marine and atmospheric chemistry

I. ASELMANN
Abt. Luftchemie
Max-Planck-Institut für Chemie
Postfach 3060
6500 Mainz, F.R. Germany

Determination of biogenic tracer sources for atmospheric studies with emphasis on CH_4 and CO_2

F. BAK
Dept. of Ecology and Genetics
University of Aarhus
Ny Munkegade
8000 Aarhus C, Denmark

Microbial ecology

D. BALDOCCHI
Atmospheric Turbulence and Diffusion Division
NOAA
P.O. Box 2456
Oak Ridge, TN 37831, U.S.A.

Canopy micrometeorology: measuring and modeling gas exchange (CO_2, H_2O, O_3, SO_2, NO_x), canopy radiative transfer and turbulence

H.G. BINGEMER
Abt. Biogeochemie
Max-Planck-Institut für Chemie
Postfach 3060
6500 Mainz, F.R. Germany

Exchange of sulfur, organic acids, and ammonia between marine and terrestrial ecosystems and the atmosphere

A.F. BOUWMAN
International Soil Reference and Information Centre (ISRIC)
P.O. Box 353
6700 AJ Wageningen, Netherlands

Role of soils in the greenhouse effect

R.J. CICERONE
National Center for Atmospheric Research
P.O. Box 3000
Boulder, CO 80307, U.S.A.

Atmospheric chemistry; sources and sinks of atmospheric methane; climatic change due to trace gases

R. CONRAD
Fakultät für Biologie
Universität Konstanz
Postfach 5560
7750 Konstanz, F.R. Germany

Microbial metabolism of trace gases (H_2, CO, CH_4, NO_x) in soil, water, and sediments

P.J. CRUTZEN
Abt. Luftchemie
Max-Planck-Institut für Chemie
Postfach 3060
6500 Mainz, F.R. Germany

Atmospheric chemistry/global climate modeling

R.A. DELMAS
Faculté des Sciences
Département de Physique
B.P. 69
Brazzaville
Congo People's Republic

Atmospheric chemistry in tropical ecosystems; sources and sinks of methane in the tropics

O.T. DENMEAD
CSIRO Centre for Environmental Mechanics
GPO Box 821
Canberra, ACT 2601, Australia

Physical ecology: atmospheric transfer processes in plant canopies; gaseous transfer in the terrestrial N cycle; gas exchange across air–water interfaces; measurement of trace gas fluxes in the field (H_2O, NH_3, N_2O, CO_2, CH_4)

R.L. DESJARDINS
Land Resource Research Center
Bldg. 74, Research Branch
Agriculture Canada
Ottawa, Ontario K1A OC6
Canada

Micrometeorology and remote sensing

J.H. DUYZER
TNO Division of Technology for Society
P.O. Box 217
2600 AE Delft, Netherlands

Dry deposition

D.H. EHHALT
Institut für Atmosphärische Chemie
Kernforschungsanlage Jülich GmbH
Postfach 1913
5170 Jülich, F.R. Germany

Atmospheric chemistry

M.K. FIRESTONE
Dept. of Plant and Soil Biology
University of California, Berkeley
108 Hilgard Hall
Berkeley, CA 94720, U.S.A.

Microbiological control of trace N-gases from soil

D. FOWLER
Institute of Terrestrial Ecology
Bush Estate
Penicuik
Midlothian EH26 0QB, Scotland

Surface/atmosphere exchange of NO, NO_2, O_3, NH_3, and HNO_3

I. FUNG
NASA Goddard Institute for Space Studies
2880 Broadway
New York, NY 10025, U.S.A.

Biogeochemical cycles; global modeling

I.E. GALBALLY
CSIRO Atmospheric Research
Private Bag No. 1
Mordialloc, Victoria 3195, Australia

Atmospheric chemistry

V.F. GALCHENKO
Institute of Microbiology
Academy of Sciences USSR
Pros. 60-letija Octjabrja 7
Moscow 117811, USSR

Microbial methane oxidation in fresh and seawater ecosystems

P.M. GROFFMAN
Dept. of Natural Resources Science
University of Rhode Island
210B Woodward Hall
Kingston, RI 02881, U.S.A.

Landscape and regional-scale biogeochemistry

R.C. HARRISS
Institute for the Study of Earth, Oceans, and Space
University of New Hampshire
Durham, NH 03824, U.S.A.

Atmospheric chemistry and biogeochemistry

B.B. HICKS
NOAA/ATDD
P.O. Box 2456
Oak Ridge, TN 37831–2456
U.S.A.

Atmospheric dispersion and deposition

C. JOHANSSON
Dept. of Meteorology
University of Stockholm
Arrhenius Laboratory
10691 Stockholm, Sweden

Atmosphere–biosphere exchange of trace gases (NO_x, O_3, terpenes); photochemical modeling

M. KANAKIDOU
Abt. Luftchemie
Max-Planck-Institut für Chemie
Postfach 3060
6500 Mainz, F.R. Germany

Atmospheric modeling

M. KELLER
National Center for Atmospheric Research
P.O. Box 3000
Boulder, CO 80307, U.S.A.

Biogeochemistry of natural and disturbed tropical environments with a special emphasis on sources and sinks of atmospheric trace gases

J. KESSELMEIER
Abt. Biogeochemie
Max-Planck-Institut für Chemie
Postfach 3060
6500 Mainz, F.R. Germany

Carbonyl sulfide metabolism in plants

L. KLEMEDTSSON
Dept. of Microbiology
Swedish University of Agricultural Sciences
Box 7025
75007 Uppsala, Sweden

N_2O emission, denitrification, and nitrification in agricultural and forest soils

P.A. MATSON
Mail Stop 239–12
NASA-Ames Research Center
Moffett Field, CA 94035, U.S.A.

Biogeochemical cycling in tropical ecosystems

F.X. MEIXNER
Fraunhofer-Institut für Atmosphärische Umweltforschung
Kreuzeckbahnstr. 19
8100 Garmisch-Partenkirchen
F.R. Germany

Biosphere–atmosphere exchange of S, N, and C trace constituents (dry deposition, biogenic emission)

J.M. MELILLO
Ecosystems Center
Marine Biological Laboratory
Woods Hole, MA 02543, U.S.A.

Terrestrial ecology, biogeochemistry

A.R. MOSIER
USDA–ARS
P.O. Box E
Fort Collins, CO 80522, U.S.A.

Soil N transformations, denitrification, trace gas production from soil–plant system

H. PAPEN
Fraunhofer-Institut für Atmosphärische Umweltforschung
Kreuzeckbahnstr. 19
8100 Garmisch-Partenkirchen
F.R. Germany

Microbiology: microbial metabolism of trace gases, especially H_2 metabolism, denitrification, nitrification, and production of trace gases during heterotrophic nitrification

W.S. REEBURGH
Institute of Marine Science
University of Alaska
Fairbanks, AK 99775–1080, U.S.A.

Methane geochemistry: tundra fluxes, soil consumption, oxidation in marine anoxic systems

G.P. ROBERTSON
W.K. Kellogg Biological Station
Michigan State University
Hickory Corners, MI 49060–9516
U.S.A.

Ecosystem ecology, biogeochemistry, microbial ecology

H. RODHE
Dept. of Meteorology
University of Stockholm
10691 Stockholm, Sweden

Chemical meteorology

T. ROSSWALL
IGBP Secretariat
The Royal Swedish Academy of Sciences
Box 50005
10405 Stockholm, Sweden

Biogenic trace gas emissions from soils (mainly nitrous oxide)

E. SANHUEZA
I.V.I.C.
Apartado 21827
Caracas 1020–A, Venezuela

Tropical atmospheric chemistry

D.S. SCHIMEL
NRC Fellow
NASA–Ames Research Center
MS 239–12
Moffett Field, CA 94035, U.S.A.

Biogeochemistry; trace gas production in terrestrial ecosystems

H. SCHÜTZ
Fraunhofer-Institut für Atmosphärische Umweltforschung
Kreuzeckbahnstr. 19
8100 Garmisch-Partenkirchen
F.R. Germany

Atmospheric chemistry

L.J. STAL
Microbiology Laboratory
University of Amsterdam
Nieuwe Achtergracht 127
1018 WS Amsterdam, Netherlands

Cyanobacteriology: ecology and physiology, microbial mats, cyanobacterial blooms

J.W.B. STEWART
Saskatchewan Institute of Pedology
University of Saskatchewan
Saskatoon, Saskatchewan S7N OWO
Canada

Nutrient cycling and interrelationships of carbon, nitrogen, sulfur, and phosphorus in natural and agroecosystems

B.H. SVENSSON
Dept. of Microbiology
Swedish University of Agricultural Sciences
Box 7025
75007 Uppsala, Sweden

Methanogenesis in wetlands and biogas reactors; denitrification; N_2O *formation*

P.M. VITOUSEK
Dept. of Biological Sciences
Stanford University
Stanford, CA 94305, U.S.A.

Tropical ecosystems

M. WAHLEN
Wadsworth Center for Laboratories and Research
New York State Dept. of Health
Empire State Plaza, Rm. D–486
Albany, NY 12201, U.S.A.

Global cycles of atmospheric trace gases; isotopic species of CH_4

R. WASSMANN
Fraunhofer-Institut für Atmosphärische Umweltforschung
Kreuzeckbahnstr. 19
8100 Garmisch-Partenkirchen
F.R. Germany

Tropical ecosystems, marine ecosystems, emission of methane- and sulfur-containing trace gases

M.J. WHITICAR
Ref. B. 433
Bundesanstalt für Geowissenschaften und Rohstoffe
Postfach 510153, Stilleweg 2
3000 Hannover 51, F.R. Germany

Stable isotopes/organic geochemistry with emphasis on H–C volatiles (hydrocarbons)

W.-X. YANG
Research Center for Eco-Environmental Sciences
Academia Sinica
P.O. Box 934
Beijing 100083, China

Atmospheric chemistry, trace gases in air

G.A. ZAVARZIN
Institute of Microbiology
Academy of Sciences USSR
Pros. 60-letija Octjabrja 7
Building 2
Moscow 117812, USSR

Microbiology

Subject Index

Author Index

Dahlem Konferenzen Workshop Reports

Physical, Chemical and Earth Science Research Reports (PC)

PC 6 **Resources and World Development: Energy and Minerals, Water and Land**
Editors: D.J. McLaren and B.J. Skinner
0 471 91568 8 958pp 1987

PC 7 **The Changing Atmosphere**
Editors: F.S. Rowland and I.S.A. Isaksen
0 471 92047 9 296pp 1988

PC 8 **The Environmental Record in Glaciers and Ice Sheets**
Editors: H. Oeschger and C.C. Langway, Jr.
0 471 92185 8 416pp 1989

WILEY

Prices on application from the Publisher

Dahlem Konferenzen Workshop Reports

Life Sciences Research Reports (LS)

LS 37 **Mechanisms of Cell Injury: Implications for Human Health**
Editor: B.A. Fowler
0 471 91629 3 480pp 1987

LS 38 **The Neural and Molecular Bases of Learning**
Editors: J.-P. Changeux and M. Konishi
0 471 91569 6 576pp 1987

LS 39 **Sexual Selection: Testing the Alternatives**
Editors: J.W. Bradbury and M. Andersson
0 471 91683 8 368pp 1987

LS 40 **Biological Perspectives of Schizophrenia**
Editors: H. Helmchen and F.A. Henn
0 471 91683 8 368pp 1987

LS 41 **Humic Substances and Their Role in the Environment**
Editors: F.H. Frimmel and R.F. Christman
0 471 91817 2 286pp 1988

LS 42 **Neurobiology of Neocortex**
Editors: P. Rakic and W. Singer
0 471 91776 1 480pp 1988

LS 43 **Etiology of Dementia of Alzheimer's Type**
Editors: A.S. Henderson and J.H. Henderson
0 471 92075 4 268pp 1988

LS 44 **Productivity of the Ocean: Present and Past**
Editors: W.H. Berger, V.S. Smetacek and G. Wefen
0 471 92246 3 492pp 1989

LS 45 **Complex Organismal Functions: Integration and Evolution in Vertebrates**
Editors: D.B. Wake and G. Roth
0 471 92375 3 Approx 458pp 1989

LS 46 **Biofilms**
Editors: W.G. Characklis and P.A. Wilderer
0 471 92480 6 Approx 408pp In Press

LS 48 **Organic Acids in Surface Waters**
Editors: E.M. Perdue and E.T. Gjessing
In Preparation.

Prices on application from the Publisher